Fundamentals of Inertial Navigation Systems and Aiding

Other volumes in this series:

Volume 1	**Optimised Radar Processors** A. Farina (Editor)
Volume 3	**Weibull Radar Clutter** M. Sekine and Y. Mao
Volume 4	**Advanced Radar Techniques and Systems** G. Galati (Editor)
Volume 7	**Ultra-Wideband Radar Measurements: Analysis and processing** L. Yu. Astanin and A.A. Kostylev
Volume 8	**Aviation Weather Surveillance Systems: Advanced radar and surface sensors for flight safety and air traffic management** P.R. Mahapatra
Volume 10	**Radar Techniques Using Array Antennas** W. Wirth
Volume 11	**Air and Spaceborne Radar Systems: An introduction** P. Lacomme (Editor)
Volume 13	**Introduction to RF Stealth** D. Lynch
Volume 14	**Applications of Space-Time Adaptive Processing** R. Klemm (Editor)
Volume 15	**Ground Penetrating Radar, 2nd Edition** D. Daniels
Volume 16	**Target Detection by Marine Radar** J. Briggs
Volume 17	**Strapdown Inertial Navigation Technology, 2nd Edition** D. Titterton and J. Weston
Volume 18	**Introduction to Radar Target Recognition** P. Tait
Volume 19	**Radar Imaging and Holography** A. Pasmurov and S. Zinovjev
Volume 20	**Sea Clutter: Scattering, the K distribution and radar performance** K. Ward, R. Tough and S. Watts
Volume 21	**Principles of Space-Time Adaptive Processing, 3rd Edition** R. Klemm
Volume 22	**Waveform Design and Diversity for Advanced Radar Systems** F. Gini, A. De Maio and L.K. Patton
Volume 23	**Tracking Filter Engineering: The Gauss-Newton and Polynomial Filters** N. Morrison
Volume 25	**Sea Clutter: Scattering, the K distribution and radar performance, 2nd Edition** K. Ward, R. Tough and S. Watts
Volume 33	**Radar Automatic Target Recognition (ATR) and Non-Cooperative Target Recognition** D. Blacknell and H. Griffiths (Editors)
Volume 26	**Radar Techniques Using Array Antennas, 2nd Edition** W. Wirth
Volume 101	**Introduction to Airborne Radar, 2nd Edition** G.W. Stimson
Volume 530	**Radar Sea Clutter: Modelling and target detection** Luke Rosenburg and Simon Watts
Volume 534	**New Methodologies for Understanding Radar Data** Amit Kumar Mishra and Stefan Brüggenwirth
Volume 537	**Ocean Remote Sensing Technologies: High frequency, marine and GNSS-based radar** Weimin Huang and Eric W. Gill (Editors)
Volume 553	**Modern Radar for Automotive Applications** Z. Peng, C. Li and F. Uysal (Editors)

Fundamentals of Inertial Navigation Systems and Aiding

Michael Braasch

The Institution of Engineering and Technology

Published by SciTech Publishing, an imprint of The Institution of Engineering and Technology, London, United Kingdom

The Institution of Engineering and Technology is registered as a Charity in England & Wales (no. 211014) and Scotland (no. SC038698).

First published 2022

The Institution of Engineering and Technology
Futures Place
Six Hills Way, Stevenage
Herts, SG1 2AU, United Kingdom

www.theiet.org

British Library Cataloguing in Publication Data
A catalogue record for this product is available from the British Library

ISBN 978-1-83953-412-6 (hardback)
ISBN 978-1-83953-413-3 (PDF)

Typeset in India by MPS Limited

Cover Image: *Flight paths illustration*: Sean Gladwell/Creative via Getty Images

Contents

About the author

Michael Braasch holds the Thomas Professorship in the Ohio University School of Electrical Engineering and Computer Science and is a principal investigator with the Ohio University Avionics Engineering Center, USA. He has been performing navigation system research for over 35 years and has taught inertial navigation on-site for numerous manufacturers of navigation-grade inertial systems. He is an instrument-rated commercial pilot and is a fellow of IEEE and the US Institute of Navigation.

Preface

The subject of inertial navigation systems and how to aid them (i.e., reduce their inherent error drift) is complex and multi-disciplinary. Mathematics and physics along with electrical, mechanical, and software engineering all are involved. I have had the privilege of teaching the subject for over 25 years and have long noted that if I placed all the necessary foundational courses as pre-requisites, the class would have few, if any, students.

This book has been written to serve as an introduction for students and those new to the field. Although I personally teach the subject strictly at the graduate level (a one-semester course on inertial systems and a one-semester course on aiding), the material is accessible to most seniors in engineering as well. Specialized topics such as rotation matrices, quaternions, and relevant stochastic processes are covered in the book. The reader is expected to have a basic understanding of vectors, matrices, matrix multiplication and Laplace transforms as well as freshman-level differential and integral calculus.

The reader will find the book has a distinctly conversational tone and style. This was done intentionally with the hope that it will prove to be more accessible than the average graduate-level textbook. When the depths of a particular topic are beyond the scope of the book, the reader is referred to select relevant references.

I have a "build a little, test a little" approach to the subject. Inertial navigation in a (conceptually) simple inertial reference frame is introduced before all the complexities associated with a spheroidal, rotating earth are covered. The key concepts of the Kalman filter are described in the context of a simple, scalar example before the reader is overwhelmed with all the matrices of the full multi-variate filter. The concept of inertial system aiding is first described in the context of a one-dimensional example (a rocket sled!). Again, it is my sincere hope that this approach will allow those new to the subject to ease into it and, along the way, gain a better conceptual understanding.

Although my name is on the title page, it should come as no surprise that there are a great many individuals who have contributed to this effort. Since 1985, I have had the distinct privilege of working in the Ohio University Avionics Engineering Center. My friends, colleagues, mentors, and supervisors are too numerous to list but it was in this research lab that I became a navigation engineer and learned the fundamentals of inertial navigation, satellite-based navigation, and navigation system integration. In 1994, I was appointed to the faculty of the School of Electrical Engineering and Computer Science at Ohio University and within a few years started to teach graduate courses in these subjects.

In the late 1990s, Dr Mikel Miller (then on the faculty of the US Air Force Institute of Technology) took my barely legible handwritten notes and converted them into a beautiful set of illustrated PowerPoint slides. That slide deck, although modified by me over the years, served as the basis for my lectures which, in turn, served as the basis for this book. I have had the privilege of teaching this material at all three US manufacturers of navigation-grade (i.e., "high end") inertial navigation systems: Honeywell, Kearfott, and Northrop Grumman. The feedback I have received when teaching at these companies has been very valuable. I must single out Mr Phil Bruner (Northrop Grumman) for the numerous occasions in which he patiently explained fundamental concepts to me. I also gratefully acknowledge the most excellent short course on the Integration of GPS with Inertial Navigation Systems presented by Dr Alan Pue (Johns Hopkins University). Of course, my most regular feedback comes from the many students who have been taking my courses in this subject area since 1997. Their feedback on early drafts of the manuscript has been very helpful indeed.

Although I had a couple of false starts (both of my sabbaticals were overcome by events, etc.), when I was approached by the Institution of Engineering and Technology (IET) in 2020 to write a book on this subject, all the pieces were finally in place to make it happen. I especially want to thank Nicki Dennis, Books Commissioning Editor, for her steadfast encouragement and support. Aside from the multiplicity of MATLAB® plots, nearly all the remaining graphics were created by Mindy Yong (mindyyongart.com). Since my own graphics skills do not even rise to the level of being called primitive, her contributions are beyond my words to express gratitude.

Over the decades, my wife and two daughters (now both grown and married) lovingly and patiently dealt with many evenings when I was at the office instead of being at home. I dedicate this book to them.

Chapter 1
Introduction and overview of inertial navigation systems

1.1 A simple thought experiment

I want you to give your imagination a little bit of a flight of fancy. Imagine that you have been kidnapped. Your home has been invaded. You have been blindfolded. Your ears have been plugged up. You have been placed into the backseat of a car. The driver of the car then proceeds to drive you somewhere. You sway side to side in the seat as the car goes around the turns. You lunge forward or backward as the car accelerates or brakes. At some point, the car finally stops. Do you know where you are? Does the amount of time that you were being driven around matter?

Before you can answer the question "where am I?" you have to address the associated problem of "determining position relative to what?" You could state that you are in the such-and-such country (e.g., France, Singapore, the United States, etc.). More generally, you could state that you are on the planet Earth or you are in the Milky Way galaxy. All of these statements are correct. There are a wide variety of places that you could use to describe your location.

So when you think about it, the question of "where am I?" really boils down to the determination of where you are relative to something else, some frame of reference. Maybe you are in the second floor of some building, or you know that you are 30 m south of your office. Notice that position is frequently given in relative terms. Someone residing in a small town will typically describe the location of their town in terms of distance and direction from a better-known larger town (e.g., Batavia is approximately 20 miles east of Cincinnati). We may express our position in terms of coordinates on a map, for example, but there is always the need to have some frame of reference in order to describe location.

In this most general sense, the term "absolute position" is a misnomer. Any position we may determine is a position relative to something. Nevertheless, for terrestrial applications (the main focus of this volume), the term "absolute position" denotes a position expressed in a reference frame that uniquely defines all points on, below or above the surface of the earth. Once an earth-fixed reference frame has been defined, a given latitude, longitude, and altitude then uniquely defines a location relative to the earth itself. "Relative positioning," in contrast, generally refers to the position of a point relative to another point. In the aforementioned example, the location of the city of Batavia is given relative to Cincinnati but the position of Cincinnati, relative to the earth, is not specified.

Returning to our kidnapping example, do you need to know where you started? Of course, clearly the answer is yes. If you are sensing accelerations and turns, then you are attempting to perform dead reckoning. However, dead reckoning does not provide you with absolute position unless you knew where you were to begin with. Thus, a dead reckoning system needs some kind of initialization.

Obviously we have to be able to measure changes. We said that if we are going around a corner, then our body sways side to side. If we are accelerating or braking, then we will lunge back or forth. Thus we have to be able to sense those changes. Within the context of an inertial system, we will have sensors to do that. Once the measurements are formed, we then need to be able to synthesize those measurements into angular orientation (referred to as attitude), velocity, and position. That requires some kind of computational capability. In this hypothetical example, you may sense that you made a turn but somehow you have to determine how much of a turn you made. Or, if you sensed that the car was accelerating, then you will need to take that information and compute subsequent changes in velocity and position.

When all of these processes are performed iteratively, position is being updated. Inertial navigation systems perform a highly technologically advanced form of dead reckoning.

Key questions (components) of a dead reckoning system:

1. Determining position relative to what? (need a frame of reference)
2. Where did I start from? What orientation did I start from? (need initialization)
3. How do I measure changes in orientation and speed? (need sensors)
4. How do I synthesize these measurements? (need computational capability)

The kidnapping example captured all of these elements. You knew where you started from (remember, you were kidnapped from your home). Thus the initialization problem was not an issue because you knew where you were to begin with. You inertially sensed your angular rotations and your linear accelerations through the swaying of your body. You had a frame of reference since, presumably, you have some mental model of the layout of your hometown (e.g., your own mental version of a street map).

But of course, if you are the least bit honest with yourself, you will also recognize that as time goes on you will become more and more uncertain of your position. In all dead reckoning systems, errors accumulate over time. This is particularly true if you are going over long stretches without much change in the acceleration or angle. It will be difficult to keep track of time, and again, whatever errors you had to begin with will just continue to grow. All inertial navigation systems, as with all dead reckoning type systems, exhibit this error-growth behavior.

1.2 Newton's first and second laws of motion

We recall from physics that a body in uniform motion will remain in that state unless acted upon by an external force (Newton's first law of motion). The application to dead reckoning is obvious. In the absence of applied forces, the current position and velocity of a body can be used to determine future position so long as one keeps accurate track of elapsed time. Newton's second law, in its commonly simplified form, relates force to induced acceleration and the mass of the object:

$$f = ma$$

If we can determine acceleration, we can start the dead reckoning process because we know that the first integral of acceleration is velocity and the first integral of velocity is position. Newton's second law states that force is equal to product of mass and acceleration. We can feel force. Force is something that pushes or pulls. Thus we know that force is something that can be measured (more details later!). If we can measure the force, then we can infer acceleration. Having determined acceleration, we can go on to determine velocity and position. The thing we have to keep in mind, though, is that Newton's second law is only valid in a so-called "inertial reference frame."

1.3 Inertial reference frames

What is an inertial reference frame? Well, rather than getting sidetracked with all the nitty-gritty details of the physics, the bottom line is that Newton's laws, and hence the first inertial reference frame, were originally derived within the context of the study of planetary motion. Newton solved the problem of determining the laws governing the motions of the planets. Prior to Newton, Tycho Brahe had collected detailed observations of the motion of the planets and Johannes Kepler subsequently poured over Brahe's data and empirically derived what are now called Kepler's laws of planetary motion. Thus, the paths of the planets were well known but, until Newton, nobody knew why the planets moved the way they did. It was Newton who formulated the law of universal gravitation and was able to show that gravity dictated the motion of the planets.

At this point, we can ask what frame of reference was used to gather the data to determine the positions of the planets in the first place? The positions of the planets were determined relative to the background stars (i.e., the celestial sphere). Although we know the universe is expanding, the stars in our own galaxy (the Milky Way) are sufficiently stationary (relative to each other) for our purposes and thus we refer to them as "fixed." Thus, the planetary motion data were collected relative to the background stars. We therefore have an existence proof. We know that the background stars serve as a good reference frame in which Newton's laws are valid. Of course, much more intricate detail may be found in physics books. For our purposes, though, suffice it to say that the background stars of our own galaxy are essentially not moving relative to each other and they serve as a valid inertial reference frame.

So, the background stars are one example of an inertial reference frame. Are there any others? Yes indeed. Any frame moving at constant velocity with respect to any other inertial reference frame is also an inertial reference frame. Note that when we are talking about reference frames it is quite acceptable simply to think of a Cartesian coordinate system. For our purposes, they are essentially the same. So imagine a 3D coordinate system that is fixed out in space or is moving with constant velocity with respect to the stars. These are inertial reference frames.

1.4 Positioning with respect to the Earth

With the aforementioned review of some elementary physics, we now need to point out that the Earth itself is not an inertial reference frame. The Earth both rotates and accelerates with respect to the fixed stars. The Earth rotates on its axis and it revolves around the Sun. Both of these, obviously, are motions that are not uniform with respect to the background stars. So we will somehow have to take this into account when we are implementing inertial positioning since Newton's laws cannot be directly applied to a non-inertial reference frame.

This is an important issue since we generally do not care where we are with respect to Alpha Centauri. We care where we are with respect to our next destination. In aviation, we care about our position relative to some desired flight path. A similar situation applies for ships crossing the ocean. Most of our navigation tasks are performed via positioning with respect to the Earth itself. As we will see later, Newton's laws cannot be applied in earth-referenced positioning without taking into account some so-called "pseudo-forces." For example, we know that circular motion involves centripetal force. However, the effects of earth rotation, when viewed from the non-inertial earth-fixed reference frame, appear as an outward or "centrifugal" force.

1.4.1 A simple example

Assume a vehicle is starting at a known location. Pick a 2D coordinate frame and an initial position (x_0, y_0) valid at an initial time (t_0). Assume the vehicle is moving at a constant velocity V at some angle measured clockwise with respect to North. At any future time, position is given by:

$$x(t) = x_0 + V(t - t_0) \sin \psi$$

$$y(t) = y_0 + V(t - t_0) \cos \psi$$

Thus, for linear motion with constant velocity and a 2D coordinate system, the positioning problem is solved. However, the solution is only valid under these very restrictive assumptions.

How do we go beyond that? How do we handle more realistic scenarios? We need to be able to accommodate accelerations, turns, and all sorts of more complex motions that a vehicle may go through. How do we do it?

1.4.2 Going beyond

First of all, we have to measure over very short intervals (e.g., milliseconds). That allows us to be able to capture the changes of vehicle angle and acceleration even during high dynamics. If we are sampling this data at very high rates, then we are able to capture the changes with high fidelity. Subsequently we will need appropriate integration schemes. Of course, along the way, we have to be able to keep track of the orientation (angle) of the vehicle in order to resolve the acceleration measurements into proper coordinates. This assumes that the accelerometers (accels) are hard mounted ("fixed") to the vehicle and that they rotate with the vehicle. We will see later that there are a couple of different ways to mechanically architect an inertial system, but if you have your sensors fixed to the vehicle itself (as is most common), then you have to resolve the measured acceleration even when the vehicle is turning and rotating.

We have already discussed the fact that the Earth is not an inertial reference frame. In addition, we have to deal with the fact that not only does the Earth rotate but it also is approximately spherical. That, of course, adds to the complication of representing position. Furthermore, it adds some complication with regard to what we think is straight-line motion with respect to the Earth. What we think of as a straight path on Earth is actually a curve in inertial space. Again, all of this has to be taken into account. Finally, we have to be able to deal with the influence of gravity on our measurements. It has a profound impact, and it is one of the key considerations in inertial navigation. We will discuss this in more detail later.

1.5 Guidance, navigation and control

Now let us consider some basic concepts of navigation. Strictly speaking, navigation has to do with the entire process of getting from where you are to where you want to go. If an airplane is flying from New York to Los Angeles, the navigation process has to do with making sure that the aircraft stays on its desired path so that it successfully gets from origin to destination. Strictly speaking, from the point of view of an airframe manufacturer (e.g., Boeing or Airbus), navigation involves all of the steps in getting from point A to point B.

Within the navigation system community, however, the term *navigation* is generally applied to the determination of position, velocity, attitude, and time (relative to some reference frame). Note the primary subject of our study here is inertial navigation systems, commonly abbreviated as INS. You can purchase a high-end INS (called navigation grade or nav grade) and you will get a device in a metal box about the size of two loaves of bread. The INS outputs position, velocity, and attitude but it is only a component in the broader navigation process. The INS, by itself, does not know what the desired flight path is nor does it compute the vehicle's deviation from that desired path. However, position/velocity/attitude determination plays a vital role in the navigation process and that is why we refer to an INS as a navigation system.

A related topic is guidance. Frequently you will see the phrase "guidance, navigation and control" or GNC or GN&C. Within the context of aviation, navigation is the determination of position, velocity, attitude, and perhaps time as well. Guidance has to do with determining the desired course and providing the necessary indications to stay on course. For example, if the vehicle is half a mile east of some desired route, then the guidance system needs to determine that and provide a course correction. The guidance system thus needs to know both the desired path as well as the current vehicle position.

The guidance task is thus the process of using the navigation information (e.g., position) in order to provide steering or maneuvering commands to maneuver the vehicle onto the desired path or desired route. The third part of guidance, navigation, and control is the control system itself. The navigation system determines where the vehicle is, the guidance system determines what changes are needed to the current trajectory, and the control system physically maneuvers the vehicle onto the desired path. The control system can be either manual (i.e., a human pilot) or automatic. Either way, navigation and guidance tasks still play vital roles.*

1.6 A self-contained system?

An INS, of course, is not the only navigation system in existence. For example, so-called radio navigation systems, also known as aids to navigation, navigation aids, or simply nav aids are systems that radiate electromagnetic signals in order to provide information to the user. Global navigation satellite systems such as the U.S. Global Positioning System (GPS) are radio-type navigation systems because they transmit electromagnetic signals through space used for the determination of position, velocity, and time.

Inertial navigation, the subject of this volume, is a system that relies almost exclusively on inertial sensors (and specifically does not rely on radio transmissions). We will see later, however, that an INS cannot determine altitude reliably on its own. It turns out that the so-called "vertical channel" in an INS is unstable. Thus, in general, some kind of external information about altitude is needed. Historically that has been provided with a barometric altimeter. More recently satellite navigation systems (e.g., GPS) are also used for this purpose. Inertial navigation is frequently referred to as "self-contained," but that is not strictly correct for the aforementioned reason.

In summary, inertial navigation systems are (mostly) self-contained systems that determine the position and velocity of a vehicle using inertial sensors. Later we will discuss how we use accelerometers to sense changes in vehicle velocity and we will discuss how we use sensors called gyroscopes (commonly referred to as "gyros") in order to determine either angular rates or angular displacements.

*The launch and space community have alternate definitions of GN&C which shall not be addressed here.

1.7 Four key processes in inertial positioning

Referring back to the key questions in dead reckoning, we can now better appreciate the four key processes in inertial positioning:

1. Instrument a reference frame
2. Measure specific force
3. Employ knowledge of the gravity field
4. Time integrate inertial acceleration to determine velocity and position

Let us consider these four processes in reverse order. All inertial navigation systems have to perform integration, over time, of acceleration to determine velocity and position. As we noted earlier, what we actually measure is force (technically, specific force).

As we will discuss in more detail later, gravity plays a critical role in our determination of acceleration since it is a component in our force measurements. We will need to eliminate the gravity component from our measurements and we need to do so effectively in real time. As a result, we will need some kind of model of gravity. Specifically, we will need some kind of software algorithm to estimate the local value of gravity in order to be able to separate it from the acceleration that we are trying to determine.

As we have discussed already, we need to have some kind of a reference frame and it has to be instrumented within the inertial system itself. Historically, this was instrumented mechanically. In more modern times, it is done virtually through software. Now, let us pause for a moment. Why do we need a reference frame? If we want to know our displacement in the north/south direction, for example, then it would be great if we had an accelerometer on our vehicle that always pointed north/south. Since a given vehicle is not always pointing north/south we have two choices. One is to mechanically isolate the accelerometer from the motion of the vehicle. We can do that using gimbals and gyroscopes. Figure 1.1 illustrates a three-degree-of-freedom

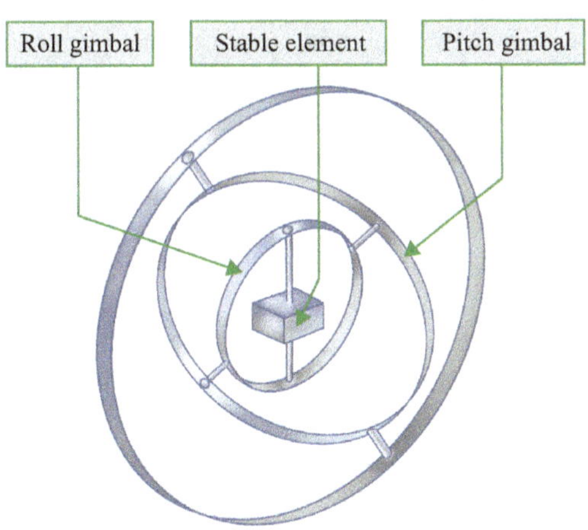

Figure 1.1 Three degree-of-freedom gimbal set

gimbal set. Although not fully illustrated in the figure, multiple mechanical gyros can be used to hold the center platform (stable element) in a fixed orientation and the gimbals allow the vehicle to rotate around it.

Thus, it is possible to mechanically isolate the sensors such that one accelerometer always points north and one accelerometer always points east. From their measurements, you can determine north and east velocity and then, hence, changes in latitude and longitude. An INS using this mechanical isolation technique is referred to as a "gimballed INS" or a "platform INS" since the "platform" is the rotationally isolated fixture inside the gimbals on which the accelerometers are mounted.

The second way to instrument a reference frame is to hard mount the accelerometers to the vehicle itself (or, rather, hard mount the accelerometers inside the case of the INS that is itself hard-mounted to the vehicle). Since the accelerometers will, in general, not be pointing in reference directions (e.g., north, east), it is necessary then to keep track of the changes in the angular orientation of the vehicle and resolve the accelerometer measurements into the directions of interest. This type of architecture is referred to as a "strapdown INS" since the sensors are physically strapped to the INS case/shell which is strapped to the vehicle. As a result, the sensors are not free to rotate independently of the vehicle.

1.7.1 A slightly less simple 2D example

Consider a ship at sea. Assume there is no current, and the ship is in calm waters (or that it has really good roll stabilization). As a result, we only need to measure acceleration along the longitudinal axis and the lateral axis (i.e., only horizontal accelerations). Now assume the ship has a directional gyro to determine its heading (i.e., where it is pointing). Are we able to perform inertial navigation with these three sensors?

We do not have to worry about the influence of gravity in this scenario since, to first order, gravity lies in the vertical direction and we do not need to determine vertical acceleration (remember, it is a ship). The accelerometers are thus orthogonal to the gravity vector so gravity is not going to play a role. We can integrate the linear accelerations from the accelerometers. We can keep track of where we are pointing through either some measurement of angular velocity from the directional gyro or, if the gyro actually keeps track of the change of angle, then heading is provided directly. As a result, yes, we can navigate with this simple sensor suite so long as we have the initial conditions (i.e., initial position, velocity, and heading).

1.7.2 More generally ...

Of course, we do not live in a simple 2D world (thank goodness!), so we need to have more information. In the most general case, we will have to determine three dimensions of position, three dimensions of velocity, and along the way, we will need to determine at least three parameters of attitude or orientation. Thus, there are at least nine parameters that have to be determined. Since we derive position from velocity, we have six basic unknowns and thus need six measurements. Typically this is achieved through the use of three orthogonal gyros and three orthogonal accelerometers.

1.8 Some historical context

Over the last century, a wide variety of "modern" navigation systems have been developed. However, mankind's quest to solve the navigation problem (getting from point A to B) dates back millennia. Let us consider a few of the key techniques, compare and contrast them, and then put them into the context of the current topic of inertial navigation. Let us start off with visual reference to the ground.

In the early days of aviation, there was no GPS and there was no inertial navigation. You had what we might call the "Mark I eyeball." The pilot simply looked out the window and kept track of where he or she was. Pilots would observe the terrain, observe the roads, observe the landmarks, and base their navigational decisions on their knowledge of the area. Certainly visual reference to landmarks is something that mankind has been using for navigation for a very long time.

Something slightly less old than the use of visual landmarks is the use of celestial bodies. For dozens of centuries, people utilized observations of the stars in order to perform navigation tasks. Ancient Polynesians were able to traverse vast regions of the Pacific Ocean with little more than observation of the stars [1]. Later, people were able to identify the pole star (Polaris, the North Star). They figured out that one particular star remained essentially stationary throughout the night whereas the rest of the star field would move over the course of the evening. Eventually people figured out that the north star was a good way to be able to determine your latitude. Specifically, all you had to do was measure the elevation angle (above the horizon) of the pole star [2].

However, both visual reference to the ground and visual reference to the stars are limited by line of sight. If you have clouds or any kind of obstruction, then you cannot make the necessary measurements. This is a large part of the reason why we, in modern times, have been developing better and better techniques to be able to navigate at all times of the day or night and in the midst of all kinds of weather conditions.

1.8.1 Dead reckoning

Dead reckoning was used for marine navigation long before inertial navigation was developed. By the way, nobody actually knows with certainty how the term "dead reckoning" came into use. The historians are conflicted over its actual origins. One theory is that it came from the term "deduced reckoning." Deduced can be shortened to "ded" and hence ded reckoning. Of course when we write it out in modern language it is "dead" instead of "ded." As a pilot, I find some of these terms rather disturbing: you use *dead* reckoning to arrive at the *terminal* area where you conduct a *final* approach. What sadist came up with these terms?†

Anyway, let us review how it works. First we measure our speed. There are a variety of techniques that can be used in aircraft. There are air speed indicators, Doppler-based systems and, of course, modern radio nav aids provide speed as well.

†As much as I would like to claim credit for this bit of humor, I must acknowledge hearing it from my former supervisor, Robert W. Lilley, Emeritus Director of the Avionics Engineering Center at Ohio University.

Then there is the need to determine heading. Historically that has been performed with a magnetic compass. In the early 1900s, the gyrocompass was developed. And finally, one also has to be able to keep track of time. Prior to modern systems, dead reckoning had fairly crude accuracy.

Heading could be estimated to within a few degrees based on a magnetic compass or early gyrocompasses. Then, the ability to measure speed was limited. In centuries past, mariners had a log. That is, a literal piece of wood. A weighted wedge-shaped piece of wood (with a knotted rope attached) was thrown over the side of the ship. That is where we get speed in terms of "knots," because the rope that was attached to the log had knots at regular intervals. The log was thrown overboard and a count was taken of the number of knots that passed in the paid out line over some interval of time. The count provided an indication of speed [3]. The accuracy was not perfect nor was the measurement of time. Thus, the computation of distance traveled (given by the product of speed and elapsed time) was approximate at best.

1.8.2 Key developments in the history of inertial navigation

Although toy spinning tops were in use since ancient times, the use of a spinning mass as the basis for an instrument did not occur until the mid-1700s. At this time, John Serson, an English sea captain, was the first person who attempted to use a spinning top (with the spin-axis oriented along the local vertical) to determine the local horizontal plane during foggy weather conditions. Unfortunately, his ship was lost during the voyage in which he was performing the first at-sea trials of his invention, but this is, nevertheless, the earliest known ancestor of the modern artificial horizon that is installed in virtually all aircraft [4].

Bohnenberger, in Germany in the early 1800s, constructed a sphere that spun on an axis and was suspended inside three rings (what we now refer to as gimbals). The device was used to demonstrate that, once spun up, the sphere would maintain its orientation regardless of how the base was moved. Other researchers developed similar devices in the early to mid-1800s, but the earliest form of the modern day mechanical instrument was developed by Leon Foucault (who also coined the term "gyroscope") and demonstrated it in 1852 [2].

The purpose of these devices was to investigate the rotation of the earth. The Foucault gyroscope could be observed to maintain its orientation while the Earth rotated (thus the spinning wheel would appear to rotate relative to the table that the device was sitting on).

The first application of the gyroscope in navigation occurred in the early 20th with the development of the gyrocompass. This was a mechanical device that was based on the idea of a spinning mass gyro. The tendency of the mechanical gyro to hold its orientation (when spinning) was exploited in order to provide a stable heading reference that was independent of the magnetic compass.

This was important for two reasons. One is that motion of the boat in rough seas can induce jitter in the compass needle. Second, as ship construction changed from wood to steel, the ferromagnetic material interfered with magnetic compass operation.

Thus a mechanical-type compass that was not based on magnetic properties was needed. The gyrocompass filled that role. This is not merely a historical note. We are still making gyrocompasses today. Additionally, they were crucial in the development of inertial navigation due to an investigation of their errors.

Specifically, when a boat would undergo lateral acceleration, that force would cause errors in the gyrocompass' output. This was a significant problem. One of the principal manufacturers of gyrocompasses was Anschütz-Kaempfe in Germany and the problem was solved by Maximillian Schuler.

We will cover this in more detail in a later chapter but Schuler discovered that the gyrocompass problem could be solved by mechanically tuning it such that it had an undamped, natural period identical to a theoretical pendulum with a length equal to the radius of the earth. Such a gyrocompass would then be insensitive to accelerations in the horizontal direction. Do not worry about understanding this completely at the moment. We will get into the details later. The important thing is that Schuler's work with regard to gyrocompasses turned out to be a very fundamental analysis used in the development and characterization of modern inertial navigation systems (recall that an inertial system needs to determine the local vertical and horizontal directions).

In parallel with Anschütz-Kaempfe, the Sperry Company developed the gyrocompass as well. There were patent disputes between the two companies and it turns out that Einstein himself was actually involved with some of those disputes since he was a patent examiner early in his career. Sperry also produced autopilot equipment for early aircraft. Gyroscopes provided the ability to measure short-term angular disturbances. Imagine you are in an aircraft and you want to fly straight and level. If the wind causes a disturbance, Sperry's gyroscopes could measure the perturbation and provide feedback through an automatic control system in order to maintain stability in the flight. This was one of the earliest uses of inertial sensing technology in aviation.

Throughout the 1920s aircraft development continued. One of the main uses of aircraft after World War I was for carrying mail. Passenger service also slowly grew. There was soon a need for instrumentation so that pilots would not have to navigate solely by visual reference to the ground. Gradually, more instrumentation was developed to make the pilot's job easier, and to enable them to be able to fly even if the weather conditions were not all that good. Rate of turn indicators were developed along with artificial horizons and directional gyroscopes. All were developed for aircraft in the 1920s and all were based on inertial sensing principles.

Open-loop accelerometers were developed at this time and north-seeking devices were used in terrestrial applications. Boykow, in Austria, conceived the use of accelerometers and gyroscopes in order to produce a full inertial navigation system but in the 1920s and early 1930s the sensor technology was not sufficiently mature [5]. Prior to and during World War II, there were tremendous technological developments. One item of note was the development of the V2 rocket by Germany. The V2 demonstrated some principles of inertial guidance but not inertial navigation.

The late 1940s and early 1950s are considered by most to be the time of the origin of modern inertial navigation. It was during this time that gyroscope technology was advanced dramatically. Prior to this time, the best gyroscopes would drift approximately 10 degrees per hour. This is perfectly acceptable if the purpose is to measure turn rate over a few seconds for stabilization applications. However, this is not nearly sufficient for navigation. It was in the early 1950s that the technology was advanced to reduce the gyro drift rate to the order of 0.01 degree per hour. Note that this is three orders of magnitude improvement! This enabled the production of inertial navigation systems capable of determining position and velocity over long periods of time (e.g., hours) and not incur hundreds of miles of error in the process.

Charles Stark Draper, affectionately known as "Doc Draper" at MIT, led the group (known as the Instrumentation Lab) that did much of the early development in inertial navigation. It was his group that eventually designed and developed the floated rate-integrating gyro. It was a particular type of mechanical gyro considered by many to be near the pinnacle of spinning mass mechanical gyro designs. At the same time, there were important developments in the accelerometer. The principle of force feedback was applied to improve the accuracy and dynamic range of accelerometers. Both the gyro and accelerometer technology were thus advanced significantly in this period of the early 1950s culminating in the first demonstration of a cross-country flight from Massachusetts to California nonstop with the aircraft being guided solely by inertial navigation [6].

After that occurred, research was aimed at making the system smaller. That first transcontinental crossing had an inertial system that was the size of a small room. In fact, they had to install it in a United States Air Force bomber because that was the only airplane big enough to be able to house it. Initially INSs were only feasible for submarines, ships, ballistic missiles, and the like. Not until the late 1950s and early 1960s did the systems become small enough for installation on small aircraft such as fighter jets.

Also in the 1960s, non-mechanical gyroscopes (now referred to as "optical gyroscopes") were under development. Research and development of the ring laser gyroscope (RLG), for example, was started in the 1960s, matured in the 1970s, and went into production in the 1980s. Work on the fiber optic gyro (FOG) also started in the 1960s, but high-precision versions (capable of supporting long-range unaided navigation) did not mature until the 1990s. In the 1970s, computational capabilities increased such that strapdown INS was feasible (strapdown processing requires much more intensive computational capability than a gimballed INS). Since strapdown INS do not have rotating mechanical components, they have significantly higher reliability and thus have largely replaced gimbal systems.

Most recently, microelectromechanical systems (MEMS) technology has been applied to the miniaturization of gyroscopes and accelerometers. These chip-scale devices are used widely in smartphones and, for example, in automobile airbag deployment systems. In the navigation area, these devices were developed for use in artillery shells (they can withstand the tens of thousands of Gs experienced when the shell is fired) but more recently have been utilized in a wider variety of applications. As of the early 2020s, the development of navigation-grade MEMS sensors was well underway.

1.9 Modern navigation systems

Over the past century, a variety of radio-based navigation systems have been developed. The VHF omnidirectional range, a radio beacon typically abbreviated VOR, is a ground-based system that has been used for domestic navigation since approximately World War II. VOR provides a bearing angle from (or to) the ground station. The distance measuring equipment (DME) was developed from early interrogation friend-or-foe (IFF) systems and provides a measure of the line-of-sight distance from the aircraft to the ground transponder. A variety of radar systems have been used for navigation as well as for surveillance. TACAN (TACtical Air Navigation system) is a system used by the military that provides both bearing and distance from the ground station. Typical accuracy for both VOR and TACAN is a few degrees. Both systems have service ranges on the order of 100 miles. Since they are VHF and UHF systems, and hence have short wavelengths, their transmissions are range-limited due to the curvature of the earth.

Historically there have been low frequency navigation aids as well. The Omega system, in operation from the 1960s until the 1990s, operated at 10 kHz. The Loran system operated at 100 kHz. Low frequency transmissions can service much larger coverage volumes (hundreds of miles in the case of Loran; thousands of miles in the case of Omega). There is a downside, though. A rough rule of thumb is that positioning accuracy is proportional to wavelength. Thus a long wavelength (low frequency) system has a relatively larger coverage volume but lower accuracy than can be achieved with a short wavelength (high frequency) system. This accuracy rule-of-thumb stems from the fact that a receiver can generally track a periodic signal with an accuracy of roughly 1% of its wavelength.

The last decade of the 20th century witnessed the maturation of satellite-based navigation. GPS was developed by the United States in the 1970s and 1980s and it was declared operational in the mid-1990s. The Russian GLONASS was developed over approximately the same time period. Over the subsequent two decades additional satellite-based navigation systems were developed by China (Beidou), Europe (Galileo), and India (NAVIC). Collectively the various systems are referred to as global navigation satellite systems (GNSS). Today, GNSS is one of the most critical elements in just about any navigation system that is currently in use or under development. The system has excellent accuracy (on the order of 5–10 m) as long as the receiver has clear line-of-sight to the satellites.

There are some limitations given the fact that the received GNSS signal is inherently weak (relative to the receiver noise floor). Interference and jamming can cause signal disruptions. In addition there are propagation uncertainties as the GNSS signals traverse the atmosphere (ionospheric and tropospheric delays). Furthermore, unwanted reflections (called multipath) also negatively impact performance. Nevertheless, GNSS provides worldwide, 24/7, all-weather navigation capability with unprecedented accuracy.

In contrast to radio navigation systems, inertial navigation does not need to receive an external signal. Even if a barometric altimeter is utilized to stabilize the vertical position/velocity channel, there is still no need to receive any external radio signals. Thus, inertial navigation is very robust due to its ability to operate in a self-contained manner. For most practical purposes it cannot be jammed. Inertial systems have limitations but they are completely independent of the error sources that affect radio navigation systems.

A typical air transport grade (nav grade) inertial navigator will drift on the order of one nautical mile per hour. Thus, if an aircraft departs Los Angeles and takes 10–12 hours to get to Tokyo, the on-board inertial system will be off by approximately 10–12 miles upon arrival. There are inertial systems with worse performance. So-called "tactical grade" systems are one to two orders of magnitude worse in performance than nav grade systems and thus can only be used over relatively short periods of time (e.g., minutes) unless aided by an external source. "Consumer grade" systems are typically a few or several orders of magnitude worse than tactical grade systems. Such systems typically are used to provide attitude and heading or are used to "coast" for a few seconds in between position/velocity fixes provided by other systems such as GNSS.

There are also inertial systems that are much better than nav grade systems. The systems installed on submarines and intercontinental ballistic missiles (referred to as "strategic grade") perform significantly better than one nautical mile per hour (although the specific performance is typically classified). In the case of the missiles, of course, the vehicle has a short flight time. In the case of submarines, the vehicle has the luxury of having sufficient space such that it can house larger instruments that have better accuracy.

1.10 INS and GNSS: made for each other

Integration of INS and GNSS is very common. The reason is because the systems were seemingly made for each other. They are almost perfectly complementary. Each one has its own advantages and disadvantages. By bringing the two together, we can minimize the disadvantages of each, and maximize the benefits that both bring to the table. For example, the inertial system, as we will see, provides very good position/velocity information over short durations (sub-second up to several minutes). This is true even in a high dynamic vehicle. You can put an appropriately engineered inertial system into a fighter aircraft that is rolling at hundreds of degrees per second and/or experiencing many G's of acceleration and the inertial system will work just fine.

Not only is the position/velocity determination of the INS not adversely affected by the vehicle dynamics, the INS also outputs this data with very high data rates and very low data latencies (both of which are important for the flight control system). Flight control systems typically need data rates on the order of 40 independent updates per second, and generally can only tolerate latencies on the order of tens of milliseconds or less. The INS does that very well. Over short periods of time, it can provide high data rates and low data latencies while undergoing high dynamics.

Contrast the INS performance with that of GNSS. In GNSS, the errors are somewhat noisy. Furthermore, it has a low data rate (generally on the order of one sample per second) and it has generally long data latencies (typically an appreciable fraction of a second). However, the advantage that GNSS has is that its errors are stable. If we set up a GNSS receiver at a surveyed location, we would find that it provides position with an accuracy on the order of five meters. If we left and came back an hour later, the accuracy would still be on the order of five meters. If we left and came back a week later, the accuracy would be about the same. The GNSS error does not drift significantly over time.

We want to be able to take advantage of the benefits of the inertial and the GPS. We want the low noise, high dynamic capability of the INS along with the high data rate and low data latency. On the other hand, we want to keep the long-term stability of the GNSS error (no drift!).

Historically, the optimum approach has been to perform the integration (fusion) of the data within a Kalman filter. INS and GNSS have certain error characteristics and they are modeled in the Kalman filter in an algorithm that is able to integrate/blend the measurements both from the INS and the GNSS to provide the optimum determination of position, velocity, and attitude.

1.10.1 The way forward ...

We have begun our study of inertial navigation. Next we will cover the various reference frames that we will use in inertial navigation and learn how to convert between various frames of reference and how we deal with rotations. Then we will discover how to determine the attitude of the vehicle based on the measurements from the gyros. Along the way, we will learn how to take our accelerometer measurements and convert them into the necessary reference frames and ultimately determine velocity and position.

We will develop the algorithms that are necessary in order to do basic inertial navigation. That will be our first step in this process. Then we will take a look at the sensors and learn how the accelerometers and gyros operate. Following this we will study their error characteristics. Once we have looked at the error characteristics, we will learn how to characterize the overall system performance.

Oddly enough, we will finish our study of inertial navigation with the topic of initialization. As we will see, knowledge of the sensors, their error characteristics and the basic inertial position/velocity/attitude determination algorithms is necessary to understand the initialization process.

Once we have covered inertial navigation system processing, we will dive into INS/GNSS integration. It is an exciting journey ...

References

[1] Thomas S. *The Last Navigator: A Young Man, An Ancient Mariner, The Secrets of the Sea*. Camden (ME): International Marine/McGraw-Hill; 1987.

[2] ION Virtual Museum [homepage on the Internet]. Manassas, VA: U.S. Institute of Navigation; c2004. [cited 2021 Aug 17]. Available from: https://www.ion.org/museum/.

[3] Bowditch N. *American Practical Navigator*. Washington (DC): U.S. Navy Hydrographic Office; 1966.

[4] Harford T. *What do Drones and GPS owe to a 1744 Shipwreck?* [homepage on the Internet]. London: BBC; 2019. [cited 2021 Aug 17]. Available from: https://www.bbc.com/news/business-47161370.

[5] MacKenzie D. *Inventing Accuracy: A Historical Sociology of Nuclear Missile Guidance*. Cambridge (MA): The MIT Press; 1990.

[6] Draper CS. Origins of Inertial Navigation. *AIAA Journal of Guidance and Control*. 1981;4(5):449–456.

Chapter 2
Reference frames and the mathematics of rotations

In most instances in inertial navigation, we are going to be concerned with multiple reference frames. For example, the accelerometers may form their measurements with respect to a set of axes that are fixed in the vehicle; but navigation is usually performed with respect to a set of axes that is fixed in the Earth. This is certainly true for applications that are on or near the surface of the Earth.

First of all, we have to understand what these various reference frames are and how they are defined and used within inertial navigation. Once we have defined the various reference frames, then we will need to discuss how to convert between them. As we have discussed previously, Newton's laws are valid in an inertial reference frame and an inertial frame is one that is moving at fixed velocity with respect to the stars. The background stars in our galaxy serve as a reference frame and any frame that is moving at constant velocity (including zero) with respect to the background stars also is an inertial frame. The true inertial frame is the only frame in which Newton's laws are valid.

2.1 Earth-centered inertial frame

It is inconvenient to use the true inertial frame in terrestrial navigation applications. We need an approximation and it is the earth-centered inertial (ECI) frame (also referred to simply as the 'inertial frame' or the i-frame). Figure 2.1 shows the inertial frame with axes labeled x^i, y^i and z^i. The x and y axes lie in the Earth's equatorial plane and the z-axis is approximately coincident with the Earth's rotational axis (the coincidence is not exact since the instantaneous location of the Earth's pole within the Earth's crust varies at the meter-level over the course of even one year).

For space applications, it can be important to define the specific orientation of the x and y axes (e.g., the x-axis points toward Aries: a particular fixed star). For inertial navigation, however, the exact orientation of the i-frame is unimportant. What is important is that the i-frame axes are centered in the Earth (i.e., the origin of the i-frame is at the center of the Earth) but they do not rotate with the Earth. The i-frame is not a truly inertial frame since it travels an elliptical path along with the Earth as the Earth orbits the Sun. The frame thus experiences non-zero acceleration and thus is not truly inertial. For our purposes, however, it is a sufficient approximation.

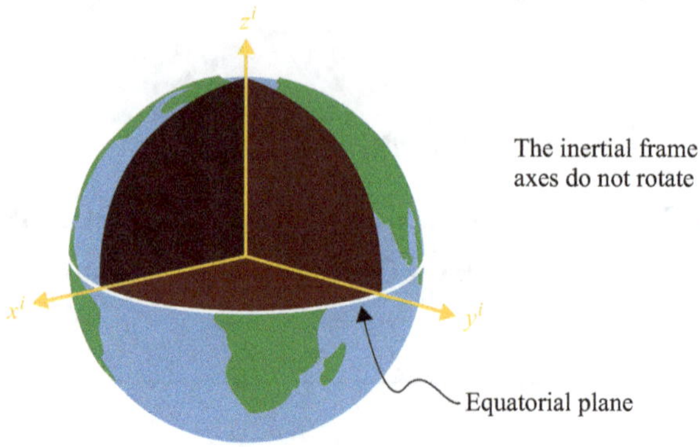

Figure 2.1 The earth-centered inertial frame (ECI or i-frame). The frame is centered in the earth but does not rotate

2.2 Earth-centered Earth-fixed frame

The next frame of interest is the so-called 'earth-centered, earth fixed' (ECEF) frame, also known as the earth frame or the e-frame. In Figure 2.2, the e-frame axes are labeled x^e, y^e and z^e. The e-frame is centered in the Earth and it is also fixed to the Earth. It *does* rotate with the Earth. Clearly the e-frame is not an inertial frame in any way. The x and y axes of the e-frame are in the equatorial plane and the positive x^e axis is defined as having its origin at the center of the Earth and it intersects the equator at the prime meridian (i.e., the line of zero degrees longitude). As with the i-frame, the e-frame z-axis is aligned approximately with the Earth's polar axis. The e-frame y-axis is simply orthogonal to x^e and z^e in order to form a right-hand coordinate system.

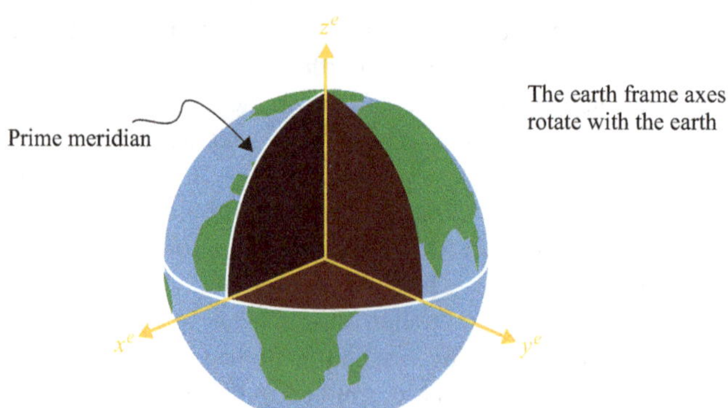

Figure 2.2 The earth-centered, earth-fixed frame (ECEF or e-frame)

2.3 Locally level frames

There are at least three so-called 'locally level' or 'local level' frames used in inertial navigation. All local level frames have two axes that lie in the local horizontal plane (i.e., the tangent plane to the Earth's reference ellipsoid at the location of interest). The origin of local level frames is the location of the vehicle in which the inertial navigation system is installed. The 'navigation frame' (also known as the nav-frame or the n-frame) is one example of a local level frame and is illustrated in Figure 2.3.

Although not a universal convention, in this book the n-frame x-axis points north, the y-axis points east and the z-axis points down (along the local vertical). Thus this n-frame is also known as the north-east-down frame or NED-frame. Another local level frame is the east-north-up (or ENU) frame. As the name implies, the x-axis points east, the y-axis points north and the z-axis points up (along the local vertical). In this book the ENU frame will sometimes be denoted as the L-frame.

As a practical matter, gimbaled inertial navigation systems physically realize a locally level frame. The accelerometers are mounted on a platform that is rotationally isolated (via gimbals and gyroscopes) from the motion of the host vehicle. Ideally, one accelerometer is pointed in a desired direction (e.g., north), another accelerometer is pointed east and the third is pointed in the vertical. However, the platform itself must be rotated, relative to the inertial frame, in order to keep the accelerometers pointed in their reference directions (note that in strapdown inertial systems, an equivalent locally level frame is mechanized in software).

This rotation of the platform (also known as "torquing") is needed to keep the platform locally level and pointed in the correct direction. There are two rotations that must be applied to the platform to keep it locally level and keep the x-axis north pointing (for the case of the n-frame). One is called 'earth rate' and the other is called 'transport rate' or 'craft rate.' Earth rate is the rotational rate that must be applied to the platform to account for the rotation of the Earth itself. To see why this is needed, Figure 2.4 illustrates what would happen if earth rate was ignored.

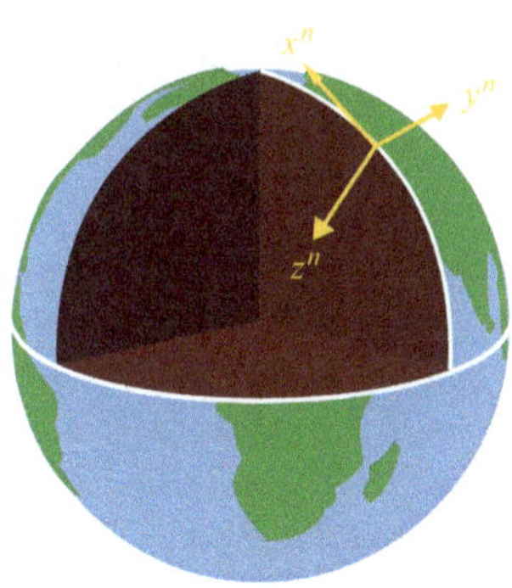

The navigation frame x-axis points north, y-axis points east and z-axis points down

Figure 2.3 The navigation frame (nav-frame or n-frame)

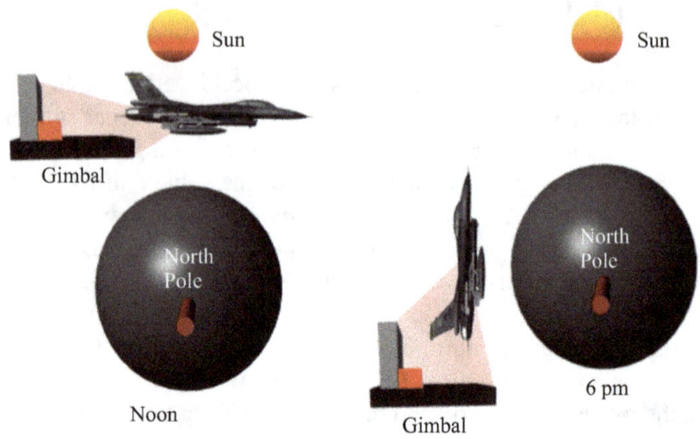

Figure 2.4 Illustration of the effect of ignoring earth rate on a gimbaled platform mounted in a stationary vehicle on the equator. At noon, the platform is locally level just as the host vehicle is. However after six hours, the Earth (and vehicle) have rotated by 90 degrees whereas the gimbaled platform has maintained its original orientation (due to the gyros and gimbals). It is thus no longer locally level

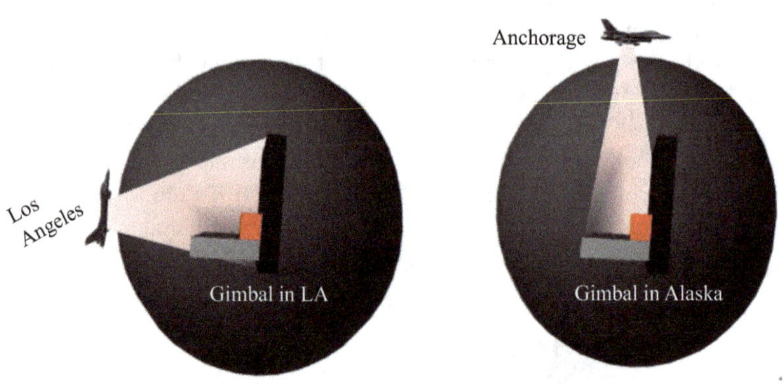

Figure 2.5 Illustration of the effects of ignoring transport rate on a gimbaled platform mounted in an aircraft flying from Los Angeles to Anchorage. Before the aircraft departs Los Angeles, the inertial sensor platform is locally level. However, by the time of arrival in Anchorage, the vehicle has rotated by 27 degrees (the illustration is not to scale) whereas the gimbaled platform has maintained its original orientation. It is thus no longer locally level (Note: the gimbaled inertial system actually resides in the aircraft itself but is shown outside of the vehicle only for the sake of visual clarity)

The second rotational rate that must be applied is transport rate (or craft rate). It is the rotation due to the movement of a vehicle over the curved surface of the Earth. Thus, in a gimbaled inertial system the platform must be physically rotated ('torqued') to account for the motion of the vehicle over the Earth. Figure 2.5 illustrates what happens if transport rate is ignored.

Thus, we see that in order to keep the frame locally level, we have to rotate it with earth rate and with transport rate. The discussion so far has been focused on keeping the frame locally level but one of the axes also must be kept pointed at north (for NED or ENU frames). As the Earth rotates, the local level frame has to be rotated around the local vertical in order to keep one of the axes north pointing. The same is needed as the vehicle traverses over the surface of the Earth.

2.4 Wander Azimuth frame

For vehicles operating in the vicinity of the Earth's poles, a different locally level frame is needed. To understand why, consider a vehicle that just barely skims past, say, the North pole. If it has a mechanical gimbaled inertial system onboard, a motor must rotate the accelerometer platform to keep the north accelerometer pointed towards north. This will not be possible because even for a reasonable finite vehicle velocity, as the vehicle just skims past the North pole, the rate at which the platform must be torqued (about the vertical) to keep the north accelerometer pointed north approaches infinity. In actual practice the rate will easily exceed the motor limits. Even in a strapdown inertial system, double-precision arithmetic in the processor will not be able to accommodate the enormous vertical rotational rate needed to mechanize, virtually, the locally level (and north-pointed) frame.

The solution is conceptually simple. If the problem is that one cannot keep a given axis pointed north, then do not even bother to try. The so-called "Wander frame" or "Wander Azimuth" frame (or w-frame) is a frame that is very similar to the n-frame. The only difference is that its x-axis is not necessarily pointed toward north. It is allowed to 'wander away' from north as the vehicle operates. More details will be provided in a later chapter.

2.5 Vehicle-fixed body frame

The body frame or the vehicle frame (referred to as the b-frame) is depicted in Figure 2.6. The x-axis lies along the longitudinal axis of the vehicle. The y-axis lies along the right wing and the z-axis is pointed down. This is normally referred to as "nose right-wing down."

A logical question would be to ask why do we pick z to be down? Why not pick z to be up? We could do that. It is just a convention, after all. We could pick right wing nose up as our right-hand vehicle coordinate frame and then the z-axis would be up. However, as we will discover later, it is more convenient to define it in terms of nose right-wing down. The reason is that positive rotations (in the right-hand sense) of the

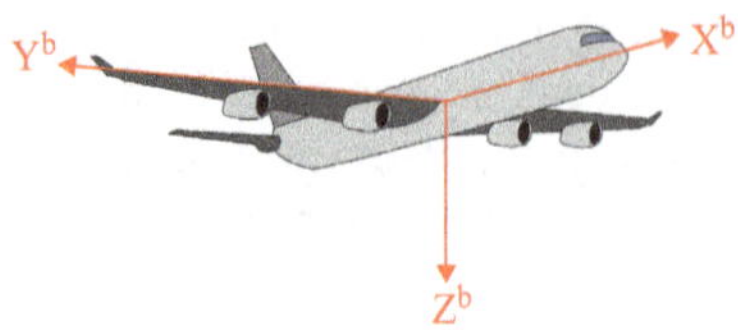

Figure 2.6 The vehicle-fixed body-frame (b-frame)

nose right-wing down frame correspond to positive Euler angles of roll, pitch, and yaw.

2.6 Sensor frames

The last set of frames that need to be considered are the sensor frames. Ideally, the accelerometer and gyro triads would be co-located at the origin of the body frame. That would be very convenient if it were possible. In our subsequent processing, we will assume so just for the sake of convenience. However, this is unrealistic for two reasons. First, the origin of the body frame is close to the center of gravity of the vehicle itself and typically it is loaded with fuel or payload. It is not practical to mount an inertial system at the vehicle origin. The second reason is, even if the space were available, it is not possible to mount all of the sensors in precisely the same spot. They physically cannot occupy the same location. We need three gyros and three accelerometers and they cannot all be mounted in one specific point on the vehicle itself.

As a result, we need to have separate coordinate frames: one for the three accelerometers and one for the three gyros. We can refer to the accelerometer frame as the a-frame and we can refer to the gyro frame as the g-frame. Ideally, we would have a simple relationship between the sensor frames and the body frame. This is similarly unrealistic since it is impossible to mount the three sensors such that they are perfectly orthogonal to each other. Manufacturing imperfections prevent a given sensor triad from being mounted perfectly 90 degrees apart. As a result, the non-orthogonality between the sensors of a given triad is an error source that can partially be compensated through calibration and external aiding. For the remainder of this book, we will assume the calibration process has already been done. Details on calibration may be found in [1]. We will also assume that the gyro and accelerometer measurements have already been converted from the sensor frames into the vehicle body frame.

2.7 The mathematics of rotations

Having defined the various reference frames utilized in inertial navigation, we now turn our attention to the development of a mathematically formal way to describe the angular relationships between different kinds of frames. Along the way, we will see that in the process of doing so, we will be describing the orientation of the vehicle. This topic is rooted in the physics of rigid body kinematics. We will boil it down to the essentials that we need for the topic at hand.

It can be derived formally, but for our purposes suffice it to say that any general motion of a rigid body can be viewed as simply a combination of translation and rotation. Figure 2.7 is an illustration of a vehicle at a particular time, t_k, and also at sometime later time, t_{k+m}. The differences in the position and orientation at the two times can be viewed simply as a translation (a change of position) and a rotation (the angular difference between the orientations at the first and second times).

It is important to understand that the translation and rotation do not necessarily represent the actual path or trajectory that the vehicle maneuvered through. We are not making any statements, yet, about *how* the vehicle got from point A to point B. We are simply describing the difference in the position and orientation *at* the second time with respect to the position and orientation *at* the first time. Translation, at least in two dimensions, is simple. It is given by the difference in the x and y coordinates:

$$\Delta x = x_{k+m} - x_k \tag{2.1}$$

$$\Delta y = y_{k+m} - y_k \tag{2.2}$$

However, rotation is another matter. How do we actually quantify a rotation? For example, if an airplane executes a pitch-up maneuver, how do we describe that mathematically? What is the framework for doing that? First, we need some kind of reference. We need to know what zero degrees of pitch is. What constitutes zero degrees of pitch? The answer is intuitive. A level attitude (level orientation) represents zero degrees of pitch (level also implies zero degrees of roll).

Considering Figures 2.3 and 2.6, if a vehicle is level, and pointed due North, then the body frame is *coincident* with the navigation frame. We can then consider the angular differences between the vehicle (in some arbitrary orientation) and the reference, which is the navigation frame. If the vehicle is rolled over and/or pitched up (or down) and/or yawed, there is an angular difference between the body frame and the navigation frame. In the course of quantifying the angular differences between the body frame and the navigation frame, we are describing the orientation of the

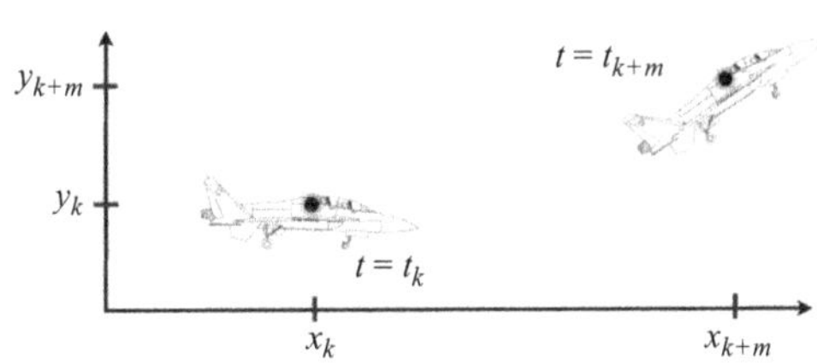

Figure 2.7 Any general motion of a rigid body can be described simply by a translation and a rotation

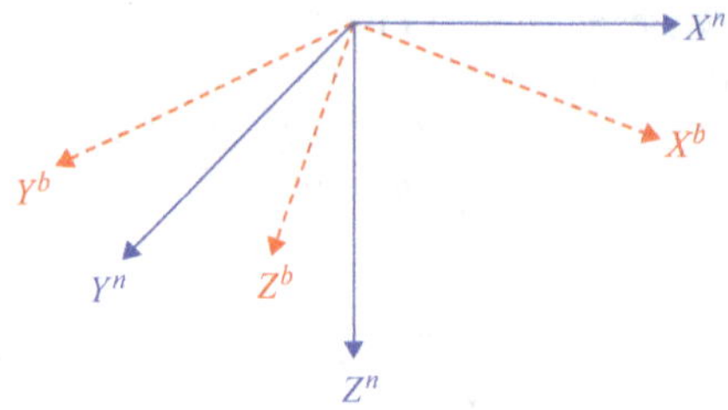

Figure 2.8 A body-frame at an arbitrary orientation with respect to a navigation frame

body. The navigation frame serves as our reference. Whatever angular differences exist between the body frame and the navigation frame, then, describe the orientation of the vehicle.

Our next task is to quantify those rotational differences in a precise manner. The orientation of a vehicle is sometimes also referred to as the vehicle 'attitude.' Sometimes the phrase "attitude and heading" is used. In that case, the term attitude is referring to pitch and roll of the vehicle. In general, the term 'attitude' can comprise not only pitch and roll, but yaw as well. Again, the orientation is defined by the rotational differences between the navigation frame and the body frame.

Consider an arbitrary navigation frame and body frame illustrated in Figure 2.8. There is some arbitrary amount of rotational difference between the two frames. It can be shown [2] that three rotations, about the principle axes, are needed to describe the rotational difference in the general case (for now, consider the 'principle' axes to be the axes of the body frame of the given object). Furthermore, each successive rotation (of the three) cannot be about the same axis. In other words, you cannot rotate around X and then rotate around X again and call that two different rotations. You can rotate about X, and then Y, and then Z. Or, you can rotate about X, then Z, and then X (in the new orientation of the x-axis). There are various ways to do it. In the aerospace community there is a particular convention that we will describe later.

2.8 Principle axis rotation matrices

First consider a rotation around just a single axis. Figure 2.9 illustrates a body frame that has been rotated about the z-axis by an amount ψ (with respect to the navigation frame). The z-axes of the two frames are coincident (and they go "into the paper" as denoted by the cross). In order to describe the rotation formally, we will first consider how an arbitrary vector V is resolved in the two frames. Assume that the components of V in the navigation frame are known: (V_x^n, V_y^n). We will derive the components of V in the body-frame and then will generalize the result.

Thus the task is to derive (V_x^b, V_y^b) given (V_x^n, V_y^n) and the rotation angle ψ. This is done by projection. Figure 2.10 illustrates how V_x^b is derived. The navigation

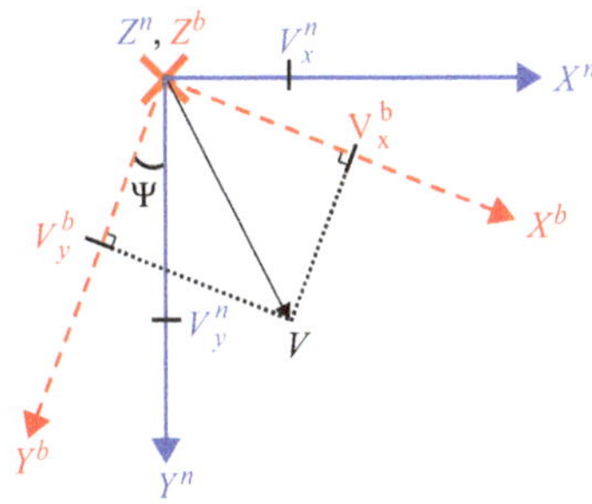

Figure 2.9 *Rotation by ψ around the z-axis of the b-frame. The vector V is defined in terms of its components in the n-frame. The projection of V onto the b-frame axes is shown*

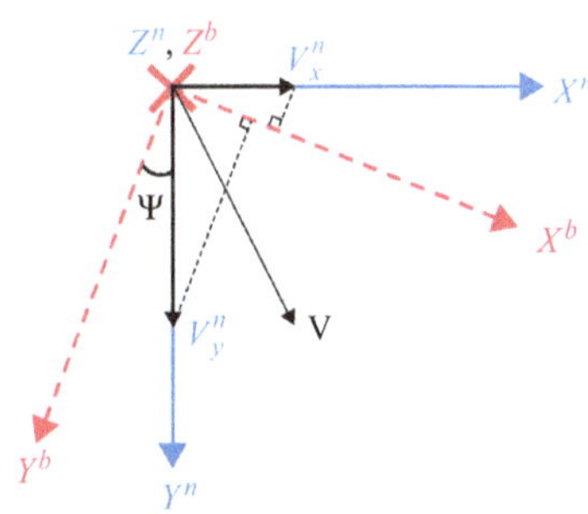

Figure 2.10 *Deriving V_x^b by projection of V_x^n and V_y^n onto the b-frame x-axis*

frame components, V_x^n and V_y^n, are projected onto the body-frame x-axis. The two components are then summed to get the final result:

$$V_x^b = V_x^n \cos\psi + V_y^n \cos\left(\frac{\pi}{2} - \psi\right) \tag{2.3}$$

$$= V_x^n \cos\psi + V_y^n \sin\psi \tag{2.4}$$

Following a similar methodology, the body-frame y-axis component of V can also be derived:

$$V_y^b = -V_x^n \sin\psi + V_y^n \cos\psi \tag{2.5}$$

For completeness, we recall that we were rotating around the z-axis (and the z-axes of the navigation and body frames were coincident). Thus, it follows:

$$V_z^b = V_z^n \tag{2.6}$$

We can bring these three equations together in a simple matrix form:

$$\begin{bmatrix} V_x^b \\ V_y^b \\ V_z^b \end{bmatrix} = \begin{bmatrix} \cos\psi & \sin\psi & 0 \\ -\sin\psi & \cos\psi & 0 \\ 0 & 0 & 1 \end{bmatrix} \begin{bmatrix} V_x^n \\ V_y^n \\ V_z^n \end{bmatrix} \tag{2.7}$$

This can be written simply as:

$$V^b = C_z(\psi)V^n \tag{2.8}$$

where $C_z(\psi)$ denotes a rotation of ψ around the z-axis:

$$C_z(\psi) = \begin{bmatrix} \cos\psi & \sin\psi & 0 \\ -\sin\psi & \cos\psi & 0 \\ 0 & 0 & 1 \end{bmatrix} \tag{2.9}$$

This matrix is known as a 'rotation matrix' since it describes the rotational difference between two frames. Following similar derivations, the rotation matrices describing rotations around the y and x axes can also be derived:

$$C_y(\theta) = \begin{bmatrix} \cos\theta & 0 & -\sin\theta \\ 0 & 1 & 0 \\ \sin\theta & 0 & \cos\theta \end{bmatrix} \tag{2.10}$$

$$C_x(\phi) = \begin{bmatrix} 1 & 0 & 0 \\ 0 & \cos\phi & \sin\phi \\ 0 & -\sin\phi & \cos\phi \end{bmatrix} \tag{2.11}$$

The letter "C" is used for these matrices since they are also known as direction *cosine* matrices (sometimes denoted 'DCM').

2.9 Multiple rotations in succession

Now we need to describe mathematically what happens when we rotate around one axis and then rotate around another axis. More generally, how do we describe the difference between two frames that cannot be handled by a single rotation about a single principle axis? It turns out that successive rotations around non-common axes can be described by simply cascading these direction cosine matrices.

For example, if we rotate first around z and then around the y-axis, mathematically it can be written as:

$$V^b = C_y(\theta)C_z(\psi)V^n \tag{2.12}$$

Note that we did not have to derive a brand new matrix for the rotate-around-z-then-rotate-around-y case. All we needed to do was cascade the individual rotation matrices. However, note the order of the matrices. The z-rotation matrix is immediately to the left of the vector since we specified that the first rotation was around z. The order is absolutely critical and this can be proven very easily with something like a Rubik's Cube (or any box that has some separately distinguishing feature on each of the six faces). Imagine a three-axis coordinate system affixed to the box. Rotate the box 90 degrees around the z-axis, then rotate it 90 degrees around the y-axis and then note the final orientation of the box. Now start all over with the box in its original orientation. This time, however, rotate around the y-axis first and then rotate around the z-axis. You will find that the box is in a completely different orientation. Thus we see that finite rotations do not commute (we will discuss infinitesimal rotations later). In fact, this is a physical proof of something you already knew: matrix multiplication does *not* commute. In equation form:

$$C_y(\theta)C_z(\psi)V^n \neq C_z(\psi)V^n C_y(\theta) \tag{2.13}$$

The fact that finite rotations do not commute implies that an order (a convention) must be specified for the aforementioned three-rotations needed to describe the orientation of one frame relative to another. We have not derived it formally, but it can be shown that these rotation matrices represent what are called 'orthogonal transformations' and the matrices are orthogonal matrices [2,3]. An orthogonal matrix is a square matrix with orthonormal columns [4]. This means that each column of the matrix (think of each column as a vector) has a magnitude of unity and is orthogonal to the other columns (i.e., their dot products are zero). This is important because of the following property of orthogonal matrices [2,3]:

$$C^{-1} = C^T \tag{2.14}$$

As we will see, rotation matrix inversions are utilized extensively in inertial navigation processing. This property, though, means that we only need to compute a transpose when we need an inverse. This is a big deal since matrix transposition is a computationally simple operation whereas matrix inversion is much more computationally intensive.

2.10 Reverse transformation

Consider the following. A certain rotation matrix transforms a vector from the navigation frame to the body frame:

$$V^b = CV^n \tag{2.15}$$

Now, what is the rotation matrix that performs the reverse operation? What matrix transforms a vector from the body frame to the navigation frame? The derivation is quite simple. Pre-multiply both sides of (2.15) by the inverse of C:

$$C^{-1}V^b = C^{-1}CV^n \tag{2.16}$$

Since we know that:

$$C^{-1}C = I \tag{2.17}$$

It thus follows that (2.16) can be written as:

$$C^{-1}V^b = C^{-1}CV^n = IV^n = V^n \tag{2.18}$$

And therefore:

$$V^n = C^{-1}V^b = C^T V^b \tag{2.19}$$

Thus, if a given rotation matrix transforms vectors from one frame to another. The opposite transformation is simply given by the transpose of the matrix.

2.11 Euler angles: the aerospace convention

So far we have discussed the fact that finite rotations do not commute and that a minimum of three rotations around principle axes are needed to describe an arbitrary orientation. Now consider a vehicle (say, an airplane) that is at some arbitrary orientation with respect to the navigation frame. What three rotations do we use to describe the airplane's attitude? Conceptually, we can think of starting with the airplane (actually the imaginary b-frame within the airplane) at coincidence with the navigation frame (i.e., level and pointing North). What rotations around the b-frame axes are needed that would result in the given orientation of the vehicle?

Within the aerospace community, the convention that is used is a rotation first around the body z-axis, then a rotation around the body y-axis and finally a rotation around the body x-axis. Why is this convention chosen? Doing so makes the rotations correspond to the aerospace Euler angles: yaw, pitch and roll. Yaw is defined as a rotation about the local vertical (when the vehicle is level). Pitch is defined as a rotation about the body y-axis (i.e., wing axis) starting with the vehicle in a level orientation. Finally, roll is defined as a rotation about the body x-axis (i.e., tail-to-nose axis). We could specify a different convention (e.g., x first, y second, and z third) but the angles would then no longer correspond to yaw, pitch and roll (it should be noted, just in passing, that physicists use other conventions depending upon what is convenient for their application).

I distinctly remember when I was first learning these principles. I was deeply troubled by, what I eventually learned was, a misconception. I said, "Look, no airplane flies like this. It does not yaw first, then pitch, and then roll. Airplanes do not do that. They have a general fluid motion as they're maneuvering." What I did not appreciate at first was that this order of rotations has *nothing* to do with *how* the vehicle got into a particular orientation. General rotation matrices (or the Euler angles) are simply a representation of a vehicle's instantaneous attitude; and that attitude is simply the rotational difference between the vehicle's body frame and a chosen frame of reference (typically, the navigation frame or some other locally level frame).

Mathematically, the aerospace rotation convention yields the following rotation matrix:

$$C_n^b = C_x(\phi)C_y(\theta)C_z(\psi) \tag{2.20}$$

This is the direction cosine matrix that transforms a vector from the navigation frame to the body frame. It is sometimes referred to as the "nav-to-body direction cosine matrix" or simply the "nav-to-body DCM."

In practice, though, what we normally will use within an inertial system is the reverse. The "body-to-nav direction cosine matrix" is needed to transform the accelerometer outputs from the body frame to the navigation frame. Although gimbaled inertial systems are convenient for describing principles conceptually, the vast majority of inertial systems produced today are strapdown. In a strapdown system, the accelerometers are hard-mounted to a case that is hard-mounted to the vehicle. In this book we assume that the accelerometer (and gyro) outputs have already been transformed from the sensor frame to the body frame.

The specific-force measured by a triad of accelerometers can be represented by:

$$f^b = \begin{bmatrix} f_x^b \\ f_y^b \\ f_z^b \end{bmatrix} \tag{2.21}$$

where the superscript refers to the frame that the vector components are being expressed in. To transform this vector to the navigation frame:

$$f^n = C_b^n f^b \tag{2.22}$$

We thus need the inverse (or transpose) of the nav-to-body DCM given in (2.20):

$$C_b^n = (C_n^b)^T = (C_x(\phi)C_y(\theta)C_z(\psi))^T = [C_z(\psi)]^T[C_y(\theta)]^T[C_x(\phi)]^T \tag{2.23}$$

Note we have utilized the matrix identity that the transpose of a product of matrices is equal to the product of the individual transposed matrices in reverse order. Substitution of (2.9), (2.10), and (2.11) into (2.23) yields (after much algebra):

$$C_b^n = \begin{bmatrix} \cos\theta\cos\psi & -\cos\phi\sin\psi + \sin\phi\sin\theta\cos\psi & \sin\phi\sin\psi + \cos\phi\sin\theta\cos\psi \\ \cos\theta\sin\psi & \cos\phi\cos\psi + \sin\phi\sin\theta\sin\psi & -\sin\phi\cos\psi + \cos\phi\sin\theta\sin\psi \\ -\sin\theta & \sin\phi\cos\theta & \cos\phi\cos\theta \end{bmatrix} \tag{2.24}$$

In practice, the body-to-nav DCM is computed/updated with an algorithm that processes the gyro outputs (we will go through the details in a subsequent chapter). Equation (2.24) thus *does not* give us the algorithm for determining the body-to-nav DCM in real time. Rather, what it does is give us the relationship between the DCM and the Euler angles. Thus, as the vehicle is moving along, the inertial software is processing the gyro outputs and is updating the body-to-nav DCM (primarily for the purpose of transforming the accelerometer measurements from the body frame to the navigation frame). If the Euler angles are desired (say, for display to the pilot), then they can be extracted as follows:

Roll:

$$\phi = arctan2(C_b^n[3,2], C_b^n[3,3]) \tag{2.25}$$

Pitch:

$$\theta = arcsin(-C_b^n[3,1]) \tag{2.26}$$

Yaw:

$$\psi = arctan2(C_b^n[2,1], C_b^n[1,1]) \tag{2.27}$$

Four-quadrant arctangent functions are needed to extract roll and yaw since these angles span from −180 deg to +180 deg (pitch, by contrast, only spans −90 deg to +90 deg).

2.12 Quaternions

At this point, we have described two different attitude representations: direction cosine matrices and Euler angles. We can think of these as two different parameter sets used to describe attitude. There is another representation that is also utilized extensively in inertial navigation. The parameter set is known as quaternions. Before we dip our toes into the messy quaternion mathematics, let us understand the big picture.

Although any arbitrary attitude can be described by three rotations around the principle axes, it can also be shown that there exists a single-axis-of-rotation, and a corresponding rotation angle about that axis, that also completely describes a given attitude [2,3] (also known as the "axis-angle" representation). The catch, of course, is finding that special axis (since it will not, in general, correspond to a principle axis) and then also finding the amount of rotation. As an example, take a Rubik's cube, imagine a body-frame affixed to it, and start with an initial orientation. Then rotate the cube 90 degrees around the body z-axis and then rotate it 90 degrees around the body x-axis. It can be shown that the equivalent single-axis-of-rotation lies along the main 'body diagonal.' That is, it is an axis that passes through the origin of the body frame and is positive in the direction of (1,1,1) in the body frame. Also, the amount of rotation about this axis can be shown to be 120 degrees (go get a box and try it!).

So, given the fact that attitude can be described by a single-axis-of-rotation and a corresponding angle of rotation, how do quaternions fit in? First, three parameters are needed to describe the orientation of the axis-of-rotation. Then a fourth parameter is needed to describe the amount of rotation. We can define a single-axis-of-rotation vector with four parameters:

$\overline{\mu}$ = single-axis-of-rotation vector

$|\overline{\mu}| = \mu$ = angle of rotation

$\hat{\mu} = \frac{\overline{\mu}}{\mu} = (\hat{\mu}_x, \hat{\mu}_y, \hat{\mu}_z)$ = unit vector in the direction of $\overline{\mu}$

Thus, the four parameters are the magnitude of the vector and the three components of the unit vector. This representation of the single-axis-of-rotation vector is accurate and complete, but it does not lend itself to cascading. That is, it is not sufficient for the purpose of representing multiple rotations about non-common axes. To do so, we need quaternions.

Strictly speaking, a quaternion is an extension of the concept of complex numbers:

$$q = q_0 + q_1 i + q_2 j + q_3 k \tag{2.28}$$

The quaternion is "q" and "i," "j," and "k" are all imaginary numbers with the following properties:

$$\begin{array}{lll} i \cdot i = -1 & j \cdot j = -1 & k \cdot k = -1 \\ i \cdot j = k & j \cdot k = i & k \cdot i = j \\ j \cdot i = -k & k \cdot j = -i & i \cdot k = -j \end{array}$$

Each of the imaginary numbers (i, j, k) behave as you would expect. The square of each is negative one. When two of them are multiplied against each other, however, the result is the third one (or its negative). This gives a hint about how quaternions are used to represent rotations in 3D space. In (2.28), q_0 is referred to as the "scalar part" of the quaternion and the remainder is referred to as the "vector part."

The coefficients in (2.28) (q_0, q_1, q_2, q_3) are all real numbers. The easiest way to understand this four-element 'number' is first to review the conventional complex number: $c = a + bi$ (where i is the square-root of -1). "a" is the real part of "c" and "b" is the imaginary part. We can plot out c by in a two-dimensional coordinate system with the x-axis being the real axis and the y-axis being the imaginary axis (this is referred to as the 'complex plane'). A conventional complex number thus has two dimensions. A quaternion is an extension of this. It is a four-dimensional number with one real axis and three orthogonal imaginary axes. Although an entire set of rules exist for quaternion mathematics, we will only need one for our use of them in inertial navigation. The product of two quaternions can be expressed in matrix form as follows [3]:

$$q * p = \begin{bmatrix} q_0 & -q_1 & -q_2 & -q_3 \\ q_1 & q_0 & -q_3 & q_2 \\ q_2 & q_3 & q_0 & -q_1 \\ q_3 & -q_2 & q_1 & q_0 \end{bmatrix} \begin{bmatrix} p_0 \\ p_1 \\ p_2 \\ p_3 \end{bmatrix} \tag{2.29}$$

where * indicates quaternion multiplication. It can be shown that the single-axis-of-rotation vector described earlier can be represented with a quaternion as follows [5]:

$$q = \begin{bmatrix} q_0 \\ q_1 \\ q_2 \\ q_3 \end{bmatrix} = \begin{bmatrix} \cos\frac{\mu}{2} \\ \hat{\mu}_x \sin\frac{\mu}{2} \\ \hat{\mu}_y \sin\frac{\mu}{2} \\ \hat{\mu}_z \sin\frac{\mu}{2} \end{bmatrix} \tag{2.30}$$

Note that, by convention, the coefficients are shown but not the imaginary numbers (i, j, and k). Note (2.30) illustrates that q_0 is the scalar part and (q_1, q_2, q_3) is the vector part.

At this point you may be thinking, "Okay, fine. There is a single-axis-of-rotation and it can be represented by a quaternion. It's another way to represent orientation.

So what? Why do I care?" The motivation for the use of quaternions lies in both their stability and their numerical efficiency. Euler angles are a three-parameter set for orientation. Direction cosine matrices appear to be a nine-parameter set for orientation. In fact, it can be shown that only six of the elements are independent and the other three can be derived from the six. Thus, a direction cosine matrix is a six-parameter set for describing orientation. Therefore you need three Euler angles or four quaternion parameters or six direction cosine matrix elements to represent attitude.

From an efficiency point of view, one is tempted to think, "Well, why do not you use Euler angles because you only have three parameters to deal with instead of four or six?" The reason that you cannot, in general, use Euler angles is because they are not always stable. One way to think about this is to notice what happens when pitch goes up to 90 degrees. If we take our airplane and we pitch it up to 90 degrees, then what happens to yaw and roll? They are meaningless. What is yaw in this orientation? Yaw, by definition, is an angle about the vertical but if you are pointing straight up then a rotation about the vertical also corresponds to roll. We thus have a singularity when pitch is equal to 90 degrees (or −90 degrees).

We can see this mathematically by considering either (2.25) or (2.27) for the case of pitch equal to 90 degrees. Let us consider the equation for roll and make the appropriate substitution from (2.24). What happens at pitch equal 90 degrees? The cosine of 90 degrees is zero and thus we are attempting to find the arctangent of 0 over 0. This, of course, is indeterminate. Note that you would get the same result with the yaw equation. In actual practice, if one represented attitude solely with Euler angles, the attitude updating algorithm in the inertial system would 'blow up' (i.e., produce erroneous results) if the vehicle pitch approached 90 degrees.

Thus, although Euler angles are efficient in terms of only needing three parameters to describe orientation, they are not stable. Although we have not yet discussed the updating algorithms, it turns out that both quaternions and direction cosine matrices can be updated throughout all attitude maneuvers without blowing up.

Of the two representations that are stable at all attitudes, quaternions are more efficient numerically. You might be thinking, "Four parameters, six parameters. Who cares? What difference does it make?" In modern times, it is certainly less of a concern, but throughout the initial development of strapdown navigation in the 1970s the computational power was limited at best so there was great motivation to be as numerically efficient as possible. Reducing the number of parameters to update from six to four is an improvement of one-third in efficiency and that is not trivial. Another point to keep in mind is that the attitude update algorithm in inertial systems, particularly in high dynamic vehicles, is performed hundreds of times per second. Even in modern processors that can be a taxing load and the use of quaternions instead of direction cosine parameters can be helpful.

Quaternions that are used to represent attitude have another nice property and that is their magnitude is unity [5]. This is convenient when the time comes to deal with things like truncation and round off errors (a non-trivial issue when the quaternion update algorithm is being run at, say, 500 times per second for hours upon hours). This property of quaternions can be used to renormalize and keep them in their proper form.

2.13 Additional quaternion equations

We will wrap up by summarizing two useful equations that relate Euler angles, quaternions, and direction cosine matrices [3,5]. Quaternions can be computed in terms of Euler angles via:

$$q_0 = \cos\frac{\phi}{2}\cos\frac{\theta}{2}\cos\frac{\psi}{2} + \sin\frac{\phi}{2}\sin\frac{\theta}{2}\sin\frac{\psi}{2} \tag{2.31}$$

$$q_1 = \sin\frac{\phi}{2}\cos\frac{\theta}{2}\cos\frac{\psi}{2} - \cos\frac{\phi}{2}\sin\frac{\theta}{2}\sin\frac{\psi}{2} \tag{2.32}$$

$$q_2 = \cos\frac{\phi}{2}\sin\frac{\theta}{2}\cos\frac{\psi}{2} + \sin\frac{\phi}{2}\cos\frac{\theta}{2}\sin\frac{\psi}{2} \tag{2.33}$$

$$q_3 = \cos\frac{\phi}{2}\cos\frac{\theta}{2}\sin\frac{\psi}{2} - \sin\frac{\phi}{2}\sin\frac{\theta}{2}\cos\frac{\psi}{2} \tag{2.34}$$

Direction cosine matrix expressed in terms of body-to-nav quaternion elements:

$$C_b^n = \begin{bmatrix} (q_0^2 + q_1^2 - q_2^2 - q_3^2) & 2(q_1q_2 - q_0q_3) & 2(q_1q_3 + q_0q_2) \\ 2(q_1q_2 + q_0q_3) & (q_0^2 - q_1^2 + q_2^2 - q_3^2) & 2(q_2q_3 - q_0q_1) \\ 2(q_1q_3 - q_0q_2) & 2(q_2q_3 + q_0q_1) & (q_0^2 - q_1^2 - q_2^2 + q_3^2) \end{bmatrix} \tag{2.35}$$

References

[1] Rogers RM. *Applied Mathematics in Integrated Navigation Systems*. 3rd ed. Reston, VA: American Institute of Aeronautics and Astronautics; 2007.

[2] Goldstein H. *Classical Mechanics*. 2nd ed. Reading, MA: Addison Wesley; 1980.

[3] Kuipers JB. *Quaternions and Rotation Sequences*. Princeton, NJ: Princeton University Press; 2002.

[4] Stewart GW. *Introduction to Matrix Computations*. Orlando, FL: Academic Press; 1973.

[5] Titterton DH, Weston JL. *Strapdown Inertial Navigation Technology*. 2nd ed. Herts, UK: The Institution of Electrical Engineers; 2004.

Chapter 3
Inertial processing in a fixed reference frame

3.1 Introduction

In the last chapter, we established a set of reference frames and studied the mathematics of rotations. We have laid the necessary groundwork for the development of inertial navigation: the method by which we take the accelerometer and gyro measurements to update position, velocity, and attitude.

As described previously, if we have a triad of orthogonal accelerometers and we can keep them in, say, a locally level orientation, one pointed north, one pointed east, one pointed vertically, then the integral of their outputs produces north, east and vertical velocity. If we integrate again, we get: (1) north–south displacement (latitude), (2) east–west displacement (longitude), and (3) vertical displacement. Recall there are two different techniques for keeping a triad of accelerometers pointed in reference directions. We can have mechanical isolation of the accelerometer platform from the vehicle itself so that, regardless of the orientation of the vehicle, the accelerometers are always pointed in the reference directions (e.g., north, east, and vertical). That is referred to as a gimbaled system or a platform system. The alternative is to have the accelerometers hard mounted ("fixed") to the vehicle itself. In so doing, the accelerometers will be outputting their values in the body frame and thus we will need to know the orientation of the vehicle so that those accelerometer values can be converted from the body frame to the reference frame of interest (e.g., the navigation frame). That is referred to as a strapdown system. Recall that either direction cosine matrices or quaternions can be used to represent all orientations (as opposed to Euler angles which have singularities).

3.2 Acceleration in an inertially fixed frame

We are going to start down the road towards developing the inertial navigation *mechanization equations*. This is a fancy term that just refers to the equations needed to update position, velocity, and attitude. We will start off with a simple case. At this point we will not worry about the curvature of the Earth and the fact that the Earth rotates. We will deal with those issues later. For now, we will develop the process whereby we take the gyro measurements and update our direction cosine matrix so that we can translate our accelerometer measurements from the body frame to the

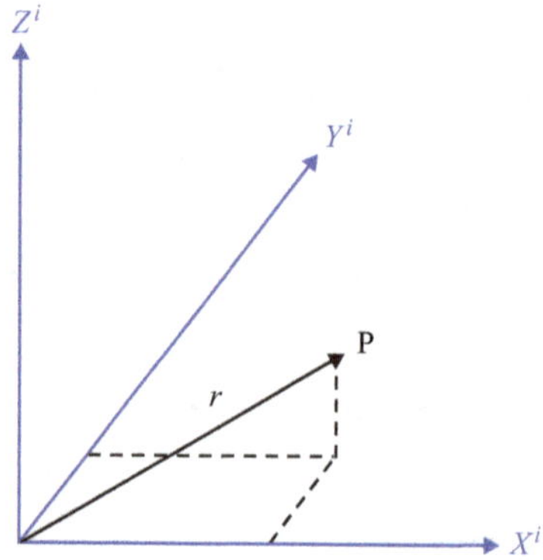

Figure 3.1 A point defined in the inertial frame by position vector **r**

navigation frame (or some locally level frame) and then use that in order to perform inertial navigation within what we call a *fixed* frame of reference. You could imagine the vehicle is in outer space some place and is navigating with respect to an inertial reference frame.

If that seems a little bit too hard to grasp, then *imagine* a flat Earth that does not rotate. If the Earth is flat, we do not have to worry about curvature and if it does not rotate, then a coordinate frame fixed to the Earth will be an inertial frame. However you want to look at it, the point is that we will start by taking the simple case of navigating with respect to a so-called inertially fixed frame or simply a 'fixed frame.'

First of all, we have to deal with the effect gravity will have on our accelerometer measurements. The output of an accelerometer is a measure of specific force but, as we will show now, gravity affects this measurement. Consider Figure 3.1 depicting a point P defined by position vector $\underline{r}$. From our study of physics, the acceleration in the inertial frame is given by the second derivative of the position vector $\underline{r}$ with respect to time:

$$\underline{a}_i = \frac{d^2}{dt^2}\underline{r}\bigg|_i \tag{3.1}$$

where the vertical bar with the subscript 'i' indicates the derivative is being evaluated in the inertial frame.

3.3 The accelerometer sensing equation

We have referred to accelerometers quite a bit up to this point without having a clue how they actually operate and how gravity plays a role. Consider the highly simplified depiction of an accelerometer shown in Figure 3.2. It consists of a so-called "proof mass" suspended between two springs (both partially in tension) that are attached to a case. The proof mass is able to displace. If you grabbed it, you could move it back

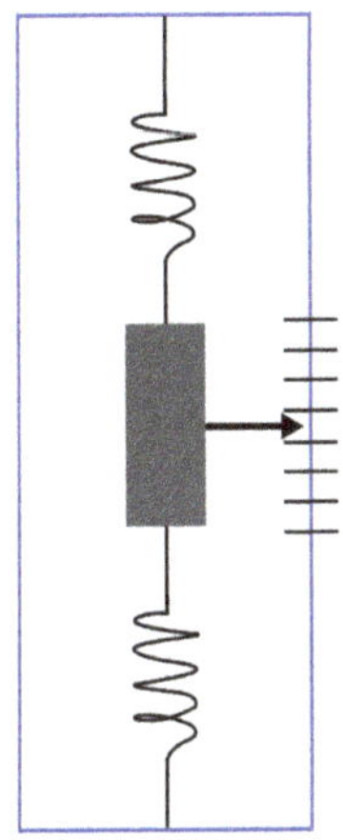

Figure 3.2 *A simplified view of an accelerometer. The central rectangle represents the "proof mass" that is suspended from the case by two springs that are partially in tension. In order to measure displacement of the proof mass relative to the case, an indicator (arrow) and gradations are provided. The so-called "sensitive axis" of the instrument lies along the line joining the two points-of-attachment of the springs*

and forth because the springs are partially in tension. Assume this accelerometer is laying on a level surface (e.g., a table). Let us also assume the springs are perfectly balanced so that the proof mass is sitting/suspended right in the middle and the arrow and gradations indicate there are no net forces acting on it.

If we push hard on the accelerometer case in the direction of the sensitive axis (i.e., the line connecting the mass and springs to the case), the mass would displace in the opposite direction, relative to the case. If the accelerometer is laying on the table and does not have any forces acting upon it, the proof mass is located in the middle (thus indicating a zero reading). If we push on it, the mass displaces in proportion to the amount of force. This sounds like exactly what we want. We have a device that can actually measure the force acting upon it. Since force is proportional to acceleration (recall $F = ma$), we thus have an accelerometer.

However, before we think our job is done, we need to consider some other cases. For example, what happens if this accelerometer was in free fall? If we took it to the top of a building and dropped it (presume that somehow it manages to fall straight down in the orientation shown in the figure), what will happen? If you remember your physics, you know that we would actually measure no force. The proof mass would be in the center of the case and the reading (output) would be zero. That is because gravity is pulling equally on the proof mass and the case so the entire device will simply fall down at the same rate. If you were able to read it while it was falling down what you would see is that the proof mass would be in the neutral position. You would measure nothing even though, relative to the earth, you would be accelerating with the force of gravity (about 9.8 m per second squared).

Let us consider a couple of other scenarios. If we had this accelerometer way out in deep space, say, and there is no gravity. Let's say it is oriented vertically (whatever that means in deep space) and we push up. The force causes the case to accelerate vertically, the proof mass displaces and the accelerometer thus acts as we want it to. However, what would happen if we set the accelerometer on its end on the surface of a table on Earth, and then walked away from it? The proof mass will displace relative to the case even though the case is not accelerating. Gravity is going to pull on the proof mass and cause it to displace. The accelerometer will give us a reading as if it was out in deep space and was being pushed up and accelerated vertically.*

The point is that the accelerometer actually measures a combination of the effects of true Newtonian acceleration and the effects of gravity. We can combine these effects in what is referred to as the "accelerometer sensing equation":

$$\underline{f} = \frac{d^2}{dt^2}\underline{r}\Big|_i - \underline{g} \tag{3.2}$$

where "f" is the measured specific force.† The specific force is equal to the Newtonian acceleration (the second derivative of the position vector with respect to time, evaluated in the inertial frame), minus gravity. Essentially the specific force is equal to the acceleration minus gravity (hold in mind the previous footnote). So how did we get this equation? The justification is as follows:

Consider the scenario where the accelerometer is lying flat on a table. In that case, the sensitive axis of the accelerometer is perpendicular to gravity so it will not be affected by gravity. Since the accelerometer is stationary, there is no applied force and thus no acceleration and therefore the measured specific force would be zero (since the acceleration is zero and the component of gravity along the sensitive axis is zero).

Now, if the accelerometer is placed on its end on the table, it is still not accelerating and thus the first term in (3.2) is zero. However, we know the proof mass will displace and the accelerometer will give a non-zero measurement. But notice the minus sign in (3.2): in this scenario where the accelerometer is sitting on the table, it measures the negative of gravity. Since gravity points in the down direction, the negative of gravity clearly is in the up direction. The proof mass is displaced downward just as it would if it had been pushed/accelerated upward. Thus, the accelerometer sensing equation covers this scenario. It handles the case where the accelerometer is laying on its side on the table and it handles the case where the accelerometer is sitting vertically on the table.

What happens in free fall? The case is accelerating downward, at +g, but this is canceled by the negative g in the equation and the total measured specific force

*Note that for the moment we are considering the classical view of gravity and are not looking at the situation relativistically.

†The measurement is specific force and not force since the measurement is independent of, or normalized by, mass (e.g., if $F = ma$, then dividing by m yields: $F/m = f = a$).

is zero. As we described earlier, in free fall the accelerometer will measure zero. At this point we must make clear that the negative g in (3.2) is actually the "reaction to gravity" and not the negative of gravity. What is the difference? If the accelerometer is sitting on its end on a table, it is actually measuring the force of the table holding the accelerometer in place. It is not measuring the force of gravity pulling the proof mass down. If the relativistic physics here are confusing, just try to remember that accelerometers measure specific force and that specific force is the sum of Newtonian acceleration and the reaction to gravity (Groves [1] notes, for example, that for an aircraft in flight, the reaction to gravity is known as lift).

In the most general scenario, of course, the accelerometer will experience both true Newtonian acceleration and the "reaction to gravity" (i.e., $-g$) and will thus measure a combination of the two. Notice in (3.2) that $\underline{f}$ and $\underline{r}$ and $\underline{g}$ are all underlined to indicate that all three are vector quantities. Thus, the vector sum will give the effective/measured specific force.

We now understand the fundamentals of accelerometer operation. However, we are not quite so interested in the specific force that we measure as we are in determining the Newtonian acceleration. We can take the accelerometer sensing equation and we can rearrange it to solve for the Newtonian acceleration:

$$\frac{d^2}{dt^2}\underline{r}\bigg|_i = \underline{a}_i = \underline{f} + \underline{g} \tag{3.3}$$

Thus, the desired acceleration is given by the sum of the measured specific force and gravity. The 3D-specific force vector is measured by the accelerometers. The gravity vector, however, is something that has to be modeled. Some kind of algorithm is needed to approximate the value of gravity at any particular location on or over the surface of the earth. Although closed-form algorithms that approximate gravity are widely utilized, ultra-high precision gravity models typically involve extensive look-up tables. This is due to the fact that the Earth is not homogeneous in its mass density. Mountains and ocean trenches, for example, induce minor variations in magnitude and direction of the gravity vector.

3.4 The navigation equation for a fixed frame

In inertial navigation, the so-called 'navigation equation' is the equation that specifies the acceleration vector in the frame of interest. In an inertially fixed frame, the navigation equation is given by (3.3). Starting with the navigation equation, we can derive discrete-time updates to determine velocity and position. Note that three orthogonal accelerometers are needed to measure the specific force vector. Virtually all accelerometers are single degree-of-freedom sensors: they only measure force in one direction.

The position/velocity/attitude updating process for the fixed-frame case is depicted in Figure 3.3. The data flow starts on the top with the accelerometers and gyroscopes. We are assuming a strapdown INS and thus these sensors are hard-mounted to the body of the vehicle itself. The diagram is shown in a continuous time format

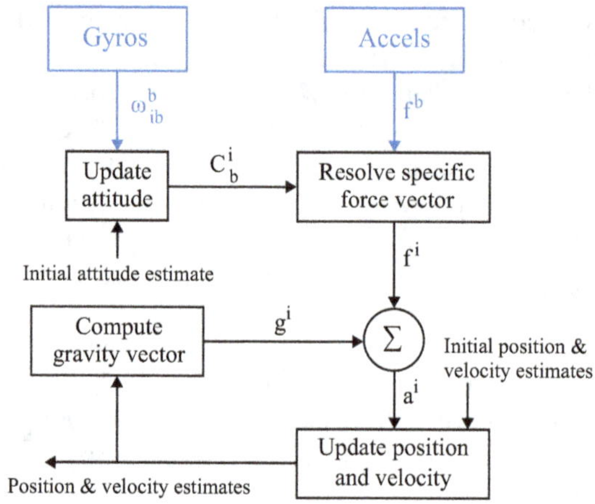

Figure 3.3 Flow diagram for inertial navigation in an inertially fixed frame

(shortly we will develop the discrete time update algorithms). The body-mounted accelerometers output a measurement of specific force and the superscript "b" indicates the measurement is expressed in the body frame. This can be a significant point of confusion and it bears further explanation. The accelerometer triad measures specific force with respect to the inertial frame but the measurements themselves are *expressed* in the body-frame. Think about it this way: any triad of accelerometers will measure the specific force vector. However, depending upon the orientation of the triad, the three particular components will differ. For example, say a triad of accelerometers is stationary with the x accel pointed up. In this case, the measured specific force vector would be:

$$\underline{f} = \begin{bmatrix} g \\ 0 \\ 0 \end{bmatrix} \tag{3.4}$$

where "g" is the magnitude of gravity. However, if we rotate the triad such that the z accel is pointed upward:

$$\underline{f} = \begin{bmatrix} 0 \\ 0 \\ g \end{bmatrix} \tag{3.5}$$

The specific force vector has not changed but the measurement vector differs depending upon the orientation. Thus, when we say a particular vector is *expressed* in a particular frame, it refers to the coordinate frame in which the vector is resolved.

The body-mounted gyroscopes measure the rotation rate of the body (vehicle). The subscript "ib" means the angular rate (ω^b_{ib}) is the rate of the body-frame *relative* to the inertial-frame. Again, the superscript "b" refers to the frame in which the measurements are expressed. Soon we will learn how to take the measurements from the gyros and update our representation of attitude. Our current challenge is to navigate with respect to the inertial frame and so ultimately we want the body-to-inertial direction cosine matrix (or quaternion, but we will address that later).

Assuming the direction cosine matrix (DCM) has been updated, resolving the specific force into the inertial frame is simply:

$$\underline{f}^i = C_b^i \underline{f}^b \tag{3.6}$$

With the specific force vector resolved into the frame of interest (the inertial frame in our current case), it then needs to be compensated for gravity in order to produce the desired acceleration vector. Notice in Figure 3.3 that gravity is a computed (modeled) quantity, not a measured quantity. Notice also that gravity is actually a function of position. Even with a first order approximation, gravity is a function of both latitude and height. Gravity decreases with height but it is also influenced by the fact that the Earth is an oblate spheroid (think of a beach ball that you've squashed on the top and bottom and it bulges out on the sides). The point is that the Earth is not perfectly spherical and that fact is one of the dominant effects on the variation of gravity with position. Thus, we have to know (or have an estimate of) our position in order to compute what the value of gravity is, which then gets fed back into the position updating process.

The sum of specific force and gravity provides the acceleration that gets integrated once to determine velocity and once again to determine position. Note that initial estimates of velocity and position are needed in the integration process. Theoretically these are the constants of integration. As a practical matter, the accelerometers allow us to determine acceleration but they do not provide initial velocity or position. As we will see later, the attitude update process also needs to be initialized.

3.5 Updating the body-to-inertial direction cosine matrix

It is important to recognize that gyros by themselves do not provide a measure of absolute attitude. The gyros only measure changes of angle. Specifically, they either measure angular rate or angular displacement over some short interval of time. Gyros thus only provide a measure of the change of orientation and not the absolute orientation itself. Broadly speaking, there are two different kinds of gyros. One is known as a "rate gyro" that obviously outputs angular rate measurements. Then there are so-called "rate-integrating gyros" that output small increments of angular change as they occur. For example, ring laser gyros (RLGs) are inherently rate-integrating gyros. Their basic measurement is a change of angle. On the other hand, a fiber-optic gyro (FOG) is inherently a rate gyro. The average navigation system engineer may not appreciate the distinction because a high-end FOG will perform the first integration in its electronics and thus still output changes of angle. Note the changes of angle output by the gyros are called "delta thetas" ($\Delta\theta$).

As discussed in the previous chapter, attitude is defined as the rotational difference between the body frame and a reference frame of interest. In an INS, the attitude representation in the computer must first be initialized and then it can be updated as the vehicle maneuvers. Initialization will be treated later and for now we will assume that the initial attitude is provided to us. Having that, we want to update the attitude using the measurements obtained from the gyros.

As we have discussed in the previous chapter, the two most robust attitude representations are direction cosine matrices and quaternions. We will start with direction cosine matrices and will consider quaternions later. The first step is to determine the derivative of the direction cosine matrix. From this, we will be able to derive a discrete-time update. Starting with the definition of the derivative:

$$\dot{C}_b^i = \lim_{\delta t \to 0} \frac{\delta C_b^i}{\delta t} = \lim_{\delta t \to 0} \frac{C_b^i(t+\delta t) - C_b^i(t)}{\delta t} \tag{3.7}$$

It can be shown that:

$$C_b^i(t+\delta t) = C_b^i(t) C_{b(t+\delta t)}^{b(t)} \tag{3.8}$$

where the second matrix is a rotation matrix relating the body-frame at time "t" to the body-frame at time "$t + \delta t$." Although this is stated without proof, it clearly is similar to the cascading of rotation matrices discussed in the previous chapter.

For very small angular changes (a reasonable assumption since the time interval is δt), the second term in (3.8) can be expressed as:

$$C_{b(t+\delta t)}^{b(t)} = [I + \delta \Lambda] \tag{3.9}$$

where:

$$I = \begin{bmatrix} 1 & 0 & 0 \\ 0 & 1 & 0 \\ 0 & 0 & 1 \end{bmatrix} \tag{3.10}$$

and

$$\delta \Lambda = \begin{bmatrix} 0 & -\delta\psi & \delta\theta \\ \delta\psi & 0 & -\delta\phi \\ -\delta\theta & \delta\phi & 0 \end{bmatrix} \tag{3.11}$$

and
$\delta\phi$ is the rotation around the body x-axis over time δt
$\delta\theta$ is the rotation around the body y-axis over time δt
$\delta\psi$ is the rotation around the body z-axis over time δt

Recall the fact that finite rotations do not commute. However, (3.11) does not indicate any order of the rotations. It can be shown that infinitesimal rotations actually *do* commute and thus if the angles are very small, then the order does not matter. That is the situation here. Over very small amounts of time the body rotates only a very small amount (in any axis). The angles in (3.11) are thus going to be very small and the order of rotation is irrelevant.

Substitution of (3.9) into (3.8) yields:

$$C_b^i(t+\delta t) = C_b^i(t)[I + \delta \Lambda] \tag{3.12}$$

and substituting (3.12) into (3.7) yields:

$$\dot{C}_b^i = \lim_{\delta t \to 0} \frac{\delta C_b^i}{\delta t} = \lim_{\delta t \to 0} \frac{C_b^i(t)[I + \delta\Lambda] - C_b^i(t)}{\delta t} = \lim_{\delta t \to 0} \frac{C_b^i(t)\delta\Lambda}{\delta t} \tag{3.13}$$

Since $C_b^i(t)$ is the DCM at the *beginning* of the time interval (see 3.8), it is thus constant in (3.13). This allows us to simplify (3.13) to:

$$\dot{C}_b^i = C_b^i(t) \lim_{\delta t \to 0} \frac{\delta\Lambda}{\delta t} = C_b^i(t)\Omega_{ib}^b \tag{3.14}$$

where:

$$\Omega_{ib}^b = \lim_{\delta t \to 0} \frac{\delta\Lambda}{\delta t} = \lim_{\delta t \to 0} \begin{bmatrix} 0 & -\frac{\delta\psi}{\delta t} & \frac{\delta\theta}{\delta t} \\ \frac{\delta\psi}{\delta t} & 0 & -\frac{\delta\phi}{\delta t} \\ -\frac{\delta\theta}{\delta t} & \frac{\delta\phi}{\delta t} & 0 \end{bmatrix} = \begin{bmatrix} 0 & -\omega_z & \omega_y \\ \omega_z & 0 & -\omega_x \\ -\omega_y & \omega_x & 0 \end{bmatrix} \tag{3.15}$$

The final term in (3.15) is called the skew-symmetric[‡] matrix form of the angular rate vector:

$$\underline{\omega}_{ib}^b = \begin{bmatrix} \omega_x & \omega_y & \omega_z \end{bmatrix}^T \tag{3.16}$$

As described earlier, this angular rate vector is measured by the gyros and provides the rotation rate of the body-frame relative to the inertial-frame, expressed in body coordinates. Since the skew-symmetric matrix form of a vector is also a cross-product operator, the final term in (3.15) is sometimes expressed as:

$$\Omega_{ib}^b = [\underline{\omega}_{ib}^b \times] \tag{3.17}$$

To see the cross-product operation, consider the cross-product of generic vectors $\underline{a}$ and $\underline{b}$:

$$\underline{a} \times \underline{b} = \begin{bmatrix} a_y b_z - a_z b_y \\ a_z b_x - a_x b_z \\ a_x b_y - a_y b_x \end{bmatrix} \tag{3.18}$$

and then note that the same result is achieved by:

$$[\underline{a}\times]\underline{b} = \begin{bmatrix} 0 & -a_z & a_y \\ a_z & 0 & -a_x \\ -a_y & a_x & 0 \end{bmatrix} \begin{bmatrix} b_x \\ b_y \\ b_z \end{bmatrix} \tag{3.19}$$

Returning now to our main goal, substitution of (3.17) into (3.14) yields:

$$\dot{C}_b^i = C_b^i(t)\Omega_{ib}^b = C_b^i(t)[\underline{\omega}_{ib}^b \times] \tag{3.20}$$

Although we derived it in the specific context of the inertial and body frames, the result is generic: the derivative of the direction cosine matrix is equal to product of the direction cosine matrix itself and the skew-symmetric matrix form of the angular rate vector (where the angular rate is the rate between the two frames):

$$\dot{C} = C\Omega = C[\underline{\omega}\times] \tag{3.21}$$

[‡]Skew-symmetric matrices are square matrices whose negative equals their transpose: $-A = A^T$.

The discrete-time solution of (3.21) is:

$$C_{k+1} = C_k e^{\int_{t_k}^{t_{k+1}} \Omega dt} \tag{3.22}$$

where the subscript "k" and "$k+1$" refer to the discrete-time points t_k and t_{k+1}. Thus, if we have the DCM at the previous time (t_k) and we have the rotations over the time interval (t_k to t_{k+1}), we can compute the DCM at the current time (t_{k+1}).

If the angular rate vector, $\underline{\omega}_{ib}^b$, is stationary, then the integral in (3.22) is simple to compute from the gyro outputs. By assuming it is stationary, we are assuming that it does not change orientation over the update interval. You may say, "Wait a minute, that sounds pretty useless." But hold on. We are not saying that the *vehicle* has to be stationary. What we are saying is that the angular rate *vector* has to be stationary. What does that mean? Let us take a very simple scenario. If our vehicle is level and we rotate around the vertical axis then we are changing yaw. The angular rate vector is oriented vertically. If, on the other hand, we rotate around the body y-axis (i.e., the axis containing the wings of an airplane), then we are changing pitch and the angular rate vector is oriented (a.k.a., lies along) the body y-axis. We could also have a more complex rotation where we are effectively rotating around the body z- and body y-axis simultaneously. This is a little harder to visualize but there exists a single axis-of-rotation for this scenario as well. In all these cases, the orientation of the angular rate vector is fixed (constant/stationary).

However, if we rotate around, say, the z-axis first and subsequently rotate around the y-axis, then the orientation of the angular rate vector changes. It starts off oriented along the z-axis and subsequently is oriented along the y-axis. This is where things can get really confusing. In the most general motions, the angular rate vector itself can change orientation continuously. If you are still struggling with this concept (and if you are, you are in good company!), consider a practical example that we have all encountered. Consider a coin that you are holding "upright" on a table (i.e., heads and tails are perpendicular to the surface of the table). Now flick the coin so that it starts to spin. After spinning a while eventually it slows down, wobbles over and finally ends up flat on the table with either heads or tails facing up. Now try to visualize the instantaneous angular rate vector of the coin throughout this entire process. It starts off oriented parallel to the faces of the coin and at the very end it is oriented perpendicular to the faces of the coin. Thus, in the body frame of the coin, the angular rate vector is changing orientation.

Although we will not dwell on the details, the changing orientation of the angular rate vector is referred to as "coning" motion. The name stems from the fact that the angular rate vector will trace out a cone in the case where the platform is experiencing angular rates about two orthogonal axes that are sinusoidal and 90 degrees out of phase (i.e., the angular rate about one axis is a sine and the angular rate about the orthogonal axis is a cosine). Although it seems to be contrived, that kind of motion is representative of some common vibration modes that have been shown to occur in real inertial platforms. Recall earlier we stated that if the angular displacements around the three orthogonal axes are very small, then the rotations commute (i.e., the order does not matter). Coning motion causes the rotations to be non-commutative and thus errors will result if the angular rate vector is blindly integrated.

The bottom line is that coning motion increases the complexity of computing the integral in (3.22). You might assume that we could eliminate the problem by using very short time intervals. This is partially correct. To accommodate high dynamic platforms (e.g., a jet fighter performing high-rate S-turns), we must process the gyro data at high rates (e.g., short time intervals) and for practical attitude update rates (e.g., hundreds/sec) coning errors still exist. In addition, coning motion in the form of high frequency vibrations also cause errors. For this reason, the so-called "coning compensation" must be applied to the gyro outputs prior to subsequent processing.

An in-depth treatment of coning error and coning compensation is given in [3]. For the remainder of this book, *we will assume that coning compensation has already been applied*. With this important caveat, the integral of the angular rate vector yields the so-called "rotation vector" (i.e., a single-axis of rotation vector):

$$\underline{\sigma} = \int_{t_k}^{t_{k+1}} \underline{\omega}_{ib}^{b}\, dt \tag{3.23}$$

The integral in (3.23) can be computed numerically if the gyro outputs angular rate. If the gyro outputs angular increments (i.e., delta-thetas), the integral is a simple matter of summing the delta-thetas. Keep in mind, though, that coning compensation must also be applied.

The rotation vector has components (magnitude and unit vector) defined in the usual way:

$$\underline{\sigma} = \begin{bmatrix} \sigma_x & \sigma_y & \sigma_z \end{bmatrix}^T \tag{3.24}$$

$$\sigma = \sqrt{\sigma_x^2 + \sigma_y^2 + \sigma_z^2} \tag{3.25}$$

$$\underline{\hat{\sigma}} = \begin{bmatrix} \frac{\sigma_x}{\sigma} & \frac{\sigma_y}{\sigma} & \frac{\sigma_z}{\sigma} \end{bmatrix}^T \tag{3.26}$$

Do not lose sight of the conceptual understanding of the rotation vector. Starting at time t_k, a rotation of the b-frame about the rotation axis $\underline{\hat{\sigma}}$ by an amount σ will yield the b-frame orientation at time t_{k+1}.

With the rotation vector obtained from (3.23), the integral in (3.22) is given by the skew-symmetric matrix form of the rotation vector:

$$\int_{t_k}^{t_{k+1}} \Omega dt = [\underline{\sigma}\times] = \begin{bmatrix} 0 & -\sigma_z & \sigma_y \\ \sigma_z & 0 & -\sigma_x \\ -\sigma_y & \sigma_x & 0 \end{bmatrix} \tag{3.27}$$

We can now complete the derivation of the discrete-time update. Substitution of (3.27) into (3.22) yields:

$$C_{k+1} = C_k e^{[\underline{\sigma}\times]} \tag{3.28}$$

We now have to evaluate this matrix exponential. In MATLAB®, you can simply use the matrix exponential function (expm) but this is computationally intensive and is not utilized in real time systems. Generally the approach taken is to perform some kind of series expansion and then truncate the series after an appropriate number of terms. We will do that and along the way we will end up deriving a more elegant form of this update equation.

The Taylor series expansion of the matrix exponential function is very similar to the Taylor series expansion for the scalar exponential:

$$e^{[\underline{\sigma}\times]} = I + [\underline{\sigma}\times] + \frac{[\underline{\sigma}\times]^2}{2!} + \frac{[\underline{\sigma}\times]^3}{3!} + \cdots \tag{3.29}$$

The square of the skew-symmetric matrix is simply the product of the matrix with itself:

$$[\underline{\sigma}\times]^2 = \begin{bmatrix} -(\sigma_y^2 + \sigma_z^2) & \sigma_x\sigma_y & \sigma_x\sigma_z \\ \sigma_x\sigma_y & -(\sigma_x^2 + \sigma_z^2) & \sigma_y\sigma_z \\ \sigma_x\sigma_z & \sigma_y\sigma_z & -(\sigma_x^2 + \sigma_y^2) \end{bmatrix} \tag{3.30}$$

However, higher values of the exponent can be shown to yield a very interesting result:

$$[\underline{\sigma}\times]^3 = -(\sigma_x^2 + \sigma_y^2 + \sigma_z^3)[\underline{\sigma}\times] = -\sigma^2[\underline{\sigma}\times] \tag{3.31}$$

$$[\underline{\sigma}\times]^4 = -(\sigma_x^2 + \sigma_y^2 + \sigma_z^3)[\underline{\sigma}\times]^2 = -\sigma^2[\underline{\sigma}\times]^2 \tag{3.32}$$

Thus all higher values of the exponent can be expressed in terms of the skew-symmetric matrix or its square. This allows (3.29) to be rewritten as:

$$e^{[\underline{\sigma}\times]} = I + \left(1 - \frac{\sigma^2}{3!} + \frac{\sigma^4}{5!} - \cdots\right)[\underline{\sigma}\times] + \left(\frac{1}{2!} - \frac{\sigma^2}{4!} + \frac{\sigma^4}{6!} - \cdots\right)[\underline{\sigma}\times]^2 \tag{3.33}$$

The terms in parentheses turn out to be, themselves, Taylor Series of trigonometric functions and thus (3.33) can be simplified to:

$$e^{[\underline{\sigma}\times]} = I + \frac{\sin\sigma}{\sigma}[\underline{\sigma}\times] + \frac{1-\cos\sigma}{\sigma^2}[\underline{\sigma}\times]^2 \tag{3.34}$$

An equivalent form of (3.34) can be written in terms of the unit vector:

$$e^{[\underline{\sigma}\times]} = I + \sin\sigma[\underline{\hat{\sigma}}\times] + (1-\cos\sigma)[\underline{\hat{\sigma}}\times]^2 \tag{3.35}$$

Equation (3.35) is a form of Rodrigues' rotation formula (Olinde Rodrigues was a nineteenth century French mathematician). It provides an elegant closed form for computing the matrix exponential and it can be used in simulations. For real time applications, however, a truncated version of a power series is preferred.

3.6 Position/velocity/attitude updating in an inertially fixed frame

Now we have the necessary mathematics to perform the position, velocity, and attitude update in an inertially fixed frame as depicted in Figure 3.3:

Step 0: Initialize position, velocity, and attitude. Initialization is a topic that we will consider later so for now we will assume that the initial values have been provided to us. For attitude, this means the initial value of the body-to-inertial direction cosine matrix.

Step 1: Form $\underline{\sigma}$ from the gyro outputs:

$\underline{\sigma} = 0$
for $k = 1$ to N
    $\underline{\sigma} = \underline{\sigma} + \underline{\Delta\theta}_k +$ (coning compensation)
end

In real applications, the gyro outputs are generated at a rate on the order of 1 kHz whereas the attitude update is performed at a rate on the order of 100 Hz. With these rates, ten sets of delta-thetas are summed to produce the rotation vector.

Step 2: Perform the attitude update

Having obtained the rotation vector, the attitude update is performed by computing (3.28) and utilizing one of the various expressions for the matrix exponential (3.29, 3.33, 3.34, or 3.35).

Step 3: Resolve the accelerometer measurements into the inertial frame

The accelerometers output specific force integrated over a short time interval. Since the integral of acceleration over a time interval provides the change of velocity, the accel outputs are referred to as delta-Vs:

$$\underline{\Delta V}_k^b = \int_{t_{k-1}}^{t_k} \underline{f}^b dt \tag{3.36}$$

Resolving the measurements into the reference frame is a simple matter of multiplying with the direction cosine matrix:

$$\underline{\Delta V}_k^i = C_b^i \underline{\Delta V}_k^b \tag{3.37}$$

As with gyros, accels typically output delta-Vs at rates significantly higher than the rate at which velocity is updated. The delta-Vs can be accumulated over the update interval but a potential error called "sculling" must be addressed. Sculling amounts to a sensed acceleration in the presence of rotation. An accel swinging back and forth at the end of some lever arm will sense a non-zero average acceleration even though its net translation is zero. As with coning compensation in gyro measurements, sculling compensation algorithms can be applied to the accel measurements [3].

A related potential error is known as "size effect." This phenomenon arises due to the fact that three accelerometers cannot all be physically mounted in the same location. The small displacement between them (typically a few centimeters) will cause each accel to sense slightly different acceleration in the presence of angular motion. Again, a size effect compensation algorithm is applied to the accel outputs to address this.

Step 4: Update velocity

Since the accelerometer itself performed the first integration of the specific-force vector, the velocity update is straightforward:

$$\underline{V}_k = \underline{V}_{k-1} + \Delta\underline{V}_k^i + \underline{g}^i \Delta t \tag{3.38}$$

Note the third term on the right hand side of the equation represents a rectangular integration of the gravity vector. This is typically acceptable since relatively large changes of latitude and/or height are needed to make an appreciable change in the gravity vector. Over a typical velocity update interval (e.g., 0.01–0.1 sec), gravity is essentially a constant. For very high-speed applications (e.g., rocket launches) and/or high precision, higher order numerical integration algorithms can be utilized.

Step 5: Update position

Having updated the velocity vector, conventional numerical integration can be used to update position. Assuming the position and velocity vectors have been defined in terms of a local Cartesian coordinate frame (that does not rotate!), simple trapezoidal integration yields:

$$\underline{P}_k = \underline{P}_{k-1} + \frac{1}{2}(\underline{V}_k + \underline{V}_{k-1})\Delta t \tag{3.39}$$

The use of a Cartesian coordinate frame to express position and velocity may be acceptable in applications involving very small regions of operation over short time intervals. Just keep in mind that our primary purpose here is to begin the gradual building up of the inertial navigation algorithms. At the moment we are determining position and velocity with respect to an inertially fixed frame of reference (e.g., an imaginary non-rotating flat earth). In subsequent chapters, we will develop the algorithms needed to extend our current results to the spheroidal, rotating earth. Nevertheless, we have completed the description of all of the processing needed for inertial position/velocity/attitude updating with respect to an inertially fixed frame. Up to this point, we have used the direction cosine matrix for the attitude representation. In the next section, we will learn how to update attitude using quaternions.

3.7 Updating the body-to-inertial quaternion

The derivative of the quaternion is given by:

$$\dot{q} = \frac{1}{2} q * \begin{bmatrix} 0 \\ \underline{\omega} \end{bmatrix} \tag{3.40}$$

where the star ($*$) indicates quaternion multiplication. The quaternion in square brackets has a zero scalar element and the vector portion is the angular rate vector. Equation (3.40) can be rewritten in matrix form as:

$$\dot{q} = \frac{1}{2}\begin{bmatrix} 0 & -\omega_x & -\omega_y & -\omega_z \\ \omega_x & 0 & \omega_z & -\omega_y \\ \omega_y & -\omega_z & 0 & \omega_x \\ \omega_z & \omega_y & -\omega_x & 0 \end{bmatrix} q = \frac{1}{2} W q \tag{3.41}$$

Assuming the orientation of the angular rate vector is fixed over the update interval, the discrete-time solution to (3.41) is given by:

$$q_{k+1} = [e^{\frac{1}{2}\int_{t_k}^{t_{k+1}} W dt}] q_k \tag{3.42}$$

We are currently considering the body-to-inertial quaternion. In this case, the matrix integral can be expressed in terms of the rotation vector $\underline{\sigma}$:

$$S_W = \int_{t_k}^{t_{k+1}} W dt = \begin{bmatrix} 0 & -\sigma_x & -\sigma_y & -\sigma_z \\ \sigma_x & 0 & \sigma_z & -\sigma_y \\ \sigma_y & -\sigma_z & 0 & \sigma_x \\ \sigma_z & \sigma_y & -\sigma_x & 0 \end{bmatrix} \tag{3.43}$$

and thus (3.42) can be rewritten as:

$$q_{k+1} = [e^{\frac{1}{2} S_W}] q_k \tag{3.44}$$

An equivalent expression in quaternion form is:

$$q_{k+1} = q_k * q_{b[k]}^{b[k+1]} = q_k * \begin{bmatrix} \cos\frac{\sigma}{2} \\ \frac{\sigma_x}{\sigma}\sin\frac{\sigma}{2} \\ \frac{\sigma_y}{\sigma}\sin\frac{\sigma}{2} \\ \frac{\sigma_z}{\sigma}\sin\frac{\sigma}{2} \end{bmatrix} \tag{3.45}$$

where, again, the star ($*$) denotes quaternion multiplication. Recall that the magnitude of the angle vector is denoted by σ as was defined in (3.25). Note also the second term in the product is a quaternion that represents the rotation of the body-frame from time t_k to t_{k+1}. For real-time applications, the trigonometric functions in (3.45) are approximated by truncated Taylor series. Keep in mind that, as with DCM updating, the rotation vector ($\underline{\sigma}$) is obtained by summing the gyro outputs ($\underline{\Delta\theta}$) over the update interval and compensating for coning.

References

[1] Groves P. *Principles of GNSS, Inertial, and Multisensor Integrated Navigation Systems*. 2nd ed. Boston, MA: Artech House; 2013.

[2] Titterton DH, Weston JL. *Strapdown Inertial Navigation Technology*. 2nd ed. Herts, UK: The Institution of Electrical Engineers; 2004.

[3] Ignagni M. *Strapdown Navigation Systems – Theory and Application*. Minneapolis, MN: Champlain Press; 2018.

Chapter 4
The impact of pseudo-forces in the velocity update

Introduction

Previously we have covered the details of inertial navigation in the context of a fixed reference frame. Now we are going to begin addressing the fact that in most applications the vehicle is actually navigating with respect to the spheroidal, rotating earth. We will deal with these two effects (earth curvature and rotation) by taking into account the additional apparent forces that come into play. This will impact our navigation equation (i.e., the expression for the derivative of velocity). In the case of the fixed frame, the expression for the derivative of velocity was very simple. In the context of a rotating reference frame, however, it is a bit more complicated.

4.1 A graphical derivation of the pseudo-forces in the Earth frame

The core theory that we will draw upon in this work is known as the Coriolis pseudo-force or Coriolis acceleration. What is Coriolis? You may have had some exposure to it during your physics classes. The usual derivation is enormously entertaining if you happen to be a mathematician. I am not and so I prefer a graphical derivation.* Let us start off by considering a simplified depiction of the earth as shown in Figure 4.1. The figure depicts the earth's rotation vector along with a 3D coordinate system and some arbitrary point P. The point P is fixed on the earth itself.

We will first derive an expression for the velocity of the point P in inertial space and then we will extend the result to take into account motion over the surface of the earth. First, how do we describe the velocity of the point with respect to the inertial frame? Figure 4.2a is an abstraction of Figure 4.1 that depicts a position vector $\underline{r}$ from the center of the earth to the point P. As the earth rotates, point P traces out a circle in space (strictly speaking, point P is actually tracing out a spiral as the earth revolves around the Sun, but this can be neglected for our purposes). Figure 4.2b depicts a

*For a more formal treatment, see [1]. Many thanks to Roger Rollins, Emeritus Professor of Physics at Ohio University, for this derivation.

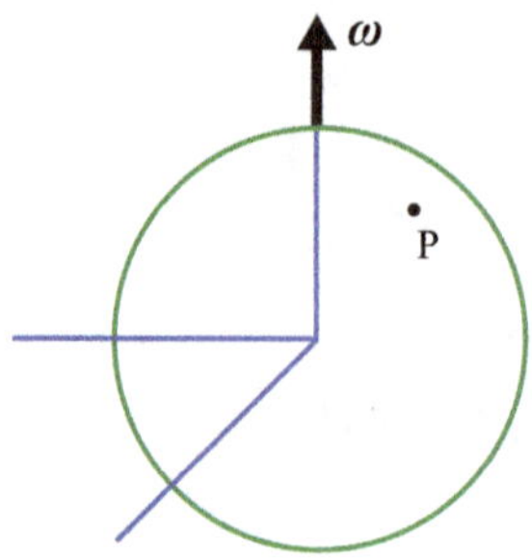

Figure 4.1 *Point "P" is fixed on the earth itself. The rotation rate of the Earth is* ω

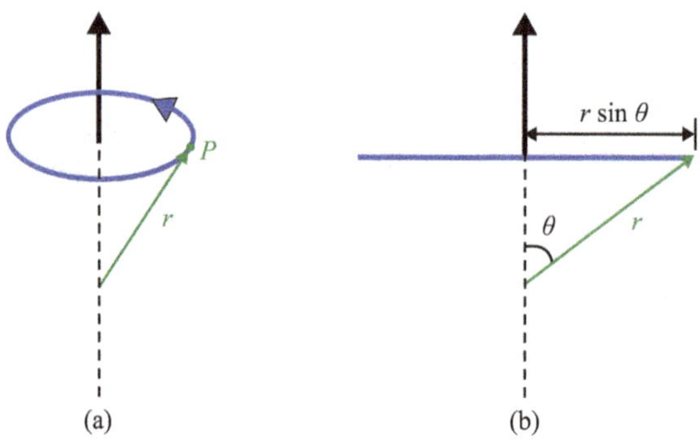

Figure 4.2 *Abstraction of Figure 4.1. The vertical arrow depicts the earth's rotation vector,* $\underline{\omega}$*. Point P is fixed on the earth and traces out a circle in inertial space (earth movement around the sun is neglected). The position vector* $\underline{r}$ *forms an angle* θ *with respect to the rotation vector.* θ *here is the co-latitude and should not be confused with other uses of* θ *elsewhere in the book (particularly for pitch)*

profile view. The horizontal line depicts the plane in which the point P is traversing and the position vector $\underline{r}$ forms an angle θ with respect to the earth's rotation axis (note that θ is the geocentric *co*-latitude since it is the complement of the geocentric latitude).

As shown in Figure 4.2b, the radius of the circle traced out by P is $r \sin\theta$. Figure 4.3 depicts the same scenario from above the earth. The earth's rotational axis, $\underline{\omega}$, is at the center of the circle and coming out of the figure (or page). The point P sweeps out an angle $\delta\Omega$ over an incremental interval of time δt.

Now consider the change in position if the time interval is infinitesimally small. In this case, the actual arc length traversed by the point P can be approximated by a straight line, specifically a vector, as depicted in Figure 4.4. This is the usual

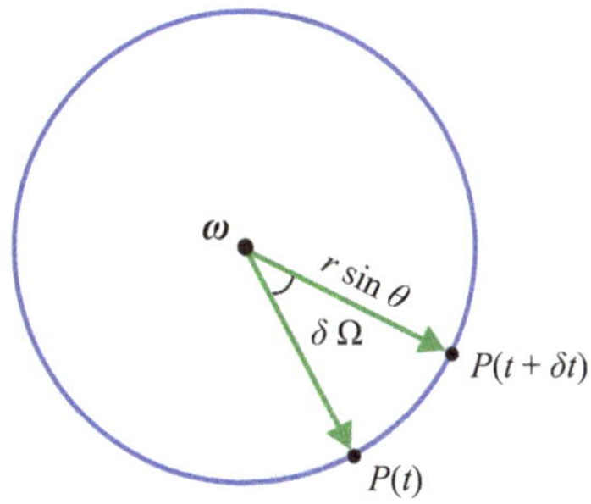

Figure 4.3 The scenario of Figures 4.1 and 4.2 depicted from above. The point P sweeps out an angle $\delta\Omega$ over an incremental interval of time δt

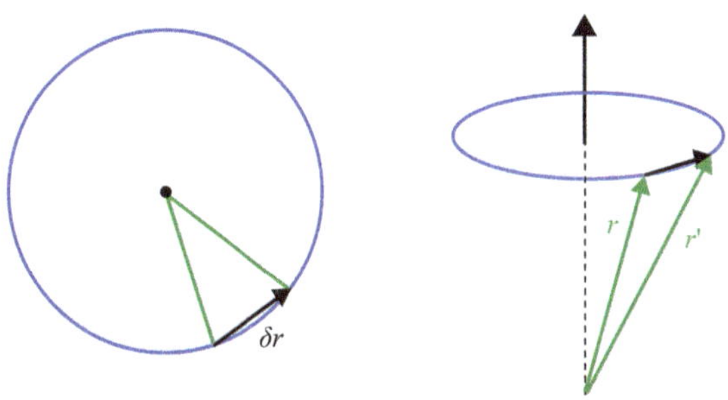

Figure 4.4 An approximation of Figure 4.2 over an infinitesimally small interval of time. The actual arc length traversed can be approximated by a straight line and $\underline{r}' = \underline{r} + \delta\underline{r}$

approximation we make in calculus. We can thus relate the position vector at the beginning and end of the time interval to the differential vector:

$$\underline{r}' = \underline{r} + \delta\underline{r} \tag{4.1}$$

By definition, the actual arc length is given by the product of the radius and the differential angle. The magnitude of the differential vector is thus:

$$|\delta\underline{r}| = (r\sin\theta)\delta\Omega \tag{4.2}$$

Since Equation (4.2) provides the magnitude of the differential vector over an infinitesimal interval of time, we can divide both sides of (4.2) by δt and get an expression for the full derivative in the limit as $\delta t \to dt$:

$$\frac{|\delta\underline{r}|}{dt} = (r\sin\theta)\frac{\delta\Omega}{dt} \tag{4.3}$$

The angular rate in (4.3) is simply earth rate:

$$\frac{\delta\Omega}{dt} = \underline{\omega} \tag{4.4}$$

We can now prove the following:

$$\frac{d\underline{r}}{dt} = \underline{\omega} \times \underline{r} \tag{4.5}$$

The proof is as follows: by definition, the magnitude of the cross-product is:

$$|\underline{\omega} \times \underline{r}| = |\underline{\omega}||\underline{r}| \sin\theta \tag{4.6}$$

Note the right hand side of (4.6) is identical to that of (4.3) substituted with (4.4). To complete the proof, we need to show the direction of the cross-product result in (4.5) is correct. Application of the right hand rule to the vectors in Figure 4.2b shows the result corresponds to the differential vector depicted in Figure 4.4. Equation (4.5) is thus proved and we can rewrite it as:

$$\underline{v}_i = \underline{\omega} \times \underline{r} \tag{4.7}$$

Points fixed on the earth thus have an inertial velocity given by (4.7). By considering the range of possible values for θ in (4.6), it follows that the inertial velocity of this fixed point will be maximum at the equator and zero at the poles. This is fully consistent with our physical intuition since a point at a pole is rotating but not translating. Since the rotation rate of the earth is essentially constant, the maximum inertial velocity for a point on the earth will occur for points that are farthest from the axis of rotation. Points on the equator satisfy this requirement.

We have derived the inertial velocity for a point fixed on the earth. We can now apply the result to a point in motion with respect to the earth. The first step is to add on the earth-referenced velocity to account for the fact that the point is no longer fixed on the earth:

$$\left.\frac{d\underline{r}}{dt}\right|_i = \left.\frac{d\underline{r}}{dt}\right|_e + \underline{\omega} \times \underline{r} \tag{4.8}$$

Note the derivative after the equal sign is evaluated with respect to the earth frame and is thus the velocity with respect to the earth. Thus, the inertially referenced velocity is equal to the sum of the earth-referenced velocity plus the velocity component due to the rotation of the earth itself. We can rewrite (4.8) as:

$$\underline{v}_i = \underline{v}_e + \underline{\omega} \times \underline{r} \tag{4.9}$$

A general form may be extracted from (4.8). The derivative of a given vector evaluated in the inertial frame is equal to the derivative of that given vector evaluated in the earth frame plus the cross product of the earth angular rate vector with the given vector:

$$\left.\frac{d\underline{k}}{dt}\right|_i = \left.\frac{d\underline{k}}{dt}\right|_e + \underline{\omega} \times \underline{k} \tag{4.10}$$

for a generic vector $\underline{k}$. Applying Equation (4.10) to the inertial velocity vector $\underline{v}_i$ yields:

$$\left.\frac{d\underline{v}_i}{dt}\right|_i = \left.\frac{d\underline{v}_i}{dt}\right|_e + \underline{\omega} \times \underline{v}_i \tag{4.11}$$

Now substitute the right-hand side of Equation (4.9) into the right-hand side of (4.11):

$$\left.\frac{d\underline{v}_i}{dt}\right|_i = \left.\frac{d}{dt}(\underline{v}_e + \omega \times \underline{r})\right|_e + \underline{\omega} \times (\underline{v}_e + \omega \times \underline{r}) \tag{4.12}$$

which can be rewritten as:

$$\left.\frac{d\underline{v}_i}{dt}\right|_i = \left.\frac{d}{dt}\underline{v}_e\right|_e + \left.\frac{d}{dt}(\omega \times \underline{r})\right|_e + \underline{\omega} \times \underline{v}_e + \underline{\omega} \times (\omega \times \underline{r}) \tag{4.13}$$

The second term on the right-hand side of (4.13) can be evaluated with the product rule:

$$\left.\frac{d}{dt}(\omega \times \underline{r})\right|_e = \left(\left.\frac{d\underline{\omega}}{dt}\right|_e\right) \times \underline{r} + \underline{\omega} \times \left.\frac{d\underline{r}}{dt}\right|_e \tag{4.14}$$

But the first term on the right-hand side of (4.14) is zero since earth rotation rate is essentially constant. Equation (4.14) thus reduces to:

$$\left.\frac{d}{dt}(\omega \times \underline{r})\right|_e = \underline{\omega} \times \underline{v}_e \tag{4.15}$$

Substituting (4.15) into (4.13) yields:

$$\left.\frac{d\underline{v}_i}{dt}\right|_i = \left.\frac{d}{dt}\underline{v}_e\right|_e + 2\underline{\omega} \times \underline{v}_e + \underline{\omega} \times (\omega \times \underline{r}) \tag{4.16}$$

Solving (4.16) for earth-referenced acceleration yields:

$$\left.\frac{d\underline{v}_e}{dt}\right|_e = \left.\frac{d}{dt}\underline{v}_i\right|_i - 2\underline{\omega} \times \underline{v}_e - \underline{\omega} \times (\omega \times \underline{r}) \tag{4.17}$$

Since the earth is a rotating frame of reference (i.e., it is not an inertial frame), the expression for acceleration is composed of true inertial acceleration plus two pseudo-acceleration terms (more precisely, pseudo-force if both sides of (4.17) are multiplied by a mass, m):

$-2\underline{\omega} \times \underline{v}_e =$ Coriolis acceleration

$-\underline{\omega} \times (\underline{\omega} \times \underline{r}) =$ Centrifugal acceleration

The Coriolis acceleration term is present only when there is relative motion between the object and the earth. Note also that it is orthogonal both to the object's earth-referenced velocity and to the earth's rotation vector. The centrifugal acceleration term is simply the negative of the true inertial centripetal acceleration due to the circular motion of the earth. When viewed from the rotating earth frame, it appears to act outward, rather than inward. Later we will study the significance of the fact that the centrifugal acceleration term is a function of position only (since the other term, earth's angular rate vector, is a constant) and is independent of the object's velocity.

Before proceeding we should point out that the aforementioned development is not limited to objects on or moving with respect to the earth frame. In fact, Equation (4.10) is not limited to the inertial and earth frames at all. The principle is applicable to any two frames that are rotating with respect to each other.

4.2 Inertial frame mechanization of the navigation equation

Although Equation (4.17) illustrates the two pseudo-forces, it is not the most convenient form for us to use since the derivatives are not both evaluated in the same frame. We ultimately want to solve for velocity in the navigation frame in order to determine local-level velocity components (i.e., north/east/vertical). As an intermediate step, we will derive the navigation equation expressed in the inertial frame. The result will subsequently be applied in the derivation of the navigation equation expressed in the navigation frame. A similar treatment of this material is presented in [2].

We start by rewriting Equation (4.8) as:

$$\left.\frac{d\underline{r}}{dt}\right|_i = \underline{v}_e + \underline{\omega}_{ie} \times \underline{r} \tag{4.18}$$

Note the subscript "*ie*" specifically denotes the angular rate vector expresses the rate of the earth frame relative to the inertial frame. Differentiating both sides of (4.18) and evaluating in the inertial frame yields:

$$\left.\frac{d^2\underline{r}}{dt^2}\right|_i = \left.\frac{d\underline{v}_e}{dt}\right|_i + \left.\frac{d}{dt}(\underline{\omega}_{ie} \times \underline{r})\right|_i \tag{4.19}$$

Following a similar development that yielded (4.15), we can reduce (4.19) to:

$$\left.\frac{d^2\underline{r}}{dt^2}\right|_i = \left.\frac{d\underline{v}_e}{dt}\right|_i + \left.\left(\underline{\omega}_{ie} \times \frac{d\underline{r}}{dt}\right)\right|_i \tag{4.20}$$

Substituting (4.18) into the second term of the cross-product in (4.20) yields:

$$\left.\frac{d^2\underline{r}}{dt^2}\right|_i = \left.\frac{d\underline{v}_e}{dt}\right|_i + \underline{\omega}_{ie} \times (\underline{v}_e + \underline{\omega}_{ie} \times \underline{r}) \tag{4.21}$$

and carrying out the initial cross-product:

$$\left.\frac{d^2\underline{r}}{dt^2}\right|_i = \left.\frac{d\underline{v}_e}{dt}\right|_i + \underline{\omega}_{ie} \times \underline{v}_e + \underline{\omega}_{ie} \times (\underline{\omega}_{ie} \times \underline{r}) \tag{4.22}$$

From the accelerometer sensing equation described in the previous chapter, we can write:

$$\left.\frac{d^2\underline{r}}{dt^2}\right|_i = \underline{f} + \underline{g} \tag{4.23}$$

Equating (4.22) to (4.23) yields:

$$\underline{f} + \underline{g} = \left.\frac{d\underline{v}_e}{dt}\right|_i + \underline{\omega}_{ie} \times \underline{v}_e + \underline{\omega}_{ie} \times (\underline{\omega}_{ie} \times \underline{r}) \tag{4.24}$$

Solving for the first term on the right-hand side of (4.24) yields:

$$\left.\frac{d\underline{v}_e}{dt}\right|_i = \underline{f} + \underline{g} - \underline{\omega}_{ie} \times \underline{v}_e - \underline{\omega}_{ie} \times (\underline{\omega}_{ie} \times \underline{r}) \tag{4.25}$$

To first order, the earth's mass attraction vector, $\underline{g}$, pulls objects towards the mass center of the earth (this ignores the fact that the earth is ellipsoidal rather than spherical). Similarly, centripetal force pulls objects towards the earth rotational axis. From the perspective of the rotating earth frame, centrifugal force is pulling the object away from the rotational axis. Both mass attraction and centripetal/centrifugal force are functions of the object position only and are not separately distinguishable. An accelerometer will measure the vector sum of the two. Thus, "effective" gravity (also known as local gravity or apparent gravity) is given by:

$$\underline{g}_{eff} = \underline{g} - \underline{\omega}_{ie} \times \underline{\omega}_{ie} \times \underline{r} \tag{4.26}$$

This is depicted graphically in Figure 4.5. The vector triple-product term in (4.26) points outward from and is orthogonal to, the earth's polar axis. Effective gravity is also known as "plumb bob" gravity since a carpenter's plumb bob will align itself with $\underline{g}_{eff}$ and not $\underline{g}$ itself. Effective gravity has historically been the definition of local vertical (note: with modern global reference datums, effective gravity is now just a first-order approximation of local vertical).

If we were being purely formal, mass attraction should be referred to as "gravitation" and the vector sum of gravitation and centrifugal acceleration is earth's "gravity." As a result the phrase "effective gravity" is actually a misnomer since it actually is gravity. Nevertheless, to be consistent with existing inertial literature, we will refer to gravitation as mass attraction and its vector sum with centrifugal acceleration as effective gravity.

Substitution of (4.26) into (4.25) yields the final form of the navigation equation mechanized in the inertial frame:

$$\left.\frac{d\underline{v}_e}{dt}\right|_i = \underline{f} - \underline{\omega}_{ie} \times \underline{v}_e + \underline{g}_{eff} \tag{4.27}$$

4.3 Navigation frame mechanization of the navigation equation

We have now laid the necessary foundation to derive the navigation equation expressed in the navigation frame. We start with a variation of (4.11) in which we relate the derivative of earth-referenced velocity evaluated in the inertial frame and the navigation frame:

$$\left.\frac{d\underline{v}_e}{dt}\right|_i = \left.\frac{d\underline{v}_e}{dt}\right|_n + \underline{\omega}_{in} \times \underline{v}_e \tag{4.28}$$

which can be rearranged:

$$\left.\frac{d\underline{v}_e}{dt}\right|_n = \left.\frac{d\underline{v}_e}{dt}\right|_i - \underline{\omega}_{in} \times \underline{v}_e \tag{4.29}$$

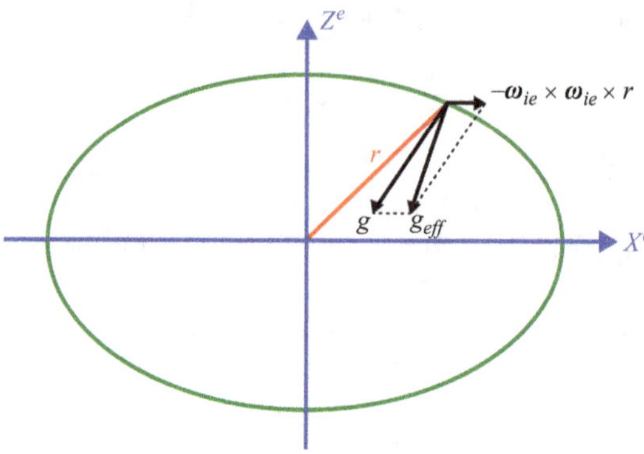

Figure 4.5 Effective gravity, $\underline{g}_{eff}$, is the vector sum of the earth's mass attraction vector, $\underline{g}$, and the centrifugal pseudo-acceleration vector. The eccentricity of the earth is grossly exaggerated for the sake of visual clarity. Note also that the mass attraction gravity vector does not point at the mass center of the earth (again, exaggerated in the figure). This is due to mass variations in the earth (e.g., mountains, deep ocean trenches, constituent variations) that perturb the local gravity vector. These perturbations are known as "gravity anomalies" and "deflection of the vertical"

The angular rate vector, $\underline{\omega}_{in}$, is the rate of the navigation frame with respect to the inertial frame and is known as "spatial" rate. Spatial rate is the sum of earth rate and transport rate (recall the description of these two rates in the Reference Frames chapter):

$$\underline{\omega}_{in} = \underline{\omega}_{ie} + \underline{\omega}_{en} \tag{4.30}$$

where:

$\underline{\omega}_{ie}$ = angular rate of the earth frame relative to the inertial frame ("earth rate")

$\underline{\omega}_{en}$ = angular rate of the navigation frame relative to the earth frame ("transport rate" or "craft rate")

The algorithms used to compute earth rate and transport rate will be discussed in a subsequent chapter. Substitution of (4.30) into (4.29) yields:

$$\left.\frac{d\underline{v}_e}{dt}\right|_n = \left.\frac{d\underline{v}_e}{dt}\right|_i - (\underline{\omega}_{ie} + \underline{\omega}_{en}) \times \underline{v}_e \tag{4.31}$$

Substitution of (4.27) into (4.31) yields:

$$\left.\frac{d\underline{v}_e}{dt}\right|_n = \underline{f} - (2\underline{\omega}_{ie} + \underline{\omega}_{en}) \times \underline{v}_e + \underline{g}_{eff} \tag{4.32}$$

The final form of the navigation equation is given by expressing all terms in the navigation frame:

$$\dot{\underline{v}}_e^n = C_b^n \underline{f}^b - (2\underline{\omega}_{ie}^n + \underline{\omega}_{en}^n) \times \underline{v}_e^n + \underline{g}_{eff}^n \tag{4.33}$$

The first term on the right-hand side of (4.33) acknowledges that the specific force vector is measured in the body frame (in a strapdown system) and multiplication by the body-to-nav DCM is necessary to convert the vector into the navigation frame. The cross-product term in (4.33) is referred to as the Coriolis term or the Coriolis correction.

The discrete-time solution of (4.33) provides the velocity update needed for inertial navigation. Before we get there, though, we have to cover several topics. Specifically, we need to be able to:

1) compute both earth rate and transport rate in the navigation frame;
2) compute effective gravity in the navigation frame;
3) extend our body-to-inertial DCM update algorithm to handle the update of the body-to-nav DCM.

For (3), note that the navigation frame rotates as well as the body frame. When updating the body-to-nav DCM, we will need to update for nav-frame rotation as well as body-frame rotation. We will address all of these issues in the next two chapters.

References

[1] Goldstein H. *Classical Mechanics*. 2nd ed. Reading, MA: Addison Wesley; 1980.

[2] Titterton DH, Weston JL. *Strapdown Inertial Navigation Technology*. 2nd ed. Herts, UK: The Institution of Electrical Engineers; 2004.

Chapter 5

Earth rate, transport rate, gravity, and the velocity update

Introduction

At the end of the previous chapter, we derived the navigation equation mechanized in the navigation frame, listed again here:

$$\dot{\underline{v}}_e^n = C_b^n \underline{f}^b - (2\underline{\omega}_{ie}^n + \underline{\omega}_{en}^n) \times \underline{v}_e^n + \underline{g}_{eff}^n \tag{5.1}$$

We need to develop algorithms to compute the earth rate, transport rate and effective gravity vectors expressed in the navigation frame.

5.1 Earth rate in the navigation frame

Strictly speaking, the instantaneous location of the earth's polar axis varies slightly over time (i.e., a few meters over the course of a year). For our purposes, however, we can assume the earth's angular rate vector is both a constant and is coincident with the z-axis of the earth frame. As a result, the most natural expression of the earth's angular rate vector is in the earth frame:

$$\underline{\omega}_{ie}^e = \begin{bmatrix} 0 \\ 0 \\ \Omega \end{bmatrix} \tag{5.2}$$

where the WGS-84 definition of the magnitude of the earth's angular rate is:

$$\Omega = 7.292115 \times 10^{-5} \qquad (\text{rad/s}) \tag{5.3}$$

What we need, however, is an expression for this vector in the navigation frame. As shown in Figures 5.1 and 5.2, we can resolve the vector into a north component and a vertical component as a function of latitude:

$$\underline{\omega}_{ie}^n = \begin{bmatrix} \Omega \cos L \\ 0 \\ -\Omega \sin L \end{bmatrix} \tag{5.4}$$

where L is the latitude. Note the z component is negative since the n-frame z-axis points down.

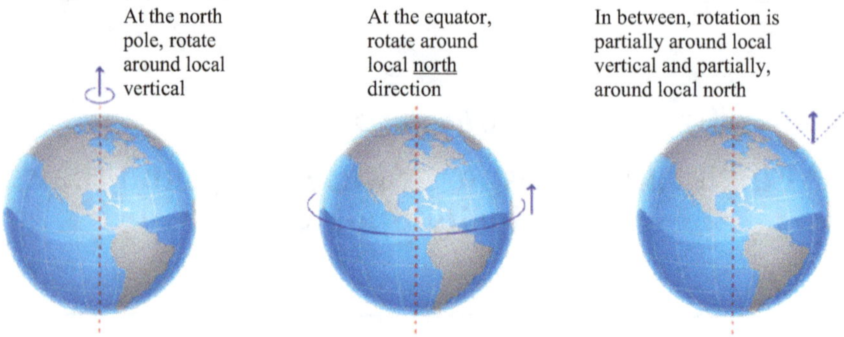

Figure 5.1 The earth's angular rate vector has north and vertical components that are functions of latitude. This can be understood by considering the two extreme cases (North pole and the equator). Note the east component of earth rate is zero

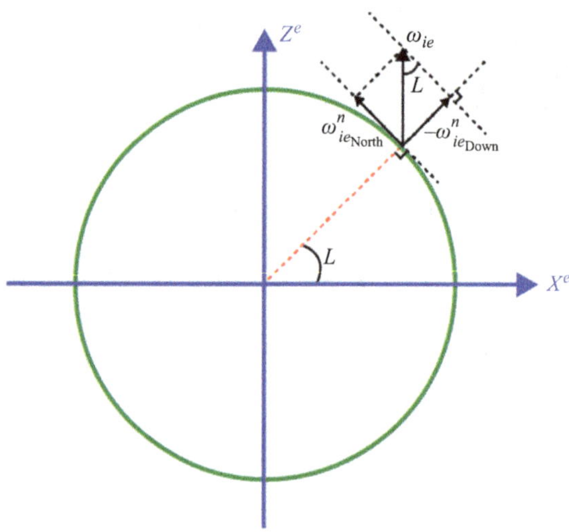

Figure 5.2 The earth's angular rate vector can be resolved into north and vertical components as a function of latitude (L)

For the sake of completeness, we also note the aforementioned result can be obtained as follows:

$$\underline{\omega}^n_{ie} = C^n_e \underline{\omega}^e_{ie} \tag{5.5}$$

where C^n_e is the direction cosine matrix relating the earth-frame to the nav-frame. This matrix will be discussed in detail in the next chapter.

5.2 Transport rate in the navigation frame

The rotation rate of the navigation frame relative to the earth frame is referred to as transport rate (or craft rate). It is governed exclusively by the motion of the body with respect to the earth. By definition, the navigation frame is locally level and north-pointing. That is, the x- and y-axes reside in the local-level plane (or, equivalently, the z-axis is aligned with local vertical) and the x-axis is also north-pointing. As the vehicle travels over the surface of the earth, the navigation frame must be rotated in order to keep it locally level and north-pointing. The transport rate vector is the vector of angular rate components that must be applied to the navigation frame in order to keep it locally level and north pointing.

In a gimbaled system, the INS platform (on which the accelerometers were mounted) is physically torqued in order to keep it locally level and north pointing. Even with modern strapdown systems, we still refer to torquing rates although the navigation frame is virtual and the torquing is performed in software.

How do we compute transport rate? Let us start with the east component. Figure 5.3 depicts the earth from the side (i.e., from the extended equatorial plane). Although it is not explicitly depicted, assume a vehicle is located on the equator at intersection of the arrows labeled V_{North} and $R_N + h$ (note: the derivation does not depend on the latitude. The equator was chosen for convenience). V_{North} is thus the instantaneous north velocity of the vehicle.

Now consider a (geometric) plane that is locally level at the location of the vehicle. As the vehicle moves in the north direction, what happens to the local level plane? Clearly it has to rotate in order to stay locally level. Furthermore, it has to rotate about an axis that is aligned in an east–west direction. For convenience, we choose an east–west axis that intersects the center of the Earth. As was demonstrated graphically in a previous chapter, the velocity of a point traversing a circle is given by:

$$\underline{v}_{tangential} = \underline{\omega} \times \underline{r} \tag{5.6}$$

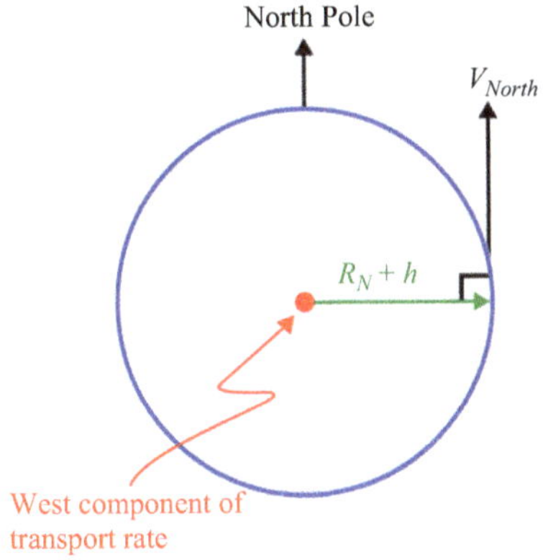

Figure 5.3 The east–west component of transport rate can be viewed as the ratio of the north velocity and the radius vector magnitude

where $\underline{\omega}$ is the angular rate of the radius vector. From the properties of the cross-product, we also know:

$$|\underline{v}_{tangential}| = |\underline{\omega}||\underline{r}| \sin\theta \tag{5.7}$$

where θ is the angle between $\underline{\omega}$ and $\underline{r}$. Since V_{North} is locally level by definition, the radius vector from the center of the Earth to the vehicle is orthogonal to V_{North}. We can thus write:

$$|V_{North}| = |\omega_{en_{East}}||\underline{r}| \sin\frac{\pi}{2} = |\omega_{en_{East}}||\underline{r}| \tag{5.8}$$

Solving for the magnitude of the angular rate:

$$|\omega_{en_{East}}| = \frac{|V_{North}|}{R_N + h} \tag{5.9}$$

In Figure 5.3, the radius is labeled "$R_N + h$." R_N is a "radius of curvature" that accounts for the fact that the Earth is an oblate spheroid rather than being purely spherical (i.e., the Earth is bulged out at the sides). The radii of curvature will be described in further detail later in this chapter. "h" is the altitude of the vehicle above the surface of the Earth. By the right hand rule, the angular rate vector is pointed out of the paper in Figure 5.3 and thus is pointed west. Since our navigation frame axes are oriented north–east-down, the negative of the angular rate vector will be the east component of transport rate:

$$\omega_{en_{East}} = \frac{-V_{North}}{R_N + h} = \frac{-V_N}{R_N + h} \tag{5.10}$$

Now we can proceed to the north component. Since the east component was a function of north velocity, we would expect that the north component of transport rate is a function of east velocity. Indeed, that is the case. Consider a top view of the

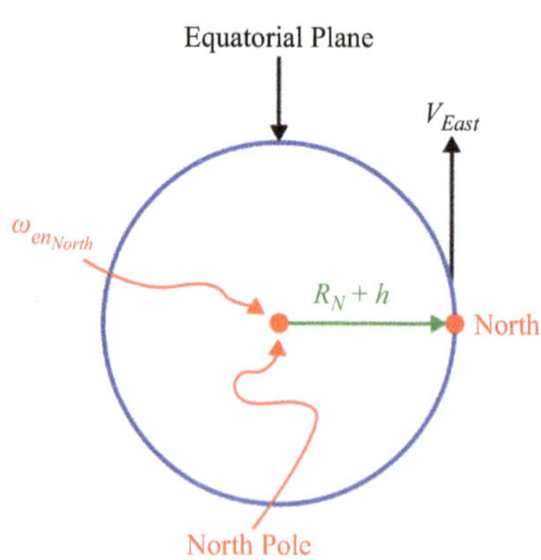

Figure 5.4 The north–south component of transport rate

Earth (i.e., from above the North pole) as shown in Figure 5.4. Following a similar development to that which yielded Equation (5.10), it can be shown that:

$$\omega_{en_{North}} = \frac{V_{East}}{R_E + h} = \frac{V_E}{R_E + h} \tag{5.11}$$

Examination of Figure 5.4 shows that by the right hand rule, the angular rate vector is aligned with North and is positive in the north direction and thus the implicit positive sign in the numerator of (5.11) is correct.

5.3 Radii of curvature

The radii of curvature used in Equations (5.10) and (5.11) are necessary since the Earth is not well modeled by a sphere and thus is not well described by a single radius. The Earth is more appropriately modeled by an ellipsoid (a.k.a., oblate spheroid). A given radius of curvature is the radius of the best-fitting circle to the ellipsoid at a given point. Of course, since a circle is a two-dimensional object, we must define its orientation before we can fit it to the three-dimensional ellipsoid. For convenience, we fit circles in an east–west plane and a north–south plane. They are illustrated in Figures 5.5 and 5.6.

The meridional radius of curvature is the radius of the best-fitting circle that lies in a north-south plane (i.e., along a meridian or line of longitude). It is given by:

$$R_N = \frac{R_0(1 - e^2)}{(1 - e^2 \sin^2 L)^{\frac{3}{2}}} \tag{5.12}$$

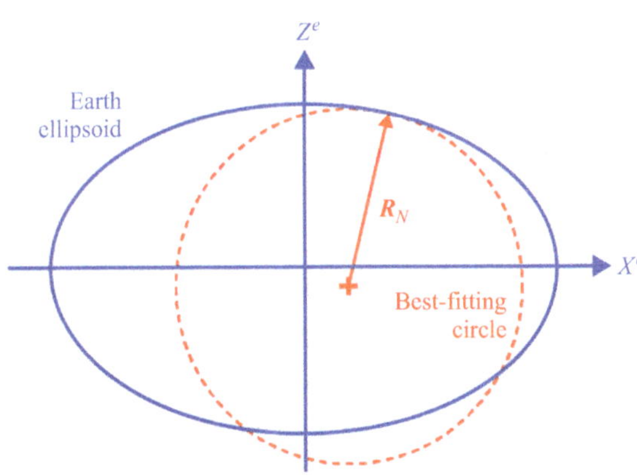

Figure 5.5 The meridional radius of curvature is the radius of the best-fitting north–south oriented circle (R_N) at the latitude of interest. The eccentricity of the earth ellipsoid is grossly exaggerated for the purposes of visual clarity

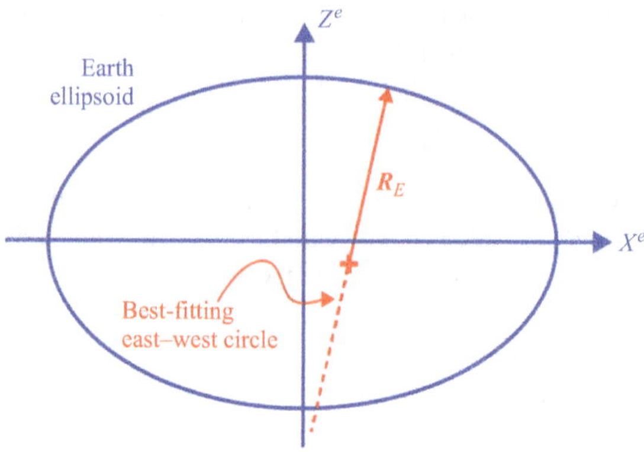

Figure 5.6 The transverse radius is the radius of the best-fitting east–west oriented circle (R_E) at the latitude of interest. The eccentricity of the earth ellipsoid is grossly exaggerated for the purposes of visual clarity

where R_0 is the semi-major axis of the earth's reference ellipsoid. This is the equatorial radius. The WGS-84 definition is 6,378,137.0 m. "e" is the eccentricity of the Earth's reference ellipsoid (it is *not* the base of the natural logarithm!). The WGS-84 definition is 0.0818191908426. L is the latitude of the point at which the radius is being computed. The subscript "N" denotes the north–south orientation of the best-fitting circle. Some authors use "R_M" where the "M" denotes meridian.

The other radius of curvature is known as the transverse or prime or principal radius of curvature, R_E. The subscript "E" denotes the east–west orientation of the best-fitting circle. Some authors use "R_P" where the "P" denotes prime or principal. As shown in Figure 5.6, it is the best-fitting circle that lies in an east–west plane containing the point of interest and the center of the ellipsoid. It is given by:

$$R_E = \frac{R_0}{(1 - e^2 \sin^2 L)^{\frac{1}{2}}} \tag{5.13}$$

Thus, the meridian radius of curvature is the radius of the best-fitting circle that passes north–south through the point of interest and the transverse radius of curvature is the radius of the best-fitting circle that passes east–west through the point. The two radii are not the same since the earth is bulged out at the equator. The Earth's eccentricity is small and thus there is not a great deal of difference between the two radii but it is enough that we have to take it into account. The so-called Gaussian radius of curvature is the radius of the best-fitting *sphere* to the ellipsoid at any particular point. It is given by the geometric mean of the two radii of curvature:

$$R_G = \sqrt{R_E R_N} \tag{5.14}$$

Additional details on these radii may be found in [1].

5.4 Relationship between transport rate and position angle rates

There is a unique relationship between transport rate and the rates of the position angles. Figure 5.7 illustrates the latitude and longitude rate vectors along with the horizontal transport rate components. A generic point of interest (e.g., the location of the INS) is at the origin of the locally level ENU coordinate frame shown (note that the "up" direction has been depicted instead of "down" so as not to clutter the diagram). The latitude rate vector lies along the east–west direction (and is positive pointing west) and the longitude rate vector lies along the Earth's rotation axis. Note these directions are consistent with the right-hand rule for positive rotations. Furthermore, by comparing Figure 5.3 with Figure 5.7, it can be concluded that latitude rate is coincident with the east–west component of transport rate and is positive in the west direction. Thus, based on (5.10), we can write:

$$\dot{L} = -\omega_{en_{East}} = \frac{V_N}{R_N + h} \tag{5.15}$$

In principle, latitude could be updated by numerically integrating Equation (5.15). However, this is not a robust technique since north-velocity is meaningless at the North and South poles. Later we will learn about a better way to update position angles.

The relationship between the longitude rate vector and the north component of transport rate is more difficult since the two are not collinear. This is illustrated in Figure 5.8. Figure 5.8 also shows that longitude rate can be decomposed into the north and "up" components of transport rate. Figure 5.9 illustrates the exact relationships.

As shown in Figure 5.9, the north component of transport rate can be expressed as:

$$\omega_{en_{North}} = \dot{\lambda} \cos L \tag{5.16}$$

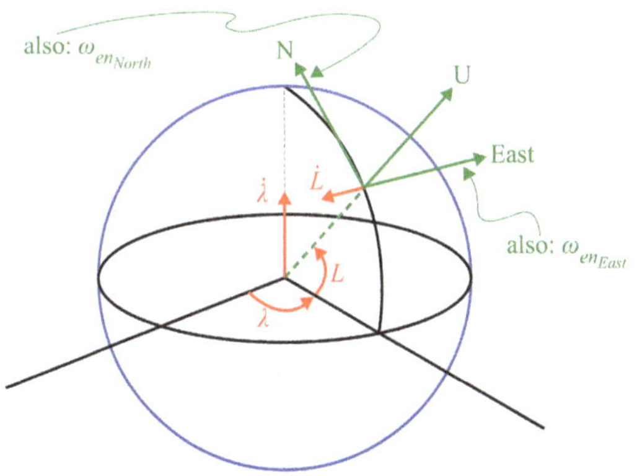

Figure 5.7 Illustration of position angles, position angle rates, and horizontal transport rates. Latitude (L) rate is positive in the west direction and longitude (λ) rate is positive along the positive z-axis of the earth-frame

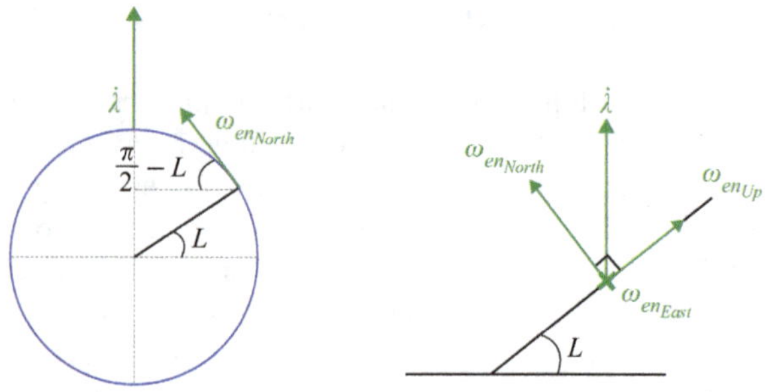

Figure 5.8 Longitude rate can be decomposed into "north" and "up" components. These components are also the north and up components of transport rate

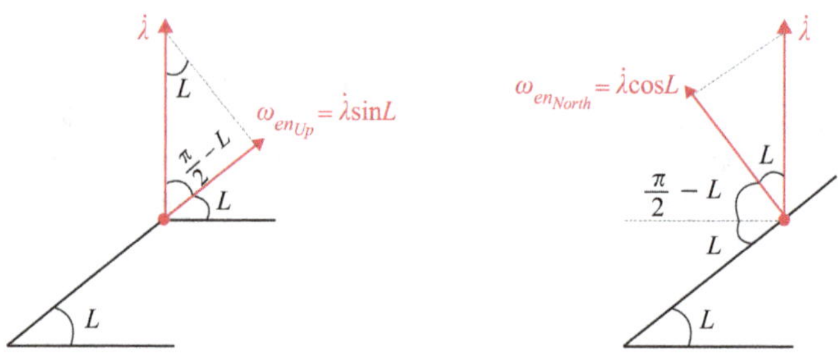

Figure 5.9 The north and "up" components of transport rate are components of longitude rate

Equation (5.16) can be rearranged to solve for longitude rate:

$$\dot{\lambda} = \frac{\omega_{en_{North}}}{\cos L} = \frac{V_E}{(R_E + h)\cos L} = \frac{V_E \sec L}{(R_E + h)} \tag{5.17}$$

This allows us to form an equation for the vertical component of transport rate in terms of the horizontal velocity. As shown in Figure 5.9, the down component of transport can be written as:

$$\omega_{en_{Down}} = -\dot{\lambda} \sin L \tag{5.18}$$

Substituting Equation (5.17) into (5.18) yields:

$$\omega_{en_{Down}} = \frac{-V_E}{(R_E + h)\cos L} \cdot \sin L = \frac{-V_E \tan L}{R_E + h} \tag{5.19}$$

Equations (5.10), (5.11), and (5.19) thus form the transport rate vector for the navigation frame:

$$\underline{\omega}^n_{en} = \left[\frac{V_E}{R_E + h} \quad \frac{-V_N}{R_N + h} \quad \frac{-V_E \tan L}{R_E + h} \right]^T \tag{5.20}$$

Now, you may say, "I understand that we needed the east and north components of transport rate in order to keep the frame locally level. But why do I need to rotate around the vertical?" We need to rotate around the vertical axis in order to keep it north-pointing. If we are going over the surface of the earth and are only maintaining local level, the x-axis of the navigation frame will drift away from North. For a north pointing mechanization, a transport rate component in the local vertical is required.

Equation (5.20) is perfectly acceptable for systems that only operate at low and/or mid-latitudes. However, there is a singularity that we must address. The navigation frame works extremely well except when you are operating in the vicinity of the earth's poles. What happens to the vertical component of transport rate as the vehicle gets close to the pole? The final expression in Equation (5.19) has the tangent of latitude in the numerator. As we get close to a pole, the tangent of the latitude approaches plus or minus infinity. That, obviously, is a problem. The north-pointing navigation frame mechanization simply does not work when you are operating in the vicinity of the poles.

Physically the problem has to do with longitude rate. Equation (5.17) also approaches infinity at the poles. The easiest way to think of it is to imagine a vehicle that is moving north along a line of longitude and is approaching the North pole. The longitude rate is zero since the vehicle is staying on a meridian. However, just as the vehicle passes over the pole, the longitude changes by 180 degrees. Thus, the longitude changes by 180 degrees instantaneously the moment the vehicle passes the pole. The 180 degrees of change in zero time is thus an infinite rate. More practically, even if the vehicle is not traveling purely along a meridian but still skims past the pole, the longitude rate will approach infinity. In a gimbaled system, the platform cannot be torqued sufficiently to keep it north-pointing. If you say, "Well, no problem. We've got computers now. We do the processing in a strapdown navigation computer." Sorry. Infinity is still a big number even in a double-precision arithmetic floating-point processor. You can simulate this scenario and prove that you will incur significant error, even when operating near the poles.

No matter how you slice it, you have to do something other than a north-pointing mechanization when operating in the vicinity of the poles. In a later chapter, we will learn about the "wander azimuth" mechanization. In the wander azimuth mechanization, we do not force one of the axes to point in the north direction. We will maintain a locally level platform but we will not force one of the axes to point north. We will do this by ignoring (specifically, setting to zero) the vertical component of transport rate. This is the most common way to enable so-called "all-latitude" operation. It works over the entire surface of the Earth. There are some additional complexities involved that we will learn about later. For the moment, we will assume that we are operating at mid-latitudes such that a north-pointing mechanization is acceptable.

5.5 Gravity models

At the surface of the Earth, the effective gravity vector, to first order, lies along the local vertical (and thus there is no horizontal component). Counterintuitively, the effective gravity vector has a *local level* component (specifically, in the north–south direction) that grows with increase in altitude. If the earth were perfectly spherical, then the gravity vector would always point “down.” However, since the earth is an ellipsoid (due to earth rotation), a strange thing happens. At the surface at, say, 45 deg North latitude, gravity is orthogonal to the ellipsoid. However, as you increase in altitude, the earth gradually becomes more like a point mass and the gravity vector gradually starts pointing at the center of the earth. This causes the gravity vector to have a non-zero “local level” component (specifically, in the north–south direction). Thus, as altitude increases, although the *magnitude* of the gravity vector decreases, the “north” component of the gravity vector actually increases from zero. If you go high enough, eventually this decreases as well.

Simple, vertical-only, gravity models are acceptable for applications operating at relatively low altitudes (i.e., less than 10 km) such as aircraft. An example is described in [2]:

$$g(0) = 9.780318(1 + 5.3024 \times 10^{-3} \sin^2 L - 5.9 \times 10^{-6} \sin^2 (2L)) \tag{5.21}$$

$$g(h) = g(0)\frac{R_G^2}{(R_G + h)^2} \tag{5.22}$$

where L is the latitude; R_G is the Gaussian radius of curvature of the Earth; h is the altitude above the Earth’s reference ellipsoid. $g(0)$ is given in units of meters-per-second-squared. R_G and h must be specified in the same units of distance (e.g., meters).

Higher fidelity gravity modeling involves determining expressions for the gravity potential (see, e.g., [3]). Theoretically this involves the computation of an infinite series with coefficients that must be determined based on data observed using satellites, the stars and ground-level gravity measurement equipment. In practice, of course, the infinite series is approximated by truncation. A commonly used low-order approximation is known as the “J2 gravity model” where “J2” is the sole coefficient. In the earth frame (with z-axis pointing out the North pole and the x-axis intersecting the equator at the prime meridian), the J2 gravity components are given by:

$$g_x = -\frac{xG_M}{r^3}\left[1 + \frac{3J_2a^2}{2r^2} - \frac{15J_2a^2z^2}{2r^4}\right] + x\Omega^2 \tag{5.23}$$

$$g_y = -\frac{yG_M}{r^3}\left[1 + \frac{3J_2a^2}{2r^2} - \frac{15J_2a^2z^2}{2r^4}\right] + y\Omega^2 \tag{5.24}$$

$$g_z = -\frac{zG_M}{r^3}\left[1 + \frac{9J_2a^2}{2r^2} - \frac{15J_2a^2z^2}{2r^4}\right] \tag{5.25}$$

where:
(x, y, z) are the earth frame coordinates of the point at which gravity is being computed;
r is the magnitude of the position vector of the point;
G_M is the WGS-84 value of the Earth’s gravitational constant: $3,986,005 \times 10^8\ \mathrm{m^3/s^2}$;

$J_2 = 108{,}263 \times 10^{-8}$;
a is the semi-major axis of the WGS-84 ellipsoid: 6,378,137 m;
Ω is the magnitude of the Earth's rotation rate (Equation (5.3))

The components can be gathered into a vector:

$$\underline{g}^e_{eff} = \begin{bmatrix} g_x & g_y & g_z \end{bmatrix}^T \tag{5.26}$$

The gravity vector can then be converted into the navigation frame with the earth-to-nav direction cosine matrix:

$$\underline{g}^n_{eff} = C^n_e \underline{g}^e_{eff} \tag{5.27}$$

where, as we mentioned earlier in the chapter, C^n_e is the earth-to-nav direction cosine matrix that we will learn how to compute in the next chapter. Although it is not obvious from a cursory examination of (5.23)–(5.25), the J2 gravity model is independent of longitude (i.e., it is only a function of latitude and altitude). It is thus symmetric about the earth's rotational axis. The north/east/down components of (5.27) can be computed for various latitudes and altitudes to show that the vector has vertical and north components only (no east component). In reality, irregularities in the earth's surface and composition cause the actual gravity vector to deviate slightly from the approximation given in (5.26) or (5.27). The magnitude of the error in the modeled gravity vector (known as "anomaly") along with the deviation of the direction of the actual vector (known as "deflection of the vertical") are only critical for high precision or high integrity applications.

5.6 Velocity update

Having derived the algorithms to compute earth rate, transport rate, and gravity, the discrete-time solution (i.e., update) of Equation (5.1) can be computed as:

$$\underline{v}^n_e[k+1] = \underline{v}^n_e[k] + C^n_b \Delta \underline{v}^b - \int_{t_k}^{t_{k+1}} [(2\underline{\omega}^n_{ie} + \underline{\omega}^n_{en}) \times \underline{v}^n_e] dt + \int_{t_k}^{t_{k+1}} \underline{g}^n_{eff} dt \tag{5.28}$$

Since the Coriolis and gravity terms change slowly with time, low-order numerical integration techniques can be employed. The velocity vector in the Coriolis term is purposely shown without a specific time-index. It can either be extrapolated from the previous velocity solution or can simply be set equal to the previous velocity solution if the vehicle does not exhibit high acceleration and/or the application does not require the highest precision.

References

[1] Grewal M, Andrews A, Bartone C. *Global Navigation Satellite Systems, Inertial Navigation, and Integration*. 3rd ed. Hoboken, NJ: Wiley; 2013.

[2] Titterton DH, Weston JL. *Strapdown Inertial Navigation Technology*. 2nd ed. Herts, UK: The Institution of Electrical Engineers; 2004.

[3] Hsu D. Comparison of Four Gravity Models. In: *Record of the IEEE Position, Location and Navigation Symposium (PLANS)*; 1996.

Chapter 6
Spatial rate, navigation frame rotation, and position update

Introduction

Previously we derived the body-frame update of the body-to-inertial DCM (C_b^i). This DCM was appropriate for use in a "fixed frame" environment (i.e., an inertial reference frame). For terrestrial, marine and aeronautical applications, however, we need to operate with respect the non-inertial earth frame. Our local-level reference frame, in many applications, is the navigation frame and thus we need to update the body-to-nav DCM (C_b^n). Since we have already derived the body-frame update, we need to derive the nav-frame update.

6.1 Navigation-frame update

In order to determine the nav-frame update, we will first consider the total body-to-nav DCM update and then will break it into two parts (body-frame update and nav-frame update). We have previously derived the equation for the derivative of the direction cosine matrix. Thus, we can write:

$$\dot{C}_b^n = C_b^n \Omega_{nb}^b \tag{6.1}$$

where $\Omega_{nb}^b = [\underline{\omega}_{nb}^b \times]$ is the skew-symmetric matrix form of the body-to-nav angular rate vector. We have not yet encountered this particular angular rate vector. It can be viewed as the difference between body-rate and spatial-rate:

$$\underline{\omega}_{ib} = \underline{\omega}_{in} + \underline{\omega}_{nb} \tag{6.2}$$

Equation (6.2) states that the angular rate of the body-frame (relative to the inertial-frame) is equal to the angular rate of the navigation-frame (relative to the inertial frame) plus the angular rate of the body-frame (relative to the navigation-frame). In other words, start with the angular rate of the navigation frame and then add on whatever extra rate there is between the body-frame and the navigation frame. Solving Equation (6.2) for the body-to-nav rate yields:

$$\underline{\omega}_{nb} = \underline{\omega}_{ib} - \underline{\omega}_{in} \tag{6.3}$$

Equation (6.1) can thus be rewritten as:

$$\dot{C}_b^n = C_b^n \Omega_{nb}^b = C_b^n [\Omega_{ib}^b - \Omega_{in}^b] = C_b^n \Omega_{ib}^b - C_b^n \Omega_{in}^b \tag{6.4}$$

The body-to-nav DCM derivative thus consists of a body-rate term and a nav-rate term. We have essentially already derived the discrete-time update for the body-rate term. It is the same as the body-to-inertial DCM update since it is a function of the body-frame angular rate vector that is measured by the gyros. The navigation-frame angular rate term, however, will take a bit more work. Consider the skew-symmetric matrix form of the navigation-frame angular rate vector in Equation (6.4):

$$\Omega_{in}^b = [\underline{\omega}_{in}^b \times] \tag{6.5}$$

Although it is not immediately obvious, the key challenge is that the navigation-frame angular rate vector is being expressed in the body-frame. Recall this angular rate is also called "spatial rate" and is the sum of earth rate and transport rate:

$$\underline{\omega}_{in} = \underline{\omega}_{ie} + \underline{\omega}_{en} \tag{6.6}$$

Previously we derived the equations for these vectors but we did so by expressing them in the navigation-frame, not the body-frame. Thus, we need to derive an alternate form of Equation (6.4) in which the spatial rate is expressed in the navigation frame:

$$\Omega_{in}^n = [\underline{\omega}_{in}^n \times] \tag{6.7}$$

Through the use of a so-called "similarity transformation," it can be shown (see Appendix A in this chapter for proof) that:

$$C_b^n \Omega_{in}^b = \Omega_{in}^n C_b^n \tag{6.8}$$

Substituting Equation (6.8) into Equation (6.4) yields:

$$\dot{C}_b^n = C_b^n \Omega_{ib}^b - \Omega_{in}^n C_b^n \tag{6.9}$$

As mentioned earlier, we have already derived the discrete-time update for the first term in Equation (6.9). The exact discrete-time solution of the second term is given by:

$$C_b^{n[k+1]} = \left[e^{-\int_{t_k}^{t_{k+1}} \Omega_{in}^n dt} \right] C_b^{n[k]} \tag{6.10}$$

If we assume the spatial rate vector is stationary over the integration interval (reasonable for all vehicles other than rockets and missiles), the integral in (6.10) can be written as:

$$\int_{t_k}^{t_{k+1}} \Omega_{in}^n dt = [\underline{\eta} \times] \tag{6.11}$$

This is the skew-symmetric matrix form of the integral of the spatial-rate vector:

$$\underline{\eta} = \int_{t_k}^{t_{k+1}} \underline{\omega}_{in}^n dt \tag{6.12}$$

The matrix exponential in (6.10) can thus be rewritten as:

$$e^{-\int_{t_k}^{t_{k+1}} \Omega_{in}^n dt} = e^{-[\underline{\eta} \times]} = e^{[-\underline{\eta} \times]} \tag{6.13}$$

where the last two expressions simply indicate the negation can be taken at the matrix level or the angle vector level. By forming the Taylor Series expansion and judiciously gathering terms, it can be shown the matrix exponential can be expressed as:

$$e^{-[\underline{\eta}\times]} = I - \frac{\sin(\eta)}{\eta}[\underline{\eta}\times] + \frac{1-\cos(\eta)}{\eta^2}[\underline{\eta}\times]^2 \tag{6.14}$$

where the magnitude of the angle vector is computed in the usual way:

$$\eta = |\underline{\eta}| = \sqrt{\eta_x^2 + \eta_y^2 + \eta_z^2} \tag{6.15}$$

Since the magnitude of the spatial rate vector is small even for high-speed aircraft, it is common to use a low-order series approximation of (6.13) or (6.14) and use low-order numerical integration (i.e., rectangular or trapezoidal) of (6.12). Trapezoidal integration with a first-order series approximation yields:

$$e^{-\int_{t_k}^{t_{k+1}} \Omega_{in}^n dt} = e^{-[\underline{\eta}\times]} \approx I - \frac{1}{2}(\Omega_{in}^n[k] + \Omega_{in}^n[k+1])\Delta t \tag{6.16}$$

where: $\Delta t = t_{k+1} - t_k$

Even more simply, rectangular integration yields:

$$e^{-\int_{t_k}^{t_{k+1}} \Omega_{in}^n dt} = e^{-[\underline{\eta}\times]} \approx I - (\Omega_{in}^n[k+1])\Delta t \tag{6.17}$$

6.2 Complete body-to-nav update

We have shown that the update of the body-to-nav DCM can be broken into two pieces. Previously we showed that the body-frame update can be expressed as:

$$C_{b[j+1]}^n = C_{b[j]}^n e^{[\underline{\sigma}\times]} = C_{b[j]}^n C_{b[j+1]}^{b[j]} \tag{6.18}$$

where $\underline{\sigma}$ is the angle vector obtained from the gyro measurements (delta-thetas) that have been summed over the update interval and compensated for coning. This is the same as the equation that we derived back in the fixed reference frame chapter except here we are explicitly showing the time update of the body frame. We are also showing that the matrix exponential can be viewed as a direction cosine matrix relating the body-frame at the end of the interval to that at the beginning of the interval.

From Equation (6.10), the navigation frame update can be expressed as:

$$C_b^{n[k+1]} = \left[e^{-\int_{t_k}^{t_{k+1}} \Omega_{in}^n dt}\right] C_b^{n[k]} = C_{n[k]}^{n[k+1]} C_b^{n[k]} \tag{6.19}$$

Equations (6.18) and (6.19) together form the complete attitude update. Since the rotation rate of the vehicle (i.e., body-frame rate) can be very high (e.g., hundreds of degrees per second for high dynamic aircraft) and the rotation rate of the navigation frame (i.e., spatial rate) is guaranteed to be low (e.g., tens of degrees per hour for everything other than rockets and missiles), it is common to perform the body update at a higher rate than the navigation frame update [1]. For example, the body-frame update (Equation (6.18)) may be executed at 100 Hz whereas the nav-frame update

(Equation (6.19)) may only be executed at 10 Hz. In this case, the body-to-nav DCM would be updated ten times for body-frame rotation before it is updated for nav-frame rotation. For the sake of completeness, it should be noted that since the gyro measurements are typically provided at an even higher rate (e.g., 1,000 Hz), there is thus a third processing rate in which the gyro measurements are summed (and coning compensated) over the body-frame update interval.

The corresponding algorithm for the navigation frame update of the body-to-nav quaternion is given in Appendix B.

6.3 Position update

So far we have developed the algorithms to update attitude and velocity. We need to develop the position update algorithm but, before we can do so, we need to derive the quantity that represents the position angles: the earth-to-nav DCM.

6.4 Derivation of the earth-to-nav DCM

The earth-to-nav DCM (C_e^n) consists of a set of rotations that rotates the earth-frame to the nav-frame. Navigation frame orientation is unique for a given point on the earth (i.e., a specific latitude and longitude). If we start with the earth-frame and rotate around the z-axis by the longitude and then around the new orientation of the y-axis by the negative of the latitude, we end up with a frame whose orientation is up-east-north (UEN). In other words, after these two rotations, the resulting frame has an x-axis pointed in the local vertical (positive up), the y-axis is pointed east and the z-axis is pointed north. This is illustrated in Figure 6.1.

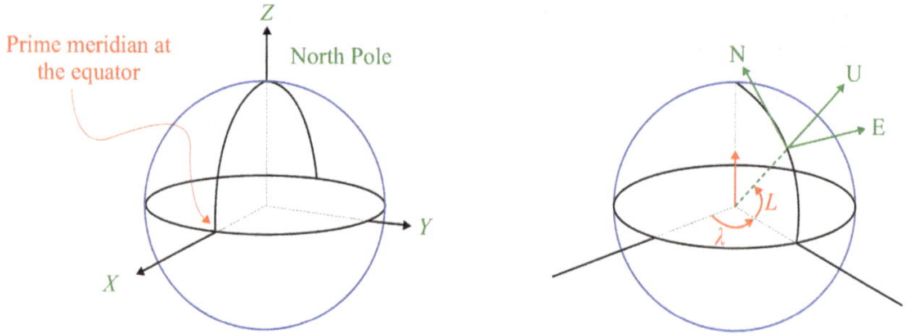

Figure 6.1 The earth-frame is depicted on the left. After a rotation about the z-axis by longitude (λ) and then a rotation about the new orientation of the y-axis by the negative of latitude (L), the resulting frame has its x-axis pointing up (U), its y-axis pointing east (E) and its z-axis pointing north (N)

The cascading of these two matrices can be written as:

$$C_e^{UEN} = C_y(-L)C_z(\lambda) \tag{6.20}$$

where L is the latitude and λ is the longitude (recall the rotation matrices were defined in the chapter on reference frames and rotations).

However, since what we want is the n-frame (i.e., the NED frame). We need to rotate about the y-axis by an additional -90 degrees. Thus:

$$C_e^n = C_y\left(-\frac{\pi}{2}\right) C_e^{UEN} = C_y\left(-\frac{\pi}{2}\right) C_y(-L)C_z(\lambda) \tag{6.21}$$

Although the two consecutive rotations about the y-axis in (6.21) could be combined into a single rotation, it is somewhat easier to compute (6.21) with the three rotations if doing so by hand. In any case, evaluation of (6.21) results in:

$$C_e^n = \begin{bmatrix} -\sin L\cos\lambda & -\sin L\sin\lambda & \cos L \\ -\sin\lambda & \cos\lambda & 0 \\ -\cos L\cos\lambda & -\cos L\sin\lambda & -\sin L \end{bmatrix} \tag{6.22}$$

Equation (6.22) allows us to initialize the DCM given initial values of latitude and longitude. We will shortly describe how to update this matrix over time through the use of the computed transport rate vector. At any given point in time, then, the position angles can be extracted from the updated C_e^n with:

$$L = \arcsin\left(-C_e^n[3,3]\right) \tag{6.23}$$

$$\lambda = \text{arctan2}(-C_e^n[3,2], -C_e^n[3,1]) \tag{6.24}$$

with the obvious exception that (6.24) cannot be evaluated when $L = \pm\frac{\pi}{2}$ (i.e., there is a singularity at the poles).

6.5 Earth-to-nav DCM update

In the chapter on fixed frame processing, we derived the derivative of the direction cosine matrix. Based on the general form of the direction cosine matrix differential equation, we can write the earth-to-nav DCM differential equation as:

$$\dot{C}_e^n = C_e^n \Omega_{ne}^e = C_e^n [\underline{\omega}_{ne}^e \times] \tag{6.25}$$

However, this is problematical since the transport rate vector in (6.25) is, first of all, reversed from our usual form ("ne" instead of "en") and, more importantly, expressed in the earth-frame instead of the navigation frame. We can make the situation significantly easier by simply writing the differential equation for the *transpose* of the earth-to-nav DCM:

$$\dot{C}_n^e = C_n^e \Omega_{en}^n = C_n^e [\underline{\omega}_{en}^n \times] \tag{6.26}$$

The discrete-time solution of (6.26) is given by:

$$C_n^e[k+1] = C_n^e[k] e^{\int_{t_k}^{t_{k+1}} \Omega_{en}^n dt} = C_n^e[k] e^{[\underline{\gamma}\times]} \tag{6.27}$$

where:

$$\underline{\gamma} = \int_{t_k}^{t_{k+1}} \underline{\omega}_{en}^n dt \tag{6.28}$$

Equations (6.27) and (6.28) are valid as long as the transport rate vector can be considered stationary over an update interval. This is a reasonable assumption for all vehicles other than rockets and missiles. The closed form expression for the matrix exponential in (6.27) is:

$$e^{[\underline{\gamma}\times]} = I + \frac{\sin\gamma}{\gamma}[\underline{\gamma}\times] + \frac{1-\cos\gamma}{\gamma^2}[\underline{\gamma}\times]^2 \tag{6.29}$$

where γ is the magnitude of $\underline{\gamma}$. Computationally simpler approximations are used in practice. Trapezoidal integration of the transport rate vector with a first-order series approximation of the matrix exponential yields:

$$e^{\int_{t_k}^{t_{k+1}} \Omega_{en}^n dt} = e^{[\underline{\gamma}\times]} \approx I + \frac{1}{2}(\Omega_{en}^n[k] + \Omega_{en}^n[k+1])\Delta t \tag{6.30}$$

where: $\Delta t = t_{k+1} - t_k$

Even more simply, rectangular integration yields:

$$e^{\int_{t_k}^{t_{k+1}} \Omega_{en}^n dt} = e^{[\underline{\gamma}\times]} \approx I + (\Omega_{en}^n[k+1])\Delta t \tag{6.31}$$

In practice, the earth-to-nav DCM is transposed in order to perform the update in Equation (6.27) and the result is transposed back so that it can be used elsewhere in the inertial processing.

6.6 Altitude

At first glance, it would appear that the inertial vertical position solution is trivial. The full velocity vector discrete-time update is given by:

$$v_e^n[k+1] = v_e^n[k] + C_b^n \underline{\Delta v}^b[k+1] + \int_{t_k}^{t_{k+1}} \left[-(2\underline{\omega}_{ie}^n + \underline{\omega}_{en}^n)\times v_e^n + \underline{g}_{eff}^n \right] dt \tag{6.32}$$

Thus the differential equation for height is simply:

$$\dot{h} = -v_e^n(z) \tag{6.33}$$

and its discrete-time update is then:

$$h[k+1] = h[k] + \int_{t_k}^{t_{k+1}} [-v_e^n(z)]dt \tag{6.34}$$

However, as we will learn later, the altitude update algorithm is *unstable*. Specifically, any error, either in initial height or in the computed vertical component of velocity, will induce a positive feedback that will cause errors both in computed vertical velocity and altitude to grow without bound. The so-called "inertial vertical channel" can thus only be used for brief periods of time (with the length dependent upon the accuracy of the accelerometers, gravity model and knowledge of initial height).

6.7 Overall strapdown inertial navigation processing

We have thus completed the development of the strapdown inertial navigation processing for position, velocity, and attitude. Figure 6.2 illustrates the overall strapdown inertial navigation processing in continuous-time form.

The inputs consist of the body-rate vector measured by the gyros ($\underline{\omega}_{ib}^b$) and the specific-force vector measured by the accelerometers ($\underline{f}^b$). In the upper left, the body-to-nav DCM is updated based on the measured body rates and the computed earth and transport rates. On the right, the body-to-nav DCM is used to resolve the specific force vector into the navigation frame as part of the velocity updating (along with the Coriolis correction and gravity compensation).

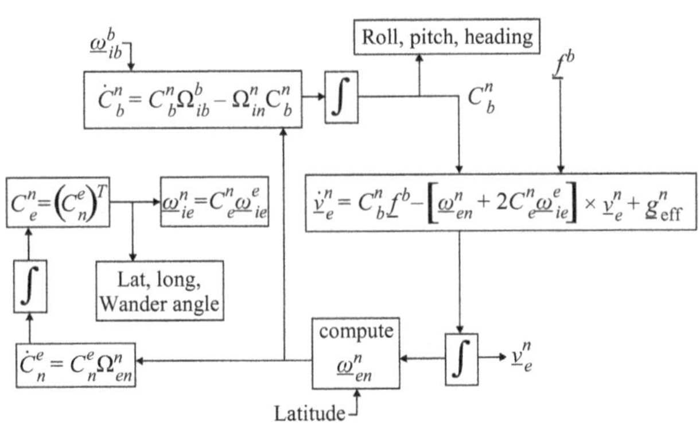

Figure 6.2 Strapdown inertial navigation processing in continuous-time form [2]

At the bottom, the velocity vector is used to compute transport rate (with recent latitude as a necessary additional input), and, on the lower left, the transport rate is used to update the nav-to-earth DCM which is then (middle left) transposed to yield the earth-to-nav DCM (from which latitude and longitude can be extracted). Also note that wander angle will be extensively treated in a later chapter.

6.8 Appendix A: Spatial-rate vector similarity transformation

As described in the main text of the chapter, we seek a transformation of the skew-symmetric matrix form of the spatial rate vector expressed in the body-frame (Ω_{in}^{b}) to an equivalent matrix with the underlying vector expressed in the navigation frame (Ω_{in}^{n}). We will start by deriving a so-called "similarity transformation" that converts a given linear transformation into an equivalent transformation expressed in a rotated coordinate frame. Equation (6.35) expresses the usual linear model:

$$\underline{y} = A\underline{x} \tag{6.35}$$

where A is a matrix that transforms vector $\underline{x}$ into vector $\underline{y}$. Now consider the same transformation viewed from a rotated coordinate frame:

$$\underline{x}' = B\underline{x} \tag{6.36}$$

$$\underline{y}' = B\underline{y} \tag{6.37}$$

where B is the rotation matrix that converts from the unprimed to the primed coordinate frame. We seek to find the transformation A' in the new (i.e., rotated) coordinate frame that relates x' to y':

$$\underline{y}' = A'\underline{x}' \tag{6.38}$$

We start by substituting (6.35) into (6.37):

$$\underline{y}' = B\underline{y} = B\,A\,\underline{x} \tag{6.39}$$

Now maintain the equality by inserting an identity matrix:

$$\underline{y}' = B\,A\,\underline{x} = B\,A\,I\,\underline{x} \tag{6.40}$$

and substitute $B^{-1}B$ for the identity matrix:

$$\underline{y}' = B\,A\,I\,\underline{x} = B\,A\,B^{-1}\,B\underline{x} \tag{6.41}$$

Finally, substitute (6.36) into (6.41):

$$\underline{y}' = B\,A\,B^{-1}\,B\underline{x} = B\,A\,B^{-1}\,\underline{x}' \tag{6.42}$$

Comparing (6.38) and (6.42) proves:

$$A' = B\,A\,B^{-1} \tag{6.43}$$

Thus, the linear transformation A in a given coordinate frame is similar to the transformation $B\,A\,B^{-1}$ in the rotated frame.

We will now exploit this relationship in order to achieve the necessary change of the skew-symmetric matrix form of the transport rate vector. In order to prove equation (6.8), we start by inserting identity matrices (which, of course, do not change the equality) around the skew-symmetric matrix on the left side of (6.8):

$$C_b^n \Omega_{in}^b = C_b^n I \Omega_{in}^b I \tag{6.44}$$

Since the transpose of the DCM is also its inverse, we know that:

$$C_n^b C_b^n = I \tag{6.45}$$

Substituting (6.45) into (6.44) yields:

$$C_b^n \Omega_{in}^b = C_b^n \{C_n^b C_b^n\} \Omega_{in}^b \{C_n^b C_b^n\} \tag{6.46}$$

which can, for convenience, be grouped as:

$$C_b^n \Omega_{in}^b = \{C_b^n C_n^b\} \{C_b^n \Omega_{in}^b C_n^b\} C_b^n \tag{6.47}$$

The first term in braces in (6.47) is the identity matrix and thus can be dropped from the equation. The second term in braces is a similarity transformation. With the relationship proven in Equation (6.43), we can conclude:

$$\Omega_{in}^n = C_b^n \Omega_{in}^b C_n^b \tag{6.48}$$

Substitution of (6.48) into (6.47) thus proves (6.8):

$$C_b^n \Omega_{in}^b = \Omega_{in}^n C_b^n \tag{6.49}$$

Appendix B: Updating the body-to-nav quaternion for navigation frame rotation

The navigation frame update for the body-to-nav quaternion is given by:

$$q_{k+1} = q_{n[k+1]}^{n[k]} * q_k = \begin{bmatrix} \cos\left(\frac{\eta}{2}\right) \\ \frac{-\eta_x}{\eta} \sin\left(\frac{\eta}{2}\right) \\ \frac{-\eta_y}{\eta} \sin\left(\frac{\eta}{2}\right) \\ \frac{-\eta_z}{\eta} \sin\left(\frac{\eta}{2}\right) \end{bmatrix} * q_k \tag{6.50}$$

where the $*$ represents quaternion multiplication. Note the first term in the product is a quaternion that represents the rotation of the navigation-frame from time t_{k+1} to t_k. It is equivalent to the quaternion of the negative of the rotation vector and is also equivalent to the quaternion conjugate of the rotation vector. The rotation vector, $\underline{\eta}$, is the integral of the spatial rate vector over the update interval as given in (6.12). As was shown in the main text, the magnitude of the rotation vector is computed in the usual way. Since spatial rate is small even for high-speed aircraft, it is common to use low-order numerical integration (i.e., rectangular or trapezoidal).

References

[1] Titterton DH, Weston JL. *Strapdown Inertial Navigation Technology*. 2nd ed. Herts, UK: The Institution of Electrical Engineers; 2004.
[2] Bose SC. *GPS/INS Multisensor Kalman Filter Navigation [Short Course Notes]*. Technalytics; 1997.

Chapter 7
The wander azimuth mechanization

7.1 Introduction

We have established the fact that maintaining a locally level reference frame that is north-pointing has some significant challenges when we are operating in the vicinity of the poles. We have discussed the fact that if we are on top of the North Pole, for example, no matter which direction we are facing: they are all south. By just barely skimming past the North Pole and trying to keep the n-frame north-pointing, the rotation rate of the local-level frame around the vertical axis approaches infinity. This occurs whether the frame is mechanical with a gimbaled system or mathematical with a strapdown system.

The primary way this problem is solved is through the so-called *wander azimuth* mechanization. We exploit the remaining degree of freedom that we have in converting from the earth-frame to the local-level frame. Previously we showed the conversion of the earth-frame to the nav-frame requires a rotation first around the earth-frame z-axis (in the amount of the longitude) followed by a rotation of $-(L + \frac{\pi}{2})$ about the (new) y-axis (where L is the latitude). In general, when converting from one frame to another we can have up to three independent rotations. In converting from the earth-frame to the nav-frame, however, we only utilize two rotations. The wander azimuth frame exploits that final degree of freedom in the conversion from the earth-frame to the locally level frame.

Specifically, the local-level frame will be allowed to rotate about the vertical axis such that it will not be forced to be north-pointing. By not forcing one axis to point north, we can derive a mechanization that does not "blow up" when operating in the vicinity of the poles.

This new locally level frame is referred to as the "wander frame" or just the w-frame. It should be noted that some references will use the term "navigation frame" but are actually working in the wander frame (because they never implement a north-pointing mechanization). The terms are thus not universal. In this book, however, the nav-frame is locally level and north-pointing whereas the wander frame is the more generalized version of nav-frame: locally level but the x-axis is not necessarily pointing north [1,2].

7.2 Derivation of the horizontal components of transport rate in the w-frame

Previously we derived the transport rate vector expressed in the navigation frame:

$$\underline{\omega}_{en}^{n} = \left[\frac{V_E}{R_E + h} \; \frac{-V_N}{R_N + h} \; \frac{-V_E \tan L}{R_E + h} \right]^T \tag{7.1}$$

As we noted, the vertical component approaches infinity for vehicle paths that go over, or skim past, one of the poles. One of the most common solutions to this problem is simply to *set* the vertical component of transport rate *to zero*. However, over time this results in the x-axis of the local-level frame "wandering away" from North. The repercussions are non-trivial: the x- and y-axes of the local-level frame are no longer pointed north and east. This impacts the entire inertial navigation mechanization. Local-level frame velocities are no longer north and east velocities and the horizontal components of transport rate will no longer be as simple as they are in Equation (7.1).

Consider Figure 7.1. In the navigation frame, we have north and east axes, north and east velocities and north and east components of transport rate. The wander frame is rotated around the z-axis by the wander angle α. The wander angle will grow over time as the vehicle moves due to the zero-ing of the vertical component of transport rate. We cannot refer to the wander-frame axes as having any particular direction since they will change over time. Thus we simply refer to the "wander-frame x and y axes," the "wander-frame x and y velocities," and the "wander-frame x and y components of transport rate."

The introduction of the wander frame forces us to derive wander-frame expressions for the horizontal components of transport rate. This will prove to be more challenging than performing a simple rotation. Later we will also need to derive the conversion between wander-frame velocities and nav-frame velocities since it is useful to know nav-frame quantities when not operating near a pole.

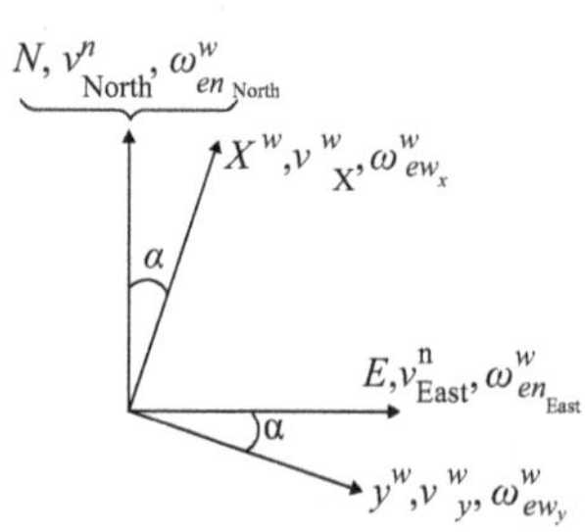

Figure 7.1 Comparison of n-frame and w-frame quantities. The angle between the two frames, α, is known as the wander angle. Note the n-frame z-axis, not shown, is "down" and thus is into the paper

To derive the horizontal components of transport rate in the wander frame, we start with the simple rotation. Figure 7.2 illustrates the geometric relationship. You can derive the result from Figure 7.2 using trigonometry or you can apply a simple rotation matrix:

$$\underline{\omega}_{ew}^{w} = C_z(\alpha)\underline{\omega}_{en}^{n} = \begin{bmatrix} \omega_{en\,North}^{n}\cos\alpha + \omega_{en\,East}^{n}\sin\alpha \\ -\omega_{en\,North}^{n}\sin\alpha + \omega_{en\,East}^{n}\cos\alpha \\ DC \end{bmatrix} \tag{7.2}$$

where C_z is the rotation matrix for rotations around the z-axis (defined in the chapter on reference frames and rotations) and "*DC*" stands for "don't care" since we will be zeroing out the vertical component anyway. Equation (7.2) is a good first step but we are far from done since the equation is a function of navigation frame transport rate components. We need to derive a form that is not dependent upon navigation frame quantities. To do so, we first substitute the horizontal components of nav-frame transport rate (from Equation (7.1) above) into Equation (7.2):

$$\underline{\omega}_{ew}^{w}(x) = \frac{V_E}{R_E + h}\cos\alpha - \frac{V_N}{R_N + h}\sin\alpha \tag{7.3}$$

$$\underline{\omega}_{ew}^{w}(y) = -\frac{V_E}{R_E + h}\sin\alpha - \frac{V_N}{R_N + h}\cos\alpha \tag{7.4}$$

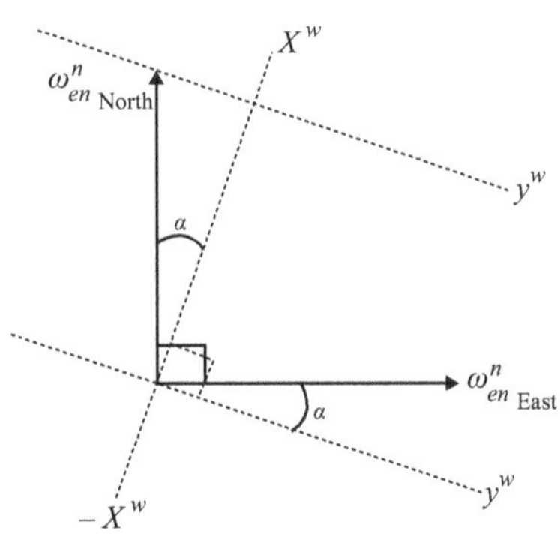

Figure 7.2 The navigation-frame transport rate components can be expressed in the wander frame either through component projections or the use of a rotation matrix

Next, we need to replace the east and north velocities with wander-frame x and y velocities. With the same rotation matrix, we can write:

$$\underline{v}^n = C_z(-\alpha)\underline{v}^w = \begin{bmatrix} v_x^w \cos\alpha - v_y^w \sin\alpha \\ v_x^w \sin\alpha + v_y^w \cos\alpha \\ v_z^w \end{bmatrix} \tag{7.5}$$

Note that we are using the reverse rotation to express navigation frame velocities in terms of wander-frame velocities. Substitution of the components of (7.5) into (7.3) and (7.4) yields a form that does not explicitly depend upon navigation frame quantities:

$$\underline{\omega}_{ew}^w(x) = \frac{v_x^w \sin\alpha + v_y^w \cos\alpha}{R_E + h}\cos\alpha - \frac{v_x^w \cos\alpha - v_y^w \sin\alpha}{R_N + h}\sin\alpha \tag{7.6}$$

$$\underline{\omega}_{ew}^w(y) = -\frac{v_x^w \sin\alpha + v_y^w \cos\alpha}{R_E + h}\sin\alpha - \frac{v_x^w \cos\alpha - v_y^w \sin\alpha}{R_N + h}\cos\alpha \tag{7.7}$$

Equations (7.6) and (7.7) can be rearranged to group the velocity terms:

$$\underline{\omega}_{ew}^w(x) = v_x^w\left[\frac{\cos\alpha\sin\alpha}{R_E + h} - \frac{\cos\alpha\sin\alpha}{R_N + h}\right] + v_y^w\left[\frac{\cos^2\alpha}{R_E + h} + \frac{\sin^2\alpha}{R_N + h}\right] \tag{7.8}$$

$$\underline{\omega}_{ew}^w(y) = -v_y^w\left[\frac{\cos\alpha\sin\alpha}{R_E + h} - \frac{\cos\alpha\sin\alpha}{R_N + h}\right] - v_x^w\left[\frac{\sin^2\alpha}{R_E + h} + \frac{\cos^2\alpha}{R_N + h}\right] \tag{7.9}$$

Two of the terms in square brackets are effectively radii of curvature in the wander frame. We can rewrite (7.8) and (7.9) as:

$$\underline{\omega}_{ew}^w(x) = \frac{v_x^w}{T} + \frac{v_y^w}{R_y} \tag{7.10}$$

$$\underline{\omega}_{ew}^w(y) = -\frac{v_x^w}{R_x} - \frac{v_y^w}{T} \tag{7.11}$$

where:

$$\frac{1}{R_x} = \left[\frac{\sin^2\alpha}{R_E + h} + \frac{\cos^2\alpha}{R_N + h}\right] \tag{7.12}$$

$$\frac{1}{R_y} = \left[\frac{\cos^2\alpha}{R_E + h} + \frac{\sin^2\alpha}{R_N + h}\right] \tag{7.13}$$

$$\frac{1}{T} = \left[\frac{\cos\alpha\sin\alpha}{R_E + h} - \frac{\cos\alpha\sin\alpha}{R_N + h}\right] \tag{7.14}$$

Equations (7.12) and (7.13) are the reciprocals of the radii of curvature in the wander frame. Equation (7.14) (where T stands for "torsional") accounts for the fact that the meridional and transverse radii of curvature are not equal.

Equations (7.10)–(7.14) provide the relationships needed to compute the horizontal components of transport rate in the wander frame. However, they raise some questions: How does the wander angle, α, get computed? Is not α the angle between north and the x-axis of the wander frame? Is not the whole reason we are doing this wander azimuth stuff precisely because keeping track of north is essentially impossible at the poles?

Yes, indeed. To untangle this knot, let us first learn how the wander angle gets computed.

7.3 The earth-to-wander direction cosine matrix

In the previous chapter, we derived the earth-to-nav direction matrix:

$$C_e^n = \begin{bmatrix} -\sin L\cos\lambda & -\sin L\sin\lambda & \cos L \\ -\sin\lambda & \cos\lambda & 0 \\ -\cos L\cos\lambda & -\cos L\sin\lambda & -\sin L \end{bmatrix} \tag{7.15}$$

Recall that L is the latitude and λ is the longitude. The wander frame differs from the navigation frame only by a rotation about the local vertical axis. The earth-to-wander DCM can thus be obtained from the earth-to-nav DCM by:

$$C_e^w = C_n^w C_e^n = C_z(\alpha)C_e^n \tag{7.16}$$

As indicated in the equation, the rotation matrix for a rotation about the z-axis by the wander angle is equivalent to the nav-to-wander DCM. Substitution of (7.15) into (7.16) yields:

$$C_e^w = \begin{bmatrix} -\cos\alpha\sin L\cos\lambda - \sin\alpha\sin\lambda & -\cos\alpha\sin L\sin\lambda + \sin\alpha\cos\lambda & \cos\alpha\cos L \\ \sin\alpha\sin L\cos\lambda - \cos\alpha\sin\lambda & \sin\alpha\sin L\sin\lambda + \cos\alpha\cos\lambda & -\sin\alpha\cos L \\ -\cos L\cos\lambda & -\cos L\sin\lambda & -\sin L \end{bmatrix} \tag{7.17}$$

Given the earth-to-wander DCM, the extraction of alpha (along with latitude and longitude) is given by:

$$L = \arcsin\left(-C_e^w[3,3]\right) \tag{7.18}$$

$$\lambda = \text{arctan2}(-C_e^w[3,2], -C_e^w[3,1]) \tag{7.19}$$

$$\alpha = \text{arctan2}(-C_e^w[2,3], C_e^w[1,3]) \tag{7.20}$$

The DCM update is performed with the same algorithm that was developed for the earth-to-nav DCM (recalling that we update the transpose rather than the DCM itself):

$$C_w^e[k+1] = C_w^e[k]e^{\int_{t_k}^{t_{k+1}}\Omega_{ew}^w dt} = C_w^e[k]e^{[\underline{\gamma}\times]} \tag{7.21}$$

where:

$$\underline{\gamma} = \int_{t_k}^{t_{k+1}} \underline{\omega}_{ew}^w dt \tag{7.22}$$

While the INS is operating and the earth-to-wander DCM has been updated, the wander angle can be extracted using (7.20), transport rate can then be computed using (7.10)–(7.14) and then the earth-to-wander DCM can be updated using (7.21)–(7.22) or the low-order numerical integration equivalent.

There is one catch. Examination of (7.17) and (7.20) shows that when latitude approaches positive or negative 90 degrees, the wander angle cannot be extracted. With modern computational power, however, it is possible to compute the wander angle up until the cosine of latitude falls below some threshold. This typically occurs within a few meters of the pole. In practice, then, transport rate can be computed until quite close to the pole and it is simply held constant for the brief moment of the pole crossing.

7.4 Alternate form of the horizontal components of transport rate in the w-frame

It is possible, however, to derive a form of the transport rate equations that do not explicitly depend on the computation of the wander angle [1,3]. The key is to reformulate the reciprocal radius terms. Specifically, by utilizing the definitions of the meridional and transverse radii of curvature, it can be shown that:

$$\frac{1}{R_x} = \frac{\sin^2\alpha}{R_E+h} + \frac{\cos^2\alpha}{R_N+h} = \frac{R_N(1+e'^2\cos^2 L\ \cos^2\alpha)+h}{(R_E+h)(R_N+h)} \tag{7.23}$$

$$\frac{1}{R_y} = \frac{\cos^2\alpha}{R_E+h} + \frac{\sin^2\alpha}{R_N+h} = \frac{R_N(1+e'^2\cos^2 L\ \sin^2\alpha)+h}{(R_E+h)(R_N+h)} \tag{7.24}$$

$$\frac{1}{T} = \left[\frac{1}{R_E+h} - \frac{1}{R_N+h}\right]\cos\alpha\sin\alpha = -\frac{R_N\, e'^2\cos^2 L}{(R_E+h)(R_N+h)}\cos\alpha\sin\alpha \tag{7.25}$$

where: $e'^2 = \frac{e^2}{1-e^2}$ = the square of the second eccentricity of the Earth ellipsoid. Substitution of (7.23)–(7.25) into (7.10)–(7.14) yields:

$$\underline{\omega}^w_{ew}(x) = \frac{-v^w_x e'^2 R_N\cos^2 L\cos\alpha\sin\alpha + v^w_y[R_N(1+e'^2\cos^2 L\sin^2\alpha)+h]}{(R_E+h)(R_N+h)} \tag{7.26}$$

$$\underline{\omega}^w_{ew}(y) = \frac{v^w_y e'^2 R_N\cos^2 L\cos\alpha\sin\alpha - v^w_x[R_N(1+e'^2\cos^2 L\cos^2\alpha)+h]}{(R_E+h)(R_N+h)} \tag{7.27}$$

Comparison of Equations (7.26) and (7.27) with the definition of the earth-to-wander direction cosine matrix given in Equation (7.17) shows that (7.26) and (7.27) can be rewritten as:

$$\underline{\omega}^w_{ew}(x) = \frac{v^w_x e'^2 R_N C^w_e[1,3]C^w_e[2,3] + v^w_y[R_N(1+e'^2\{C^w_e[2,3]\}^2)+h]}{(R_E+h)(R_N+h)} \tag{7.28}$$

$$\underline{\omega}^w_{ew}(y) = \frac{-v^w_y e'^2 R_N\, C^w_e[1,3]C^w_e[2,3] - v^w_x[R_N(1+e'^2\{C^w_e[1,3]\}^2)+h]}{(R_E+h)(R_N+h)} \tag{7.29}$$

Although they clearly are messy, Equations (7.28) and (7.29) eliminate the need to compute the wander angle (α) explicitly. Transport rate is used in the update of the earth-to-wander DCM and, in turn, components of the earth-to-wander DCM are used in the computation of transport rate.

7.5 Computation of heading and navigation-frame velocity components

Although the INS can perform perfectly well solely in the wander frame, navigation-frame quantities are typically needed by other systems on the vehicle outside of the INS (e.g., cockpit displays, flight control systems, etc.). Thus, there is a need to be able to compute heading and navigation-frame velocity components. The relationship between heading, yaw and wander angle is illustrated in Figure 7.3.

Heading is thus given by:

$$H = \alpha + \psi \tag{7.30}$$

where α is the wander angle and ψ is the yaw angle.

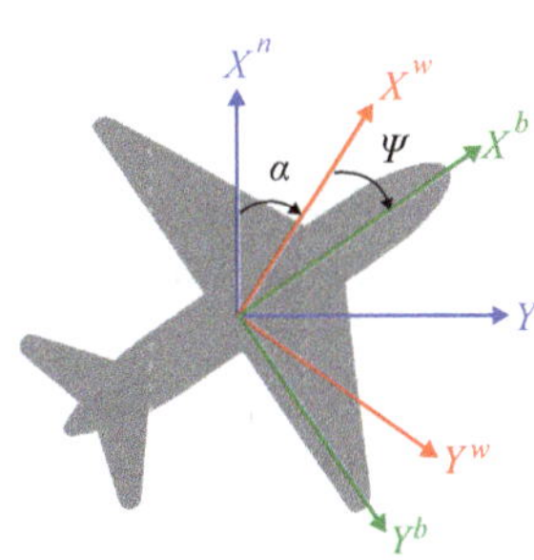

Figure 7.3 In a wander azimuth mechanization, the Euler angles (yaw, pitch, and roll) are defined between the b-frame and the w-frame. Heading is thus given by the sum of yaw (ψ) and the wander angle (α)

Although you may not have noticed, we slipped in the wander-frame to nav-frame conversion of velocity components in the context of the transport rate derivation. The conversion is given in Equation (7.5). Just make sure not to forget that the vertical velocity component is the same in both frames.

Neither heading nor navigation frame velocity can be computed near the poles (within the radius in which the wander angle cannot be extracted). The core inertial processing, however, continues unabated.

7.6 Earth rate converted to the wander frame

In an earlier chapter, we noted the expression for earth rate in the navigation frame could be obtained by:

$$\underline{\omega}_{ie}^{n} = C_e^n \underline{\omega}_{ie}^{e} = \begin{bmatrix} \Omega \cos L \\ 0 \\ -\Omega \sin L \end{bmatrix} \tag{7.31}$$

where Ω is the magnitude of earth rate (7.292115×10^{-5} (rad/s)), L is the latitude and where earth rate in the earth frame is given by:

$$\underline{\omega}_{ie}^{e} = \begin{bmatrix} 0 \\ 0 \\ \Omega \end{bmatrix} \tag{7.32}$$

The expression for earth rate in the wander frame can be computed by:

$$\underline{\omega}_{ie}^{w} = C_e^w \underline{\omega}_{ie}^{e} = C_e^w \begin{bmatrix} 0 \\ 0 \\ \Omega \end{bmatrix} = \begin{bmatrix} \Omega C_e^w[1,3] \\ \Omega C_e^w[2,3] \\ \Omega C_e^w[3,3] \end{bmatrix} = \begin{bmatrix} \Omega \cos\alpha \cos L \\ -\Omega \sin\alpha \cos L \\ -\Omega \sin L \end{bmatrix} \tag{7.33}$$

In actual practice, this vector is computed using the next-to-last expression in (7.33) since it is a simple multiplication of the third column of C_e^w by Ω.

Appendix: Wander-angle rate: the designer's choice

As previously discussed, the most common wander azimuth implementation involves the setting of the vertical component of transport rate to zero. However, the practicing inertial navigation engineer should at least be aware that this is a design choice. Any implementation that keeps the vertical transport rate component within finite bounds will be acceptable. For example, consider the following. From Equation (7.33), the vertical component of earth rate is:

$$\omega_{ie}^{w}(z) = -\Omega \sin L \tag{7.34}$$

If we purposely set the vertical component of transport rate equal to the negative of this:

$$\omega_{ew}^{w}(z) = \Omega \sin L \tag{7.35}$$

then the vertical component of spatial rate is zero:

$$\omega_{iw}^{w}(z) = \omega_{ie}^{w}(z) + \omega_{ew}^{w}(z) = 0 \tag{7.36}$$

In a gimbaled platform, this would have the advantage of not needing to torque about the vertical axis at all. Both the vertical components of transport rate and spatial rate are finite and thus there is no singularity. This is a perfectly acceptable implementation. A more detailed treatment of the design choices is given in [1]. The takeaway is simply that the wander azimuth mechanization requires a design choice. The vast majority of implementations in the world have employed the choice of setting the vertical component of transport rate equal to zero.

References

[1] Brockstein A, Kouba J. *Derivation of Free Inertial, General Wander Azimuth Mechanization Equations*. Woodland Hills, CA: Litton Systems, Inc., Guidance and Control Systems Division; 1981.

[2] Bose SC. *GPS/INS Multisensor Kalman Filter Navigation [Short Course Notes]*. Technalytics; 1997.

[3] Jekeli C. *Inertial Navigation Systems with Geodetic Applications*. Berlin, DE: De Gruyter; 2001.

[4] Titterton DH, Weston JL. *Strapdown Inertial Navigation Technology*. 2nd ed. Herts, UK: The Institution of Electrical Engineers; 2004.

Chapter 8
Navigation-grade inertial sensors

8.1 Introduction

Up to this point, we have been treating accelerometers and gyroscopes as ideal instruments and have focused almost exclusively on the "strapdown processing" algorithms also known as the "free inertial mechanization" equations. Having accomplished that goal, we now turn our attention to the actual sensors that are used in real inertial systems. Prior to the proliferation of inertial measurement units (IMUs) based on micro-electro-mechanical systems (MEMS) fabrication techniques, the term "inertial navigation system" (INS) universally referred to inertial systems built with sensors capable of supporting stand-alone (i.e., unaided) navigation with low drift rates (e.g., a few nautical miles per hour).

However, the aforementioned explosion of low-cost, MEMS-based, IMUs (found, for example, in every consumer drone), has given rise to a far broader use of the phrase "inertial navigation system." A quick web search will reveal authors referring to low-cost units as an INS despite the fact that the units could not possibly be used for stand-alone positioning for more than a handful of seconds. Very low cost gyroscopes (i.e., priced well under one U.S. dollar in volume production) may have drift rates on the order of several degrees *per second* and, as we shall see in the next chapter, are completely unsuitable for long-duration stand-alone positioning. At the other end of the spectrum are so-called "navigation-grade" and "strategic-grade" sensors that have extremely low drift rates (i.e., less than 0.01 degree *per hour*).

This chapter will provide an overview of the principles of operation of the two technologies that have been the mainstay of navigation-grade gyro offerings since the 1980s as well as one of the most popular techniques used in the design of navigation-grade accelerometers. We will then discuss the major sensor errors.

8.2 Sagnac interferometer

Although mechanical gyros were used exclusively in INSs from their very beginning in the 1950s, the move to the so-called "optical gyros" started in the 1980s and by the turn of the century was essentially complete. The main principle in optical gyros (both "ring laser gyros" and "fiber optic gyros") is the so-called Sagnac interferometer as depicted in Figure 8.1. The device consists of an optical light path, a light source and a beam splitter. After light enters the device, the beam splitter splits the beam and light propagates in opposite directions around the path and then back to the splitter itself.

If the device is stationary and we input light, the light splits, counter-propagates in opposite directions around the device, and both light beams will come back to the splitter at the same time. However, if we rotate the device, the counter-propagating beams do not return to the splitter at the same time. The figure illustrates the case of clockwise rotation. As the clockwise propagating beam travels around the device, it has to travel all the way around the circle and then has to keep going until it finally catches up to the splitter. On the other hand, the counterclockwise-propagating beam does not even have to travel all the way around the circle before it gets back to the splitter.

The key concept with the Sagnac interferometer is the measurement of some quantity that is proportional to the time difference between the arrivals of the two counter-propagating beams back at the splitter. For the example shown in Figure 8.1, the counterclockwise beam gets to the splitter before the clockwise beam arrives. Again, the idea is that we want to measure something that is proportional to that time difference since the time difference is proportional to the applied angular rate.

For a stationary device, the time needed to propagate around the full path is equal to the circumference divided by the speed of propagation:

$$t = \frac{2\pi R}{c} \tag{8.1}$$

In order to calculate the time difference (between arrival of the two beams back at the splitter) duration rotation, we write the equations for the path lengths traveled by

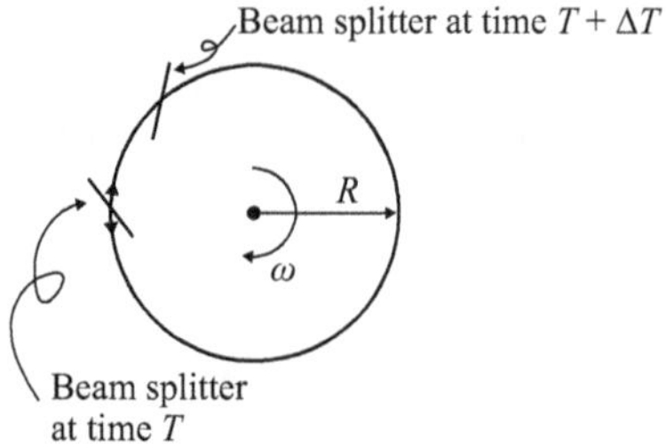

Figure 8.1 A circular Sagnac interferometer with radius "R" rotating at rate ω. A source of light (not shown) is injected into the device at the splitter and the light is thus split in two different directions. If the device is stationary, the two counter-propagating light beams arrive back at the splitter at the same time. If the device is rotating, however, one of the beams arrives back at the splitter before the other. Optical gyros exploit this phenomenon to measure angular rate

each of the two beams. The beam traveling the longer path (clockwise in Figure 8.1) traverses a total path length of:

$$ct_+ = 2\pi R + R\omega t_+ \tag{8.2}$$

The variable t_+ represents the time it takes the beam to leave the splitter, travel all the way around the device and then go a bit more in order to catch up to the splitter. Obviously t_+ is larger than t in Equation (8.1). The first term in Equation (8.2) is simply the circumference of the device. The second term is the product of the tangential velocity ($R\omega$) and the total time. In other words, the second term is the extra length that the beam must travel to catch up with the splitter. Solving (8.2) for t_+ yields:

$$t_+ = \frac{2\pi R}{c - R\omega} \tag{8.3}$$

Similarly, the path length and travel time for the counterclockwise beam are:

$$ct_- = 2\pi R - R\omega t_- \tag{8.4}$$

$$t_- = \frac{2\pi R}{c + R\omega} \tag{8.5}$$

With a bit of algebra, the time difference (of arrivals back at the splitter) between the two beams can be shown to be:

$$\Delta t = t_+ - t_- = \frac{4\pi R^2\omega}{c^2 - R^2\omega^2} \tag{8.6}$$

For all practical applications:

$$c^2 >> R^2\omega^2 \tag{8.7}$$

Thus, (8.6) can be well approximated by:

$$\Delta t \approx \frac{4\pi R^2\omega}{c^2} \tag{8.8}$$

Since the area enclosed by the circular light path is given by:

$$A = \pi R^2 \tag{8.9}$$

Equation (8.8) can thus be rewritten as:

$$\Delta t \approx \frac{4A\omega}{c^2} \tag{8.10}$$

The equivalent path length difference between the two beams is thus:

$$\Delta L = c\Delta t = \frac{4A\omega}{c} \tag{8.11}$$

For gyros of practical size (e.g., palm-sized), neither the time difference nor the path length difference can be measured directly for rotational rates of interest (Earth rate up through the roll rate of an aerobatic aircraft). For example, given a gyro with an area of 0.00283 m^2 (e.g., a circular path with a radius of 3 cm), an applied angular rate of 360 deg/sec yields a time difference of approximately 8×10^{-19} seconds and a path length difference of approximately 2.4×10^{-10} meters. The numbers corresponding to Earth rate (15 deg/hour) would obviously be even lower.

Two primary techniques have been developed to exploit the concept of the Sagnac interferometer in a practical device. One is implemented in the ring laser gyro (RLG). The other is implemented in the fiber optic gyro (FOG).

8.3 Ring laser gyro

A RLG exploits the Sagnac interferometer by taking advantage of the properties of lasers. Laser light propagates in straight lines and thus the only practical way to change its direction of travel is with mirrors. Most RLGs are either three-sided, as shown in Figure 8.2, or four-sided.

A typical laser used in an RLG is made with helium-neon with an optical wavelength of 632.8 nanometers. The key physical property in an RLG is the fact that the path length change (under rotation) induces an effective change on the frequencies of the counter-propagating beams. Before discussing the significance of the frequency change, let us remember that a laser produces monochromatic light. What does that mean? Single frequency. It produces a single frequency and thus it is highly coherent. By contrast, a typical flashlight produces white light that has a broad spectrum. A laser produces one frequency. It is essentially just one narrow portion of the visible light band. Fine, you say, but what does that have to do with the RLG?

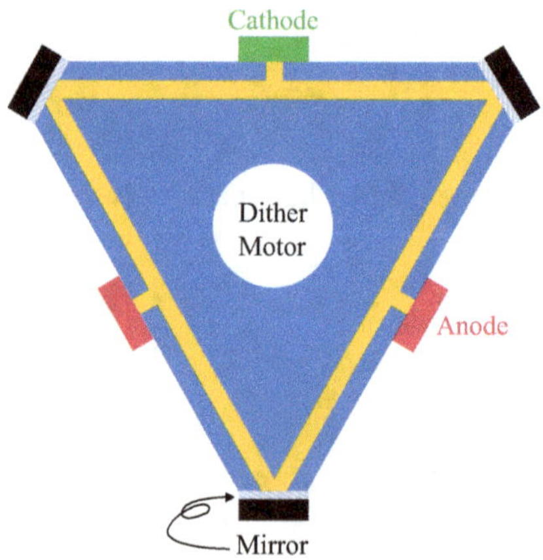

Figure 8.2 Conventional three-sided ring laser gyro. The two anodes are used to set up the counter-propagating beams. The dither motor is used to prevent a phenomenon known as lock-in

The idea is that as the laser beam is propagating around the light cavity, it has to close upon itself *in phase*. If the beam has a certain phase (peak, trough or something in between) at a certain point in the cavity, it has to have that same phase after it propagates all the way around the cavity and gets back to the same point. Another way to think about it is that the complete light path has to correspond to an *integer* number of wavelengths. If not, then the beam would not close upon itself in phase. Within the laser community, there is a verb "to lase" and it, at least partially, refers to the production of a coherent beam. An RLG is lasing properly if it maintains phase coherence.

When the RLG is rotating, this phase coherence has an important implication. As we have discussed, the rotation effectively induces a path length change that is positive for one beam and negative for the other. In order to keep the lasers lasing, one of the beams needs to *decrease* in frequency and the other beam has to *increase* in frequency. The frequency changes are essential so that the new effective path lengths still correspond to an integer number of wavelengths. Although the path length difference is so small that we cannot measure it, the frequency change that it induces on the counter-propagating beams is something that we *can* measure.

In practice, the beams are kept inside the cavity but a small amount of energy is allowed to pass through one of the mirrors (i.e., it is mostly reflecting but is partially transmitting). The two beams are mixed to create a fringe pattern. The fringes move across a detector at the rate of the difference frequency. We will now show that the difference frequency is proportional to the applied angular rate.

The frequency difference can be derived from the path length difference as follows. The ratio of path length difference to nominal path length is equal to the ratio of the frequency difference to the nominal frequency:

$$\frac{\Delta L}{L} = \frac{\Delta f}{f} \tag{8.12}$$

Solving for the frequency difference:

$$\Delta f = f\frac{\Delta L}{L} = \Delta L\frac{f}{L} \tag{8.13}$$

Substituting (8.11) into (8.13):

$$\Delta f = \left(\frac{4A\omega}{c}\right)\left(\frac{f}{L}\right) = \left(\frac{4A\omega}{\frac{c}{f}}\right)\left(\frac{1}{L}\right) = \frac{4A\omega}{\lambda L} \tag{8.14}$$

For a three-sided gyro with 4 cm sides, an input angular rate of Earth rate yields a difference frequency of approximately 0.4 Hz. Clearly this is a measurable quantity.

The basic resolution, or sensitivity, of an RLG can also be determined from Equation (8.14). Although the normal units of frequency difference are radians/second, as discussed earlier, Equation (8.14) is also the rate at which the fringe pattern moves past the detector. Thus, the units of Equation (8.14) can also be thought of as fringes/second. Since the applied angular rate ω is in units of radians/second, the units of the coefficient thus have to be fringes/radian:

$$\Delta f = \left(\frac{4A}{\lambda L}\right)\omega \Rightarrow \left[\frac{fringes}{radian}\right]\left[\frac{radians}{sec}\right] \Rightarrow \left[\frac{fringes}{sec}\right] \tag{8.15}$$

An additional verification of these units is given in the appendix. From (8.15), we can see that the number of fringes per radian is:

$$FpR = \frac{4A}{\lambda L} \tag{8.16}$$

The basic measurement resolution of the device is thus obtained by the reciprocal of (8.16) which is the number of radians per fringe:

$$\Delta\theta_{resolution} = \frac{\lambda L}{4A} \tag{8.17}$$

Thus, when the detector counts the passage of a single fringe, the device has rotated by an amount equal to $\Delta\theta_{resolution}$ as given in Equation (8.17). In the simplest implementation where the gyro only counts integer numbers of fringes, the quantity in (8.17) represents the smallest angular rotation that the device can measure and it thus represents the resolution (or sensitivity) of the device. Modern RLG designs have been enhanced to measure fractions of a fringe and thus increase resolution, but (8.17) still provides a good estimate of the sensitivity of an RLG as a function of laser wavelength, nominal optical path length, and area enclosed by the path.

For the case of three-sided RLGs, Equation (8.17) can be simplified. Assuming the triangular light path (with a total length of L) has equal length sides, the area enclosed is:

$$A = \frac{\sqrt{3}}{4}\left(\frac{L}{3}\right)^2 \tag{8.18}$$

Substitution of (8.18) into (8.17) yields:

$$\Delta\theta_{resolution} = \frac{\lambda 3\sqrt{3}}{L} \tag{8.19}$$

As an example, with a 12 cm path length and a wavelength of 632.8 nm, the basic resolution is approximately 0.0016 degree (or 5.65 arc-seconds). Equation (8.19) shows that RLG resolution is inversely proportional to nominal optical path length. To put it simply: Bigger is better.

All RLGs exhibit a phenomenon called “lock-in” that must be addressed for the unit to operate successfully. At very low input angular rates, the counter-propagating beams “lock on” to each other (in frequency) rather than maintain the ideal, slightly different, frequencies. The primary cause of lock-in is the backscatter of energy due to non-ideal mirrors at the ends of each straight segment in the light path. Without some kind of compensation, the gyro would have no output within the range of input angular rates known as the “lock-in region.”

A few different techniques have been developed in order to deal with this. The most common technique is called "dither." The gyro is mounted onto a piezoelectric wheel that vibrates when a voltage is applied to it. The idea is that if the gyro is being dithered back and forth, it spends very little time in the lock-in region. The dither motion must, of course, subsequently be removed from the output data. The process is known as "dither stripping" [1,2]. Over the history of ring laser gyros, dither mechanisms have proven to be highly reliable.

The one down side of the dither technique, though, is an effective increase in angular rate noise that, when integrated, becomes a so-called "random walk" error at the angular displacement (i.e., delta-theta) level. As a result, non-mechanical techniques have been developed to address lock-in. One example is the so-called multi-oscillator concept that typically involves a square light path with two pairs of counter-propagating beams (one is right-hand polarized and the other is left-hand polarized) thus creating four modes [3]. The modes are biased away from the lock-in region through the use of a Faraday rotator and a non-planar light path. In essence, these devices are implementing electro-magneto-optical dithering instead of mechanical dithering. They achieve lower noise performance but are penalized by an increase in size, weight, and complexity that translates into higher manufacturing cost.

8.4 Fiber optic gyro

The FOG implements the Sagnac interferometric technique by using a coil of many loops of optical fiber (imagine a spool of fishing line). Thus the propagating light travels around the circle again and again and again. The time difference of arrival (back at the detector) between the counter-propagating beams increases with each loop of the coil. Sensitivity is thus increased as a function of coil diameter and the number of loops. Instead of measuring time difference, though, the detector acts on the phase difference (which, of course, is directly proportional to time difference).

This technique was originally developed at the Naval Research Labs in the 1960s. By the 1970s, they had shown that a practical device could be built with optical fiber as the medium through which the light would propagate. Fiber optic gyros have several distinct advantages over ring laser gyros. FOGs do not need highly polished mirrors. The mirror polishing technology in RLGs is extremely complicated and very expensive. FOGs do not need mirrors at all and can be built as solid-state sensors. When the development was actively pursued in the 1980s and 1990s, the FOG architecture offered the promise of much lower manufacturing costs than the RLG. In practice, the basic FOG design could not offer nav-grade performance and thus became a high-end tactical grade instrument. In the 1990s, a variation known as the closed-loop interferometric FOG (IFOG) was developed that did provide nav-grade performance. However, the design enhancements needed to achieve this performance effectively eliminated the lower-cost nature of the basic FOG architecture.

The principle of basic FOG operation, as alluded to earlier, is that a broadband light source (no laser needed) is split and injected into a coil of optical fiber thus creating two counter-propagating beams. After the two beams have traveled around the coil and arrive back at the starting point, they are combined in order to form an interference pattern. The resultant intensity is observed with a photo detector. If the device is stationary, both beams will arrive at the same time (thus in-phase) and the detector will observe a maximum amplitude of the combination. When the device is being rotated, the differential path length difference will induce a phase difference between the two counter-propagating beams. That will change the amplitude of the sum when the two beams are combined.

We can calculate the phase difference as a function of the path length difference:

$$\Delta\Phi = 2\pi\frac{\Delta L}{\lambda} \tag{8.20}$$

Substituting Equation (8.11) into (8.20) yields the phase difference for a single loop:

$$\Delta\Phi = \frac{8\pi A\omega}{c\lambda} \tag{8.21}$$

For a coil of N turns, then:

$$\Delta\Phi = \frac{8\pi AN\omega}{c\lambda} \tag{8.22}$$

Since the total length of the fiber is given by

$$L = 2\pi RN \tag{8.23}$$

we can substitute (8.9) and (8.23) into (8.22) to get the phase difference as a function of the length of the fiber and the radius of the coil:

$$\Delta\Phi = \frac{8\pi\left(\pi R^2\right)N\omega}{c\lambda} = \frac{4\pi R\left(2\pi RN\right)\omega}{c\lambda} = \frac{4\pi RL\omega}{c\lambda} \tag{8.24}$$

For a 3 cm radius with 200 m of fiber and an optical wavelength of 850 nm, the phase difference is approximately 0.0012 degree with an applied angular rate of 15 deg/hour and is 118 degrees with an input rate of 400 deg/sec.

Thus, high angular rates are easily detected but low angular rates are not. These results are for the so-called "open loop" configuration. It has fairly poor sensitivity and limited dynamic range. Phase-biased versions have been developed that have good sensitivity at low rates, but still limited dynamic range. The next level up in design is the closed loop version that has good sensitivity at both low and high rates and has an extended dynamic range. The reader is referred to [4] for more details.

8.5 Force-feedback pendulous accelerometers

One of the most common architectures for navigation-grade accelerometers is based, unsurprisingly, on the principle of inertia. Namely, that a body at rest tends to stay at rest. Figure 8.3 is a notional illustration of a pendulum in two different states. On the left it is at rest. On the right, the *pivot*, not the mass, is being accelerated to the left. Due to the property of inertia, the mass tends to stay at rest even though the pivot is undergoing an acceleration. Thus, we could construct an accelerometer by measuring the displacement of the mass (relative to the pivot) and accounting for the scale factor between displacement and the applied force.

Such an architecture would be referred to as "open loop" for reasons that will become evident shortly. Although conceptually simple, this open loop architecture suffers from limited dynamic range. Refer back to Figure 8.3. Let us measure the displacement as the angle that the pendulum makes with respect to the vertical (i.e., the resting state). With a small amount of force, the pendulum displaces a small angle. With increasing force the angular displacement increases. At some amount of force the pendulum will displace 90 degrees. However, any force greater than this will also displace the pendulum by 90 degrees. Thus, the dynamic range of this accelerometer is limited to the range from zero up to the minimum force that displaces the pendulum by 90 degrees.

Now we can consider the commonly utilized "closed loop" architecture. Rather than let the pendulum swing freely, a feedback control system is implemented as follows. A sensing mechanism is used to measure the displacement of the pendulum and then a restoring mechanism is used to force the pendulum back into the null/rest position. If a small force is applied to the accelerometer, a small amount of displacement is sensed and a small amount of restoring force is applied to hold the pendulum in place. The larger the applied force, the larger the required restoring force. The measurement of the applied force is a scale factor times the required restoring force.

The basic elements of the so-called "force-feedback pendulous accelerometer" are shown in Figure 8.4. This notional figure shows the accelerometer from the "side" in that the hinge axis is into the paper. Although a pickoff coil is depicted for sensing displacement of the proof mass, a variety of sensing techniques have been developed and utilized (optical, inductive, capacitive [5]).

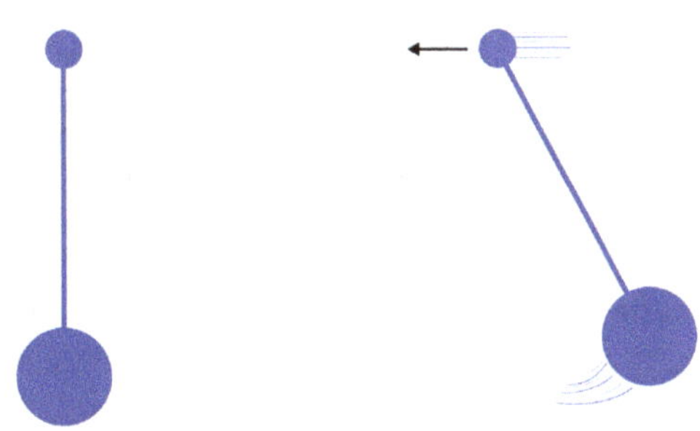

Figure 8.3 Notional pendulum with a pivot point at the top, an arm and the mass at the bottom. On the left the pendulum is at rest. On the right, the pivot is undergoing acceleration to the left. Due to the property of inertia, the mass initially stays at rest even though the pivot is accelerating

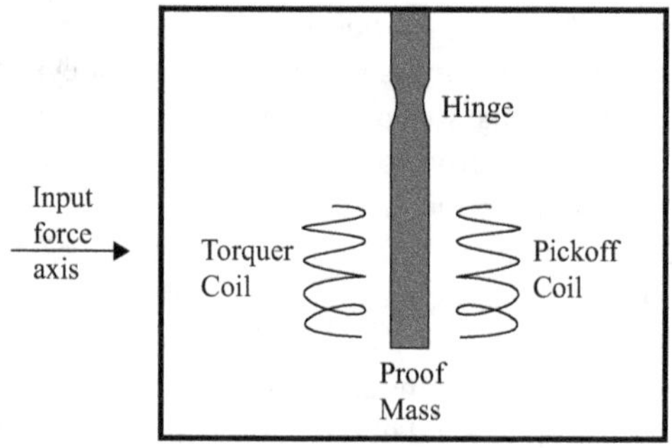

Figure 8.4 The main components of a force-feedback pendulous accelerometer. A proof mass is suspended from a hinge. An applied input force induces a displacement of the proof mass which is sensed by the pickoff coils. A restoring force is applied by the torquer coils

8.6 Primary inertial sensor errors

Inertial sensors have a wide variety of errors. We will focus on the primary ones in this section. The most significant of all the inertial sensor errors are the so-called "biases." The word is put in quotes to emphasize the fact that the actual error is not perfectly constant but, rather, changes slowly enough that the effect can be considered as a bias over short time periods (e.g., minutes). Throughout the remainder of this book the word *bias* will be used in this sense. It is also important to emphasize that there are several components of the total bias error. First, the bias can be separated into two components: repeatable and non-repeatable. The repeatable portion of the bias is, obviously, the component that *is* the same every time a given set of operating conditions are encountered (temperature being the most critical operating condition). The repeatable portion of the bias is estimated during laboratory calibration. Calibration values are loaded into the embedded processor in the INS and thus calibration corrections are applied to the raw sensor data prior to subsequent processing for attitude, velocity, and position updating. For details on calibration procedures, the reader is referred to [6].

At the system performance level, it is the non-repeatable portion of the error that is of concern. In the system error analyses presented in the next chapter (as well as the remainder of the book), the word "bias" will refer to the bias component that remains after calibration corrections have been applied. There are two primary components of this remaining, non-repeatable, portion of the bias. One is the change in error that occurs each time the INS is powered on. This is sometimes referred to as "turn-on to turn-on bias" sometimes shortened simply to "turn-on bias." With well-designed sensors, and good calibration, this component is zero-mean over an ensemble of power-ups. Manufacturers may provide "one-sigma" values in the specification sheets. What they are actually providing is a measure of the standard deviation.

Equating standard deviation to 1σ assumes a Gaussian population which, in practice, is an approximation at best (this is generally true for most of the error values listed in the spec sheets).

The second primary component changes (slowly) over the course of a given run of the system. This component is frequently referred to as "in-run bias stability" even though it is referring to the *instability* of the error. To make the matter even more confusing, some references will explicitly refer to this error as "bias instability." The context of the usage of the term typically makes the meaning clear but those who are new to the literature are urged to be careful.

Over sufficiently short intervals of time, the actual sensor measurement can simply be modeled as a constant added to the true value:

$$s_{meas} = s_{true} + bias \tag{8.25}$$

Navigation-grade accelerometers typically exhibit bias errors (after calibration) less than 100 μg. For navigation-grade gyros, the bias errors are typically less than 0.01 deg/hour.

Following bias, the next most significant sensor error typically is scale-factor error. Ideally there should be a perfectly linear relationship between the input to the sensor (i.e., linear acceleration or angular rotation) and the output measurements. In practice, however, there are two reasons the actual input–output curve deviates from the ideal. First, the slope of the actual curve may not be unity. That is, a given change in the input may result in an output (measurement) of:

$$s_{meas} = (1 + SF_{error})s_{true} \tag{8.26}$$

where SF_{error} is the scale-factor error. If SF_{error} was constant, it could be eliminated through factory calibration. However, as with bias error, real sensors exhibit some amount of scale-factor instability. That is, SF_{error} varies from run to run and within a given run. The constant portion of SF_{error} will be calibrated out but the variable portion remains. This component of scale-factor error is sometimes modeled and estimated in aided-inertial systems (a simple example is given in a later chapter). The other scale-factor error component is scale-factor non-linearity. This component generally cannot be estimated in an aided system and thus designers of navigation-grade sensors seek to minimize this error component. For navigation-grade sensors, the scale-factor stability and non-linearity is less than 10^{-4} (i.e., 100 ppm).

The next major sensor error source is cross-coupling errors. That is, a given sensor may erroneously sense a portion of the input from an orthogonal axis. For example, an accelerometer installed along the x-axis may sense a portion of the specific force in the y and z directions. This can occur because of error in the determination of the sensitive axis of a given sensor. This can also occur because a given sensor is not mounted perfectly orthogonal to the other two sensors in the triad. Some references in the literature refer to this phenomenon as "non-orthogonality" error and some references use the term "misalignment" error. The latter term, regrettably, has become quite prevalent despite the fact that "alignment" is part of the inertial initialization procedure (details provided in a subsequent chapter) and errors in that procedure are separate and distinct from non-orthogonality sensor error. Nevertheless, for consistency with the bulk of the literature, we will abbreviate non-orthogonality error as "MA" (for misalignment). With the reasonable assumption that the non-orthogonality is constant, the erroneous measurement can be modeled with a matrix equation:

$$\begin{bmatrix} s_x \\ s_y \\ s_z \end{bmatrix}_{meas} = \begin{bmatrix} 1 & MA_{xy} & MA_{xz} \\ MA_{yx} & 1 & MA_{yz} \\ MA_{zx} & MA_{zy} & 1 \end{bmatrix} \begin{bmatrix} s_x \\ s_y \\ s_z \end{bmatrix}_{true} \tag{8.27}$$

where, for example, MA_{xy} is the fraction of the input along the y-axis that is erroneously being sensed by the x sensor. For navigation-grade inertial measurement units (IMUs), the cross-coupling error is typically less than 10^{-4} (i.e., less than 0.1 milli-radian of axis non-orthogonality).

Equations (8.25), (8.26), and (8.27) can be combined to model the total actual sensor measurement including all three effects:

$$\begin{bmatrix} s_x \\ s_y \\ s_z \end{bmatrix}_{meas} = \begin{bmatrix} bias_x \\ bias_y \\ bias_z \end{bmatrix} + \begin{bmatrix} SF_x & MA_{xy} & MA_{xz} \\ MA_{yx} & SF_y & MA_{yz} \\ MA_{zx} & MA_{zy} & SF_z \end{bmatrix} \begin{bmatrix} s_x \\ s_y \\ s_z \end{bmatrix}_{true} \tag{8.28}$$

8.7 Noise and other effects

We conclude this chapter with some comments about sensor noise and other error effects. Although legacy spinning-mass gyros exhibited negligible amounts of noise, modern optical-based sensors are not so fortunate. Thermal noise inherent in the electronics is one source and, for ring-laser gyros, the mechanisms used to alleviate the lock-in problem are another source. For those working at the system-level, the bottom line is that in both accelerometers and gyros, the bulk of the noise effect manifests itself at the *rate* level (i.e., acceleration or angular rate). Since most nav-grade sensors output velocity-change (ΔV) or angular displacement ($\Delta\theta$), it is the integrated noise which is experienced in the measurements. As described in an appendix at the end of this book, the integral of white noise is called "random walk" and its standard deviation grows proportionately with the square-root of time. Thus, in a spec sheet, the units of gyro noise (specified as "angle random walk") are typically given as deg/$\sqrt{\text{hour}}$. For accelerometers, the units of velocity random walk are a bit less intuitive and are typically given as $\mu g/\sqrt{\text{Hz}}$. This form is derived from the power spectral density representation of the accelerometer noise. To see how this is also a random walk, alternative units can be derived as follows:

$$1\left[\frac{\mu g}{\sqrt{Hz}}\right] \approx X_g\left[\frac{\frac{m}{s^2}}{\sqrt{\frac{1}{s}}}\right] = X_g\left[\frac{m}{s^2 \cdot s^{-\frac{1}{2}}}\right] = X_g\left[\frac{m}{s^{\frac{3}{2}}}\right] = X_g\left[\frac{\frac{m}{s}}{\sqrt{s}}\right] \tag{8.29}$$

where $X_g \approx 9.81 \times 10^{-6}$. Note that, by convention, the "cycles" in Hz ("cycles/sec") have been ignored. The final form shows that, indeed, the velocity error grows in proportion to the square root of time. For nav-grade sensors, accelerometer noise is on the order of 10 $\mu g/\sqrt{Hz}$ and for gyros is on the order of 0.002 $deg/\sqrt{hour}$.

In simulations, the analyst typically needs to know the standard deviation of the noise that is being simulated. For the accelerometer, start by assuming the velocity random walk is given as V_{rw} $\mu g/\sqrt{Hz}$. As shown above, this can be converted to more convenient units quite simply:

$$V_{rw}\left[\frac{\mu g}{\sqrt{Hz}}\right] \approx V_{rw}X_g\left[\frac{\frac{m}{s}}{\sqrt{s}}\right]$$

where, again, $X_g \approx 9.81 \times 10^{-6}$. We can determine the standard deviation of the specific-force noise by multiplying the noise spec by the square root of the sampling rate:

$$\sigma_f = V_{rw}X_g\left[\frac{\frac{m}{s}}{\sqrt{s}}\right]\sqrt{f_s\left[\frac{1}{s}\right]} = V_{rw}X_g\sqrt{f_s}\left[\frac{\frac{m}{s}\sqrt{\frac{1}{s}}}{\sqrt{s}}\right] = V_{rw}X_g\sqrt{f_s}\left[\frac{m}{s^2}\right] \quad (8.30)$$

where f_s is the sampling rate. To determine the standard deviation of the delta-V noise, multiply the noise spec by the square-root of the sampling interval:

$$\sigma_{\Delta V} = V_{rw}X_g\left[\frac{\frac{m}{s}}{\sqrt{s}}\right]\sqrt{T_s\,[s]} = V_{rw}X_g\sqrt{T_s}\left[\frac{\frac{m}{s}\sqrt{s}}{\sqrt{s}}\right] = V_{rw}X_g\sqrt{T_s}\left[\frac{m}{s}\right] \quad (8.31)$$

where T_s is the sampling interval.

A similar procedure works for gyro noise. It is convenient first to convert the noise spec from $deg/\sqrt{hour}$ to $rad/\sqrt{s}$. Following this, again, multiplication by the square-root of the sampling rate or sampling interval yields the standard deviation of angular rate noise or delta-theta noise.

8.7.1 Other error sources

Some sensors, such as spinning mass gyros and some FOG designs, have errors that result from applied specific force. These are known as *g-dependent bias* or *g-sensitive bias* [5,7]. Some sensors also have errors when in vibration environments. The vibratory motion excites both the scale factor error and cross-coupling error in the sensor. If these two errors were perfectly linear and symmetrical, a zero-mean sinusoidal error would manifest in the sensor output. In reality, though, both the scale factor error and the cross-coupling error exhibit some degree of asymmetry and non-linearity and the vibration environment produces a biased sensor output that is proportional to the amplitude of the vibration [7]. These are known as *g^2-sensitive bias*.

Appendix: Verification of units in the RLG difference frequency equation

The coefficient of the difference frequency Equation (8.15) is repeated here:

$$\frac{4A}{\lambda L} \tag{8.32}$$

Substituting (8.9) into (8.32) yields:

$$\frac{4\pi R^2}{\lambda L} \tag{8.33}$$

The relationship between the circular path length (i.e., the circumference of the circle) and the radius is simply:

$$L = 2\pi R \Rightarrow R = \frac{L}{2\pi} \tag{8.34}$$

Substituting (8.34) into (8.33) yields:

$$\frac{4\pi \left(\frac{L}{2\pi}\right)^2}{\lambda L} = \frac{L}{\lambda \pi} \tag{8.35}$$

which we will rewrite (for reasons that will soon be apparent) as:

$$\frac{L}{\lambda \pi} = \frac{L}{\left(\frac{\lambda}{2}\right) 2\pi} \tag{8.36}$$

Now set Equation (8.36) aside for a moment. Since each fringe in the pattern is a half wavelength long, the total number of fringes around the circular path is simply the path length divided by a half-wavelength:

$$\frac{L}{\left(\frac{\lambda}{2}\right)} \tag{8.37}$$

Dividing (8.37) by 2π gives an expression for total number of fringes per 2π radians:

$$\frac{L}{\left(\frac{\lambda}{2}\right) 2\pi} \tag{8.38}$$

The quantity in (8.38) thus has units of number of fringes per radian. Since (8.38) is the same as (8.36), and (8.36) has been shown to be equal to (8.32), the units in (8.15) are verified.

References

[1] Killpatrick JE, inventor; Honeywell, assignee. Laser angular rate sensor. United States Patent US 3373650; 1968 Mar 19.

[2] Killpatrick JE, Berndt DF, Fritze KR, *et al.*, inventors; Honeywell, assignee. Laser gyro dither stripper gain correction method and apparatus. United States Patent US 5486920; 1996 Jan 23.
[3] Kondo N. Application of zero-lock laser gyro technology to high accuracy stellar-inertial systems. In: *Proceedings of the 1992 National Technical Meeting of The Institute of Navigation*; 1992.
[4] Lefevre HC. *The Fiber-Optic Gyroscope*. 2nd ed. Boston, MA: Artech House; 2014.
[5] Titterton DH, Weston JL. *Strapdown Inertial Navigation Technology*. 2nd ed. Herts, UK: The Institution of Electrical Engineers; 2004.
[6] Rogers RM. *Applied Mathematics in Integrated Navigation Systems*. 3rd ed. Reston, VA: American Institute of Aeronautics and Astronautics; 2007.
[7] Groves P. *Principles of GNSS, Inertial, and Multisensor Integrated Navigation Systems*. 2nd ed. Boston, MA: Artech House; 2013.

Chapter 9
Inertial error analysis

9.1 Introduction

We have studied inertial sensors and their error characteristics. We have learned how to process the measurements in order to determine position, velocity and attitude. We now turn our attention to performance. For example, how does an accelerometer bias impact the position, velocity and attitude solution? What about a gyro bias? How do we go about analyzing the performance of the overall inertial navigation system? The first step is to introduce the various classes of errors found in inertial navigation. We will learn about the dominant horizontal error mode known as Schuler oscillations and then we will also study the instability of the vertical channel. We will close with some comments about noise.

The accuracy in an inertial navigation system is limited by a variety of errors in the data that it is processing. This includes any errors in the initialization of position, velocity and attitude. There are, of course, imperfections in the system components (the gyros and accelerometers) but there are also imperfections in the mounting of the components. Up to this point, we have implicitly assumed that the accelerometer triad consists of three orthogonal accelerometers. What if they are not perfectly orthogonal? What if we do not know exactly how they are mounted relative to the body frame? The same goes for the gyros. Finally you have computational inaccuracies. We have discrete time updates. We are not implementing perfect analog integration and as a result there is a potential for errors in the computation as well.

As we understand from our development of the position/velocity/attitude updating equations, we know that any inaccuracies at any step in the process will be passed on to the next step. The errors will simply build upon each other and thus, as is the case in any kind of dead reckoning system, the errors will grow. As a result, we say an inertial navigation system "drifts" over time. A typical nav-grade inertial system is characterized as being about a "one nautical-mile-per-hour navigator." What this means is that the position error drifts off at approximately one nautical mile per hour.

This is a gross over-generalization. As we will see shortly, inertial errors are not linear over time periods of hours. Nevertheless, average drift rates are estimated and used to characterize the overall system. If an inertial system is characterized as a nautical-mile-per-hour navigator, the implication is that a certain level of sensors, a certain level of mounting accuracy and a certain performance on the initialization is inherent.

We will start by characterizing the dominant horizontal and vertical errors and then will derive expressions to characterize the impact of the dominant sensor errors (accelerometer and gyro biases).

9.2 The Schuler pendulum

The Schuler pendulum is named for its inventor, Dr Maximilian Schuler. Schuler was a German engineer who worked on a problem that plagued gyrocompasses in the early 1900s. Put simply, a gyrocompass involves a mechanical gyro with a spin axis that is pointed in a reference direction (e.g., North). If the vehicle, typically a ship (on which the gyrocompass was mounted), turned, the gyro would maintain its orientation and the heading of the ship could be determined from the angular difference between the ship's longitudinal axis and the spin axis of the gyro. Sounds good in theory but in practice there were problems. Lateral acceleration caused the gyro to process. This caused the gyro's spin axis to rotate out of the horizontal plane that, accordingly, resulted in large errors. Thus there was a need to design a system that would maintain a local-level reference (or equivalently, a local-vertical reference) despite the presence of lateral acceleration. Schuler's ingenious solution to the problem was to build a mechanical system that had the *characteristics* of a theoretical pendulum with a length equal to the radius of the Earth.

9.2.1 Pendulum basics

Let us first briefly review some pendulum basics. Figure 9.1 depicts the three components of a pendulum. A weight is suspended from a pivot point with an arm length "l" (lower case L). To make the math easier, it is usually assumed that the arm has no mass and thus the center-of-gravity (c.g.) of the pendulum is collocated with the weight (known as the "bob").

In our undergraduate physics, we learned that when the bob is displaced, the period of oscillation is proportional to the square-root of the ratio of the arm length to the restoring acceleration (due to gravity):

$$T = 2\pi\sqrt{\frac{l}{g}} \tag{9.1}$$

Figure 9.1 The theoretical pendulum of length l. On the left, the pendulum is displaced and will oscillate. On the right, the pendulum is at rest and will not oscillate

As shown in Figure 9.1, if the pendulum is at rest, the center of gravity and the pivot lie along the same line as the restoring force (the gravity vector). When the pendulum is displaced, the force of gravity can be resolved into two components as shown in Figure 9.2. The restoring force is the gravity component that is perpendicular to the pendulum arm: $g \sin \theta$

Now consider the situation in which the pendulum is mounted in a vehicle that is accelerating horizontally. The pendulum is attached to the vehicle at the pivot point. As a result, when the vehicle is accelerating, the pivot is accelerating too. However, as depicted in Figure 9.3, the "resting" orientation of the pendulum lies along the *vector sum* of gravity and the horizontal acceleration.

9.2.2 Back to Schuler

We are now ready to discuss the Schuler pendulum. Imagine that we have a pendulum with the pivot at the *surface* of the earth and the bob is at the *center* of the earth. As shown in Figure 9.4, such a pendulum is *always* at rest even if it is being accelerated horizontally. No matter how the pivot is accelerated horizontally, the arm of the pendulum is always aligned with gravity.

At this point, it is understandable if you are wondering how this imaginary Earth-radius-length pendulum has anything to do with gyrocompasses much less inertial navigation. What Schuler realized was that a man-made, mechanical system (small

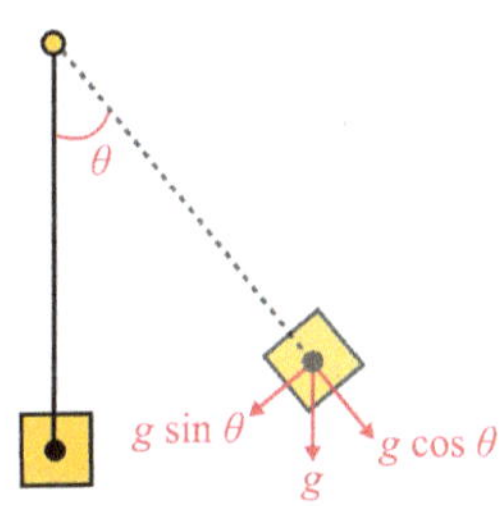

Figure 9.2 The force of gravity acting upon the pendulum can be resolved into components that are parallel and perpendicular to the pendulum arm. The perpendicular component is the restoring force

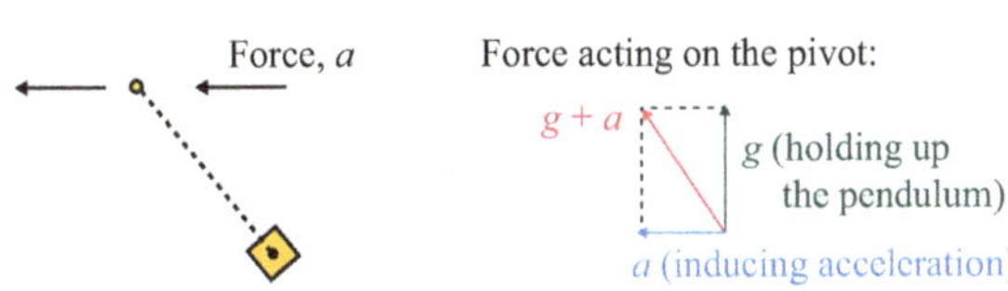

Figure 9.3 When undergoing horizontal acceleration, the pendulum aligns itself with the vector sum of the reaction to gravity and the horizontal acceleration

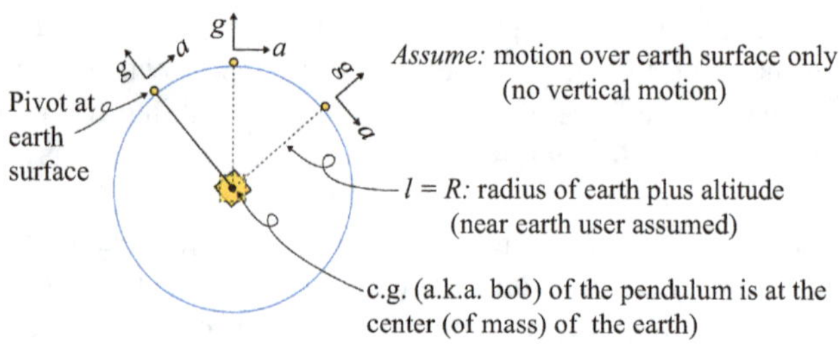

Figure 9.4 The Schuler pendulum. Regardless of horizontal acceleration of the pivot, the pendulum is always aligned with gravity and thus is always at rest

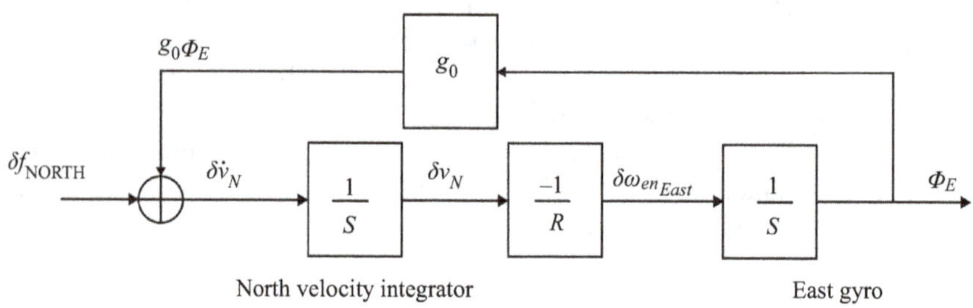

Figure 9.5 Simplified error analysis for the north channel. The vehicle in which the INS is mounted is assumed to be stationary, located on the surface of the Earth, level and north-pointing. A spherical Earth is assumed as is perfect knowledge of gravity

enough to mount and use on a ship) specifically designed with a natural oscillation period (i.e., the inverse of its natural frequency) equal to the period of an Earth-radius-length pendulum would *behave* in the desired manner. Specifically, despite horizontal acceleration such a mechanical system would maintain its orientation (specifically, local-level and vertical).

In the case of a gyrocompass, the spin axis of the gyro needed to be maintained in the local horizontal. However, recall that in inertial navigation we mechanize a locally level platform too (physical in the case of a gimbaled system and virtual/mathematical in the case of a strapdown system). In the early days of inertial navigation, the process of mechanizing a platform to maintain local-level was known as "Schuler tuning." The period of a Schuler pendulum is:

$$T = 2\pi\sqrt{\frac{l}{g}} = 2\pi\sqrt{\frac{R}{g_0}} \approx 2\pi\sqrt{\frac{6.37 \times 10^6\,\text{m}}{9.81\,\frac{\text{m}}{\text{s}^2}}} \approx 5{,}063\,\text{s} \approx 84.4\,\text{min} \qquad (9.2)$$

where g_0 is the surface level gravity and R is the mean radius of the Earth. As we will see shortly, not only is the so-called "Schuler period" the natural period of a system mechanized to maintain local level, it is also the period of some of the dominant errors in such a system.*

9.3 Schuler oscillations in the horizontal inertial channels

Figure 9.5 is a block diagram for a simplified error analysis in the north channel of an inertial navigation system. In the figure, it is assumed that the vehicle (in which the INS is mounted) is stationary, level, and north pointing. Starting in the lower left hand corner, δf_{NORTH} is the error of the north accelerometer output (e.g., an accel bias). This gets summed with the component of gravity that is coupled into the horizontal channel (this will be discussed further shortly) and the result is $\delta \dot{v}_N$, the error in computed north acceleration. The error in computed north acceleration is then integrated to produce δv_N, the error in computed north velocity.

Recalling that the east component of transport rate is the negative of north velocity divided by the earth radius, the *error* in the east component of transport rate, $\delta \omega_{\text{en}_{\text{EAST}}}$, is the negative of the error in north velocity divided by the Earth radius (note that for this simplified analysis a spherical earth is being assumed and thus we do not have to worry about radii of curvature). The integral of the transport rate error, ϕ_E, is then the angle about the east axis that the platform (or nav-frame in the case of strapdown) has rotated off of level. One way to think about this is to note that since the INS has developed an erroneous value for velocity, it responds by rotating the local-level platform (or reference frame) in an attempt to keep it locally level. Of course, what actually happens is that the platform rotates off of level since the actual vehicle is not moving. In any case, since the vehicle is north-pointing, the erroneous rotation about the East axis represents an error in pitch, ϕ_E. Any error in the East-pointing gyro would be added to this value but we are assuming perfect gyros in this particular analysis.

Finally, there is a feedback path. As we just described, the erroneously computed velocity causes the INS to rotate the platform off of level. This is illustrated in Figure 9.6. The easiest way to think about this is in terms of a gimbaled platform. One of the accelerometers on the platform is, ideally, pointed north. When the platform is erroneously tilted off of level, that accelerometer will now start sensing a component of the reaction to gravity. That component is $g_0 \sin \phi_E$ but since the tilt (pitch) error is very small for a navigation-grade inertial system, it can be approximated simply by $g_0 \phi_E$. Although it is not obvious, the sensed component of the reaction to gravity

*For the nitpickers: at any given position on the earth, a more precise value for the Schuler period can be calculated by using the geometric mean of the local values of the two earth radii of curvature and the local value of gravity. Latitude-alone is sufficient for the earth radii and is also sufficient for simple models of gravity. If you have an application that needs better gravity accuracy then you probably are already an expert in Schuler periods.

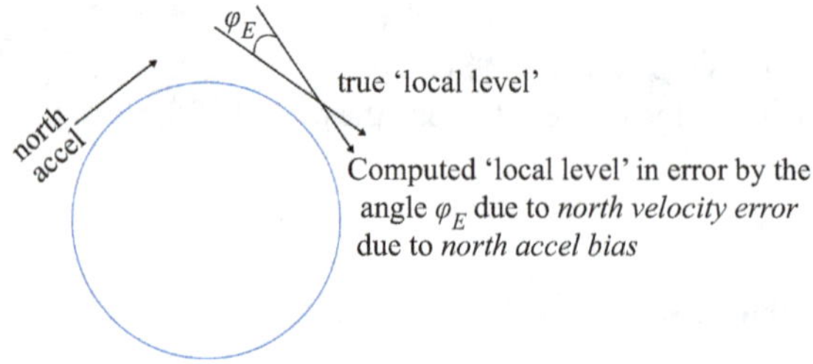

Figure 9.6 A north accelerometer bias leads to erroneous north velocity and a tilting of the platform about the east axis. Since the north axis of the platform is no longer level, a component of the reaction to gravity will be sensed

is always *opposite* to the direction of the accelerometer bias. There is, therefore, a *negative* feedback loop which we will now show produces Schuler oscillations.

The generic characteristic equation of a feedback loop is given by:

$$1 - \text{loop gain} = 0 \tag{9.3}$$

where the loop gain is simply the product of the gain blocks around the loop. For Figure 9.5, then, the characteristic equation is:

$$1 - \left(\frac{1}{s}\right)\left(\frac{-1}{R}\right)\left(\frac{1}{s}\right)g_0 = 0 \tag{9.4}$$

The gain terms can be gathered:

$$1 + \frac{g_0}{s^2 R} = 0 \tag{9.5}$$

which can be rewritten as:

$$s^2 + \frac{g_0}{R} = 0 \tag{9.6}$$

The stability of the feedback loop is determined by solving for the roots (i.e., poles) of the characteristic equation:

$$s = \pm j\sqrt{\frac{g_0}{R}} \tag{9.7}$$

where j is the square root of -1. The roots thus consist of a conjugate pole pair on the imaginary axis (Figure 9.7). The loop is thus marginally stable. That is, any disturbance will cause the loop to oscillate. The frequency of oscillation is given by the square-root term in Equation (9.7). Comparing this term with Equation (9.2), we see that the frequency of oscillation is the inverse of the Schuler period. It is thus referred to as the Schuler frequency:

$$\omega_s = \frac{2\pi}{T} = \sqrt{\frac{g_0}{R}} \tag{9.8}$$

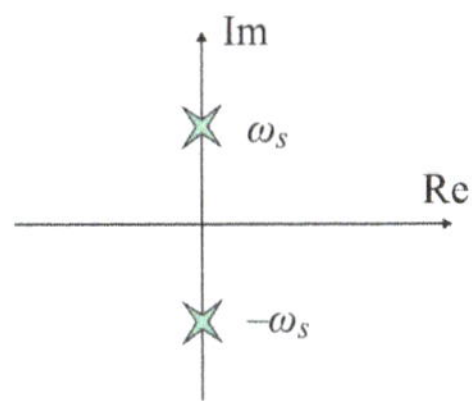

Figure 9.7 *The poles of the characteristic equation of the north inertial error channel negative feedback loop consist of a conjugate pair on the imaginary axis*

Although we have studied the north channel, a similar analysis on the east channel would reveal the Schuler oscillation there as well. Furthermore, similar analyses for initial velocity errors, initial tilt errors or gyro errors (instead of accel bias) would also show the presence of Schuler oscillations. We will investigate some of these now.

9.4 Horizontal error analyses

We have shown that perturbations in the horizontal channel lead to Schuler oscillations. We can extend this analysis to derive closed form expressions for position and velocity error for a variety of sensor or initialization errors. Figure 9.8 is a variation of the block diagram of Figure 9.5. Since the North channel is considered here and the vehicle is assumed to be level and north-pointing, the body-x accelerometer is pointed

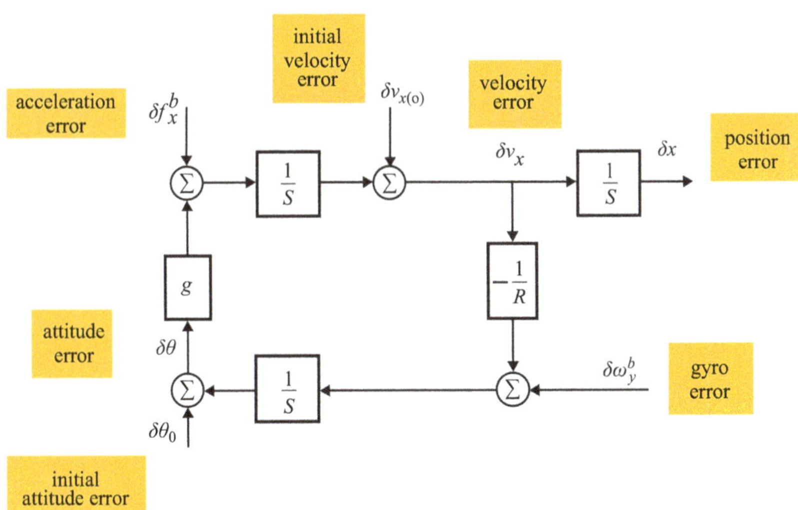

Figure 9.8 *Single channel (North) Schuler error loop. The vehicle is assumed to be stationary, level and north-pointing. The attitude error is about the pitch axis*

north, the body-y gyro is pointed east and the attitude error ($\delta\theta$) is the pitch error about the body y-axis. We will go through two examples in detail: a north accelerometer bias and an east gyro bias. These analyses are so-called "single channel" analyses because they ignore the cross-coupling between axes that occurs in reality (e.g., a north accelerometer error eventually leads to east-west velocity and position errors). As we will see in the next chapter, the cross-coupling is due to the effects of Earth rate.

9.4.1 Analysis of an accelerometer bias

To analyze the effect of a north accelerometer bias using Figure 9.8, we assume initial velocity and attitude errors are zero and that the gyro error is zero also. In order to apply standard linear system/Laplace transform theory, we define the system input to be the accelerometer error and we define the system output to be the velocity error (later we will integrate the velocity error to determine the position error):

$$x(t) = \delta f_x^b(t) \Rightarrow X(s) = \mathscr{L}\{x(t)\} \tag{9.9}$$

$$y(t) = \delta v_x(t) \Rightarrow Y(s) = \mathscr{L}\{y(t)\} \tag{9.10}$$

where $X(s)$ and $Y(s)$ are the Laplace transforms of $x(t)$ and $y(t)$, respectively. Applying this and focusing on just the accelerometer bias, the loop can thus be redrawn as shown in Figure 9.9. We seek first to describe the relationship between the input and the output of this system. We will then derive a transfer function from which we can determine the velocity error for a particular accelerometer error.

From Figure 9.9, we see that $Y(s)$ is the output of an integration block and the input to that block consists of the input, $X(s)$, summed with the result of a feedback loop:

$$Y(s) = \frac{1}{s}\left[X(s) + \left(\frac{-1}{R}\right)\left(\frac{1}{s}\right)(g)\,Y(s)\right] \tag{9.11}$$

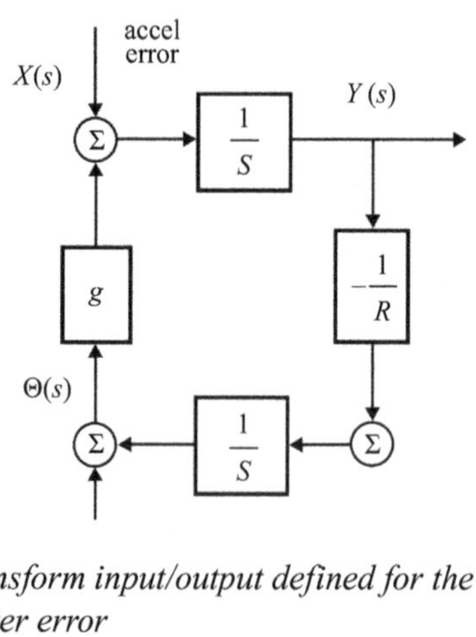

Figure 9.9 Laplace transform input/output defined for the case of an accelerometer error

With a bit of algebra, we can solve for the transfer function (the ratio of the output to the input):

$$H(s) = \frac{Y(s)}{X(s)} = \frac{s}{s^2 + \frac{g}{R}} \tag{9.12}$$

We seek to determine the errors that result from an accelerometer bias. Without loss of generality, we assume the bias begins at $t = 0$ and thus can be modeled by a scaled unit step function:

$$x(t) = \delta f_x^b u(t) \Rightarrow X(s) = \delta f_x^b \left(\frac{1}{s}\right) \tag{9.13}$$

where δf_x^b is the magnitude of the accelerometer bias. The output for this particular input is thus:

$$Y(s) = X(s)H(s) = \frac{\delta f_x^b}{s^2 + \frac{g}{R}} \tag{9.14}$$

If, like most engineers, you solve inverse Laplace transforms by using tables, you will want to modify the form of (9.14) as follows:

$$Y(s) = \frac{\delta f_x^b}{s^2 + \frac{g}{R}} = \delta f_x^b \left(\sqrt{\frac{R}{g}}\right) \frac{\sqrt{\frac{g}{R}}}{s^2 + \frac{g}{R}} \tag{9.15}$$

This allows us to take advantage of the following entry in a table of standard Laplace transforms:

$$\mathcal{L}^{-1}\left\{\frac{\omega}{\omega^2 + s^2}\right\} = \sin(\omega t)u(t) \tag{9.16}$$

Therefore:

$$\mathcal{L}^{-1}\{Y(s)\} = y(t) = \delta f_x^b \left(\sqrt{\frac{R}{g}}\right) \sin\left(\sqrt{\frac{g}{R}}\, t\right) u(t) \tag{9.17}$$

Recalling Equation (9.8), we can write the expression for north velocity error as a function of north accelerometer bias in terms of the Schuler frequency:

$$y(t) = \delta v_x(t) = \frac{\delta f_x^b}{\omega_s} \sin(\omega_s t)u(t) \tag{9.18}$$

The integral of (9.18) provides the position error (in linear units):

$$\delta p_x(t) = \int_0^t \delta v_x(\lambda)d\lambda = \int_0^t \frac{\delta f_x^b}{\omega_s} \sin(\omega_s \lambda)d\lambda \tag{9.19}$$

The solution of this integral yields:

$$\delta p_x(t) = \frac{\delta f_x^b}{\omega_s^2}[1 - \cos(\omega_s t)]u(t) \tag{9.20}$$

Let us now consider some implications of these equations. From (9.18), we can see that the maximum velocity error is:

$$\max\{\delta v_x(t)\} = \frac{\delta f_x^b}{\omega_s} \tag{9.21}$$

For an accelerometer bias, B_a, specified in the typical units of micro-gs (i.e., millionths of the nominal magnitude of gravity), the maximum velocity error is:

$$\frac{\delta f_x^b}{\omega_s} \approx \frac{9.81 \times 10^{-6} B_a}{\sqrt{\frac{g}{R}}} \approx \frac{9.81 \times 10^{-6} B_a}{1.241 \times 10^{-3}} \approx 0.008 B_a \tag{9.22}$$

Thus, we conclude that maximum velocity error is approximately 8 mm/sec per μg. A 100 μg accelerometer bias thus induces a maximum velocity error of 0.8 m/sec. Following a similar analysis, we can show that the maximum position error is approximately 12.74 m per μg. A 100 μg accelerometer bias thus induces a maximum position error of approximately 1,274 m, or roughly 0.7 nautical mile. Nav-grade accelerometers typically exhibit so-called "bias instabilities" (i.e., the residual bias after calibration) on the order of tens of micro-gs. The position and velocity errors ((9.18) and (9.20)) over the course of 4 hours are depicted in Figure 9.10. Although

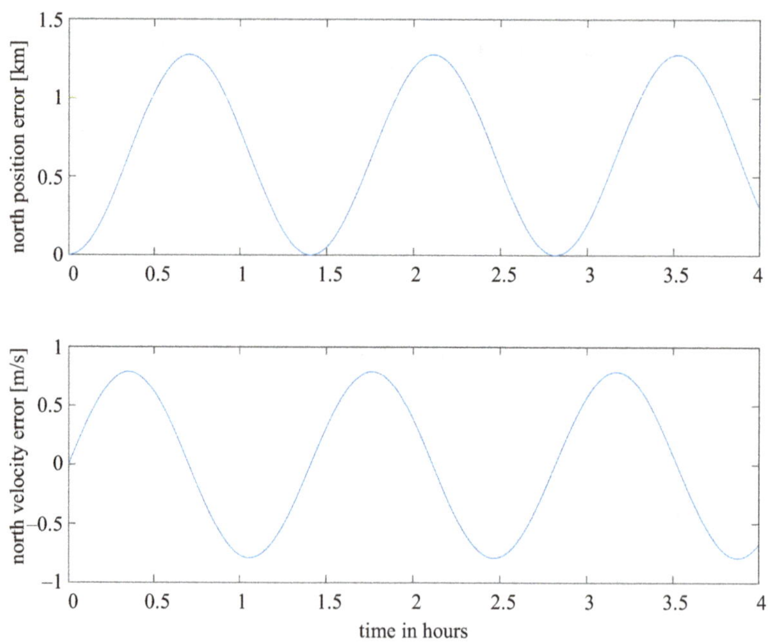

Figure 9.10 Single-channel north error results for a 100 μg accelerometer bias

we have not shown it here, a closed form expression for the pitch error can also be derived. Like the velocity error, it consists of a sinusoid at the Schuler frequency.

9.4.2 Analysis of a gyro bias

Starting with Figure 9.8 and considering a gyro error as the only error, we can follow a similar analysis as for the accelerometer case. This time the gyro error is the system input and we will keep the velocity error as the system output. The relationship between the two is:

$$Y(s) = \left[X(s) + \left(\frac{-1}{R}\right)Y(s)\right]\left(\frac{1}{s}\right)g\left(\frac{1}{s}\right) \tag{9.23}$$

After some algebra, the transfer function is:

$$H(s) = \frac{Y(s)}{X(s)} = \frac{g}{s^2 + \frac{g}{R}} \tag{9.24}$$

As with the accelerometer, we assume the gyro bias begins at $t = 0$ and thus can be modeled by a scaled unit step function:

$$x(t) = \delta\omega_y^b u(t) \Rightarrow X(s) = \delta\omega_y^b \left(\frac{1}{s}\right) \tag{9.25}$$

The output for this particular input is:

$$Y(s) = X(s)H(s) = \frac{\delta\omega_y^b}{s}\left(\frac{g}{s^2 + \frac{g}{R}}\right) \tag{9.26}$$

In this case, it can be easier to get the inverse transform of the term inside the parentheses and then find the integral of the result (recall that division by s in the Laplace domain is the equivalent of integration in the time domain). In order to take advantage of the previously discussed table entry (9.16), we write the term in parentheses as:

$$\begin{aligned} \mathscr{L}^{-1}\left\{\frac{g}{s^2 + \frac{g}{R}}\right\} &= \mathscr{L}^{-1}\left\{\frac{g\sqrt{\frac{g}{R}}\sqrt{\frac{R}{g}}}{s^2 + \frac{g}{R}}\right\} \\ &= g\sqrt{\frac{g}{R}}\mathscr{L}^{-1}\left\{\frac{\sqrt{\frac{R}{g}}}{s^2 + \frac{g}{R}}\right\} \\ &= \sqrt{gR}\sin\left(\sqrt{\frac{g}{R}}t\right)u(t) \end{aligned} \tag{9.27}$$

Utilizing (9.27), the inverse Laplace transform of (9.26) is:

$$\begin{aligned} \mathscr{L}^{-1}\{Y(s)\} = y(t) &= \delta\omega_y^b \int_0^t \mathscr{L}^{-1}\left\{\frac{g}{s^2 + \frac{g}{R}}\right\} d\lambda \\ &= \delta\omega_y^b \int_0^t \sqrt{gR}\sin\left(\sqrt{\frac{g}{R}}\lambda\right) d\lambda \end{aligned} \tag{9.28}$$

that after a bit of freshman calculus yields:

$$y(t) = \delta v_x(t) = -\delta\omega_y^b R\left[\cos\left(\sqrt{\frac{g}{R}}t\right) - 1\right] = -\delta\omega_y^b R\left[\cos\left(\omega_s t\right) - 1\right] \quad (9.29)$$

where again we have used (9.8) for the Schuler frequency. More freshman calculus applied to (9.29) yields the position error expression:

$$\delta p(t) = \delta\omega_y^b R\left[t - \frac{\sin\omega_s t}{\omega_s}\right] \quad (9.30)$$

Equation (9.29) indicates the velocity error is a biased sinusoid. It can be shown that the average velocity error is approximately 30.88 m/s per deg/hour of gyro bias. The position error (9.30) is a linearly increasing trend (first term in square brackets) with a sinusoid riding on it (second term). It can be shown that the linear term has a slope of approximately 60 nautical miles per hour per deg/hour of gyro bias. In other words, if the gyro bias is 1 deg/hour, after 1 hour of operation, the inertial navigation position error will be 60 nautical miles! It is easy to see, then, that nav-grade gyros have bias instabilities on the order of 0.01 deg/hour. The position and velocity errors (Equations (9.29) and (9.30)) over the course of 4 hours are depicted in Figure 9.11. As we will learn in the next chapter, the linear trend in (9.30) and Figure 9.11 *does not* continue forever. When the effects of cross-coupling (due to Earth rate) are taken

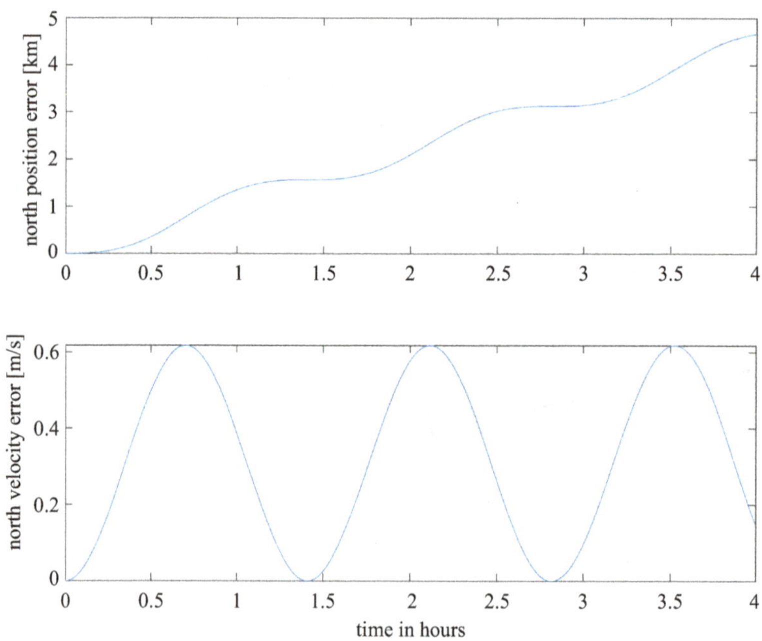

Figure 9.11 Single-channel north error results for a 0.01 deg/hour gyro bias

into account, there is a long period oscillation for which the linear trend shown here is merely the initial portion.

9.5 Instability in the vertical inertial channel

In the horizontal, most error sources in an inertial system will initiate oscillations with a Schuler period of approximately 84.4 min. As we will now show, the vertical channel is characterized by a positive feedback loop in which errors grow without bound. This behavior in the vertical channel stems from incorrect correction of gravity in the vertical accelerometer measurements. Thus we start the analysis by modeling gravity. Gravity decreases with altitude and this altitude dependency is well approximated by:

$$g(h) = g_0 \left(\frac{R}{R+h} \right)^2 \tag{9.31}$$

where g_0 is the surface level gravity, R is the mean radius of the Earth, and h is altitude above the Earth. Equation (9.31) can be written as a Taylor series expansion at $h = 0$:

$$g(h) = g_0 \left(\frac{R}{R+h} \right)^2 = g_0 \left(1 - \frac{2h}{R} + \frac{3h^2}{R^2} - \frac{4h^3}{R^3} \cdots \right) \tag{9.32}$$

Since $h << R$ for terrestrial and aeronautical platforms, it is sufficient to approximate gravity by truncating the series after the first order term:

$$g(h) \approx g_0 \left(1 - \frac{2h}{R} \right) \tag{9.33}$$

We can now derive an expression for the error in computed gravity as a function of error in computed altitude. We write the erroneously computed altitude as:

$$h_{computed} = h + \delta h \tag{9.34}$$

where the altitude error is δh. The computed value of gravity at this computed altitude is thus:

$$g(h_{computed}) = g(h + \delta h) = g_0 \left(1 - \frac{2(h + \delta h)}{R} \right) = g_0 \left(1 - \frac{2h}{R} - \frac{2(\delta h)}{R} \right) \tag{9.35}$$

Now gather the true and error terms:

$$g(h_{computed}) = g_0 \left(1 - \frac{2h}{R} \right) - \left(\frac{2g_0}{R} \right) \delta h \tag{9.36}$$

Substitution of Equation (9.33) into (9.36) yields:

$$g(h_{computed}) = g(h) - \left(\frac{2g_0}{R} \right) \delta h \tag{9.37}$$

The error in the computed gravity at the erroneously computed altitude is thus:

$$\delta g(\delta h) = g(h_{computed}) - g(h) = - \left(\frac{2g_0}{R} \right) \delta h \tag{9.38}$$

As expected, Equation (9.38) shows that for positive altitude errors (i.e., the computed height is greater than the true height), the computed gravity is less than true gravity and thus the error is negative. The vertical channel error block diagram is depicted in Figure 9.12.

In the upper left corner of the diagram, we see the gravity model error has the same magnitude as Equation (9.38) but is opposite in sign. This is necessary since a negative gravity modeling error results in a positive vertical velocity error (when the up direction is taken as positive). The remainder of the block diagram is straightforward with the sum of specific force measurement error and gravity modeling error being the vertical acceleration error. The integral of this is vertical velocity error and the next integral yields vertical position error.

The diagram can be used to analyze the effects of initial height error or initial vertical velocity error but the vertical channel error characterization will be the same as for the case of a vertical accelerometer error as depicted in the figure. The characteristic equation for the feedback loop is:

$$s^2 - \frac{2g_0}{R} = 0 \tag{9.39}$$

The solution of (9.39) yields two real roots (poles):

$$s = \pm\sqrt{\frac{2g_0}{R}} = \pm\sqrt{2}\omega_s \tag{9.40}$$

where the Schuler frequency, ω_s, was defined in Equation (9.8). As we know from linear system theory, the presence of a real pole in the right half plane indicates the system is unstable. Specifically, any perturbation of the system will yield a response that grows without bound.

Let us conceptualize this instability. If a stationary INS is on the surface of the Earth but has been erroneously initialized with a higher value of altitude, the gravity compensation algorithm will compute a value of gravity that is too small. Since the INS is stationary, the only force being measured by the vertical accelerometer is the reaction to gravity. Thus the specific force measured by the accelerometer triad, converted to the navigation frame is:

$$\underline{f}_b^n = \begin{bmatrix} 0 \\ 0 \\ -g_0 \end{bmatrix} \tag{9.41}$$

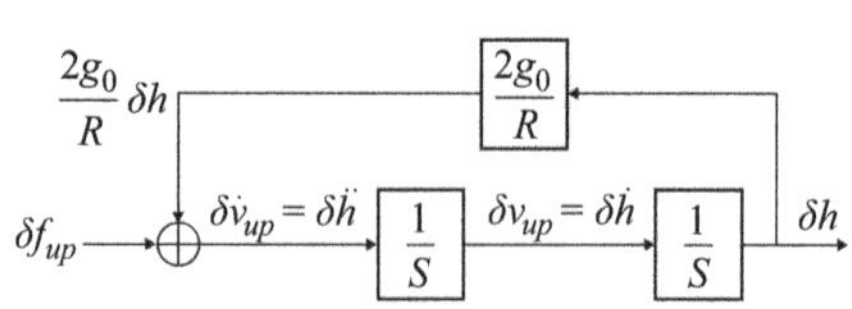

Figure 9.12 Inertial vertical channel error analysis

For the sake of illustration, let us assume that the height error is such that the computed gravity is too small by a factor of 10^{-4}. As a result, the computed gravity vector would be:

$$\underline{g}_{computed} = \begin{bmatrix} 0 \\ 0 \\ 0.9999g_0 \end{bmatrix} \tag{9.42}$$

The value of vertical acceleration computed by the accelerometer sensing equation is thus:

$$\underline{f}_b^n + \underline{g}_{computed} = \begin{bmatrix} 0 \\ 0 \\ -0.0001g_0 \end{bmatrix} \tag{9.43}$$

Since the nav-frame z-axis points down, the computed acceleration is in the up direction. The computed acceleration is small, but it is non-zero. As a result, it will get integrated into non-zero vertical velocity (in the up direction) and this will get integrated into an increasing value of altitude. Thus, the erroneous initialization of height (to a value that is too high) eventually produces a computed value of height that is even higher. The next time gravity is computed, an even smaller value is obtained and an even larger computed vertical acceleration results. This positive feedback loop continues as computed vertical velocity and altitude grow toward infinity.

To quantify this phenomenon more precisely, we can perform error analyses using the block diagram illustrated in Figure 9.12. For example, if the vertical channel is initialized with an incorrect altitude, we can exploit the following Laplace Transform:

$$\mathscr{L}\left\{\frac{d}{dt}y(t)\right\} = sY(s) - y(0) \tag{9.44}$$

First, let:

$$y(t) = \delta h(t) \tag{9.45}$$

Then, from Figure 9.12, we can see that the derivative of altitude error is vertical velocity error and is the integral of $\frac{2g_0}{R}\delta h$ (recall we are assuming the accelerometer error is zero). Thus:

$$\delta v_{up}(t) = \int_0^t \frac{2g_0}{R} y(\lambda) d\lambda \tag{9.46}$$

Taking the Laplace transform of both sides of Equation (9.46) yields:

$$sY(s) - y(0) = \left(\frac{1}{s}\right)\frac{2g_0}{R}Y(s) \tag{9.47}$$

Grouping terms:

$$\left(s - \frac{\frac{2g_0}{R}}{s}\right)Y(s) = y(0) \tag{9.48}$$

Solving for $Y(s)$:

$$Y(s) = y(0)\frac{1}{s - \frac{\frac{2g_0}{R}}{s}} = y(0)\frac{s}{s^2 - \frac{2g_0}{R}} \tag{9.49}$$

A perusal of the Laplace transform tables shows the inverse Laplace transform yields:

$$y(t) = y(0)\cosh\left(t\sqrt{\frac{2g_0}{R}}\right) \tag{9.50}$$

Substituting (9.45) into (9.50) yields:

$$\delta h(t) = \delta h(0)\cosh\left(t\sqrt{\frac{2g_0}{R}}\right) \tag{9.51}$$

After 1 hour, the hyperbolic cosine term is approximately 276. Thus if, for example, the initial height error is 1 meter, the altitude error will grow to 276 m after an hour (see Figure 9.13). Note this occurs even with a *perfect* accelerometer and a *perfect* gravity model.

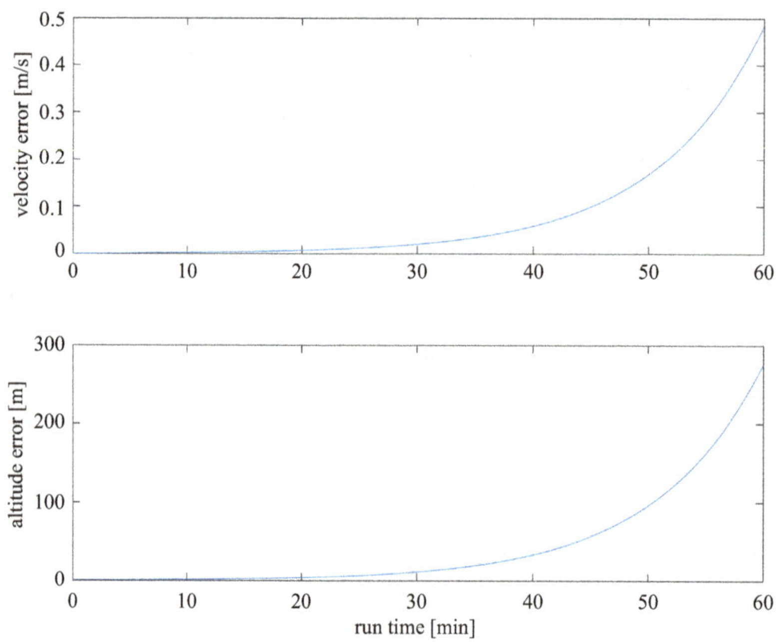

Figure 9.13 Vertical velocity and altitude errors for the case where altitude is erroneously initialized to a value 1 m higher than true altitude. All other error sources have been set to zero

Following a similar analysis, it can be shown that an initial vertical velocity error yields an altitude error given by:

$$\delta h(t) = \delta v_{up}(0)\sqrt{\frac{R}{2g_0}}\sinh\left(t\sqrt{\frac{2g_0}{R}}\right) \tag{9.52}$$

Following an analysis similar to that of the horizontal channel, it can be shown that a vertical accelerometer bias yields an altitude error given by:

$$\delta h(t) = \delta f_{up}\left(\frac{R}{2g_0}\right)\left(\cosh\left(t\sqrt{\frac{2g_0}{R}}\right) - 1\right) \tag{9.53}$$

Note that in all three cases, the error growth is hyperbolic. This is the fundamental difference between the horizontal channel and the vertical channel. Perturbations in the horizontal channel yield Schuler oscillations. Perturbations in the vertical channel, however, yield errors that head off toward infinity. First-generation inertial navigation systems handled this problem by simply not mechanizing the vertical channel. These were gimbaled platforms and they did not even have a vertical accelerometer. The system did not compute either vertical velocity or vertical position (altitude). This was acceptable in aircraft for enroute navigation applications since height was determined independently by a barometric altimeter. Although unmodeled atmospheric pressure variations caused bias errors in the measured altitude, it was nevertheless stable (i.e., the measured altitude would never run off towards infinity).

However, the barometric altimeter had one particularly undesirable characteristic. Lags inherent in the measurement of pressure caused lags in both the indicated altitude and vertical velocity. The need to determine accurate, and low latency, vertical velocity led to the development of techniques to implement and stabilize the inertial vertical channel. Specifically, altitude measurements from the barometric altimeter were combined with the vertical accelerometer measurements to provide vertical velocity and altitude outputs that had the low latency characteristics of the inertial solution and the stability of the barometric altimeter measurements. Early implementations involved fairly simple analog complementary filters [1] but later implementations made compensations for atmospheric variations [2]. Eventually, Kalman filters were developed not only to output the vertical velocity and altitude but also estimate altimeter and vertical accelerometer errors [3]. We will cover baro stabilization later in this book. Since the advent of GPS, it has also been used to stabilize the inertial vertical channel.

9.6 Sensor noise and other errors

Although scale factor error and non-orthogonality/misalignment along with g-dependent and g-squared dependent errors can be significant, their dependency on vehicle maneuvers (dynamics) makes it impossible to derive general equations such as for sensor biases. Before closing this chapter, however, we should briefly discuss the impact of sensor noise.

Since the error models described in this chapter are linear and time-invariant (recall we used Laplace transforms to characterize the transfer functions), random signal processing theory can be applied to the analysis of random inputs. Since the noise at the rate level (specific force or angular rate) can be considered zero-mean and white, the only necessary statistical characterization is the variance (or standard deviation). Over very short periods of time (e.g., less than 5 or 6 min), the effects of earth radius, gravity feedback, earth-rate, and Coriolis can be ignored. For accelerometer noise with a standard deviation of σ_{acc} (given in units of meters-per-second per root second), the position error standard deviation can be approximated by [4]:

$$\sigma_{pos_{acc}} = \sqrt{\frac{1}{3}\sigma_{acc}^2 \, t^3} \tag{9.54}$$

where t is the elapsed time in seconds. For gyro noise with a standard deviation of σ_{gyr} (given in units of radians-per-root-second), the position error standard deviation can be approximated by:

$$\sigma_{pos_{gyr}} = \frac{g}{2}\sqrt{\frac{1}{5}\sigma_{gyr}^2 \, t^5} \tag{9.55}$$

where g is the nominal magnitude of gravity. Figure 9.14 depicts the standard deviation of position error due to accelerometer noise of 10 $\frac{\mu g}{\sqrt{Hz}}$ (approximately 10^{-4} meters-per-second per root second). Figure 9.15 similarly depicts the position error standard

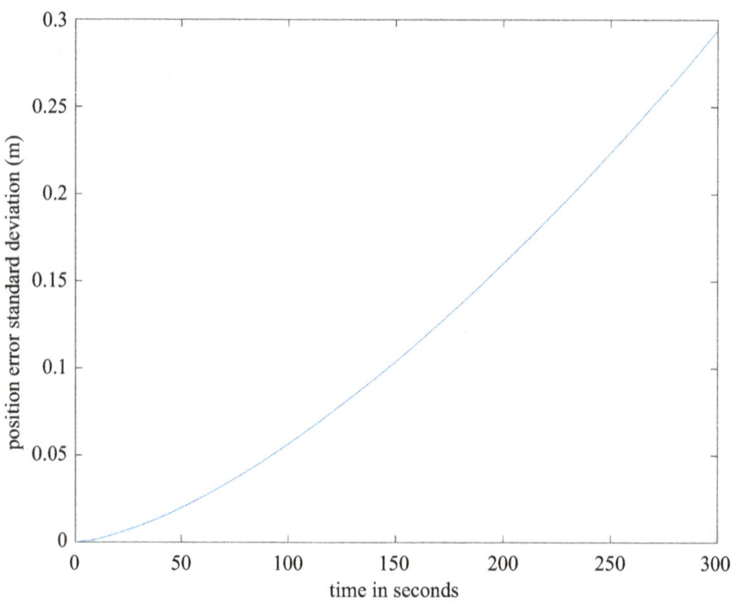

Figure 9.14 Standard deviation of position error due to accelerometer noise of 10 $\frac{\mu g}{\sqrt{\text{Hz}}}$ as given by Equation (9.54)

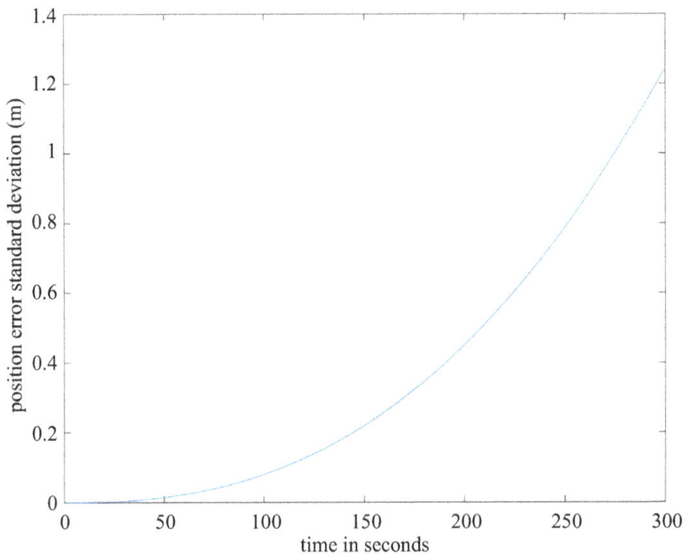

Figure 9.15 Standard deviation of position error due to gyro noise of 0.00125 $\frac{\text{deg}}{\sqrt{\text{hour}}}$ as given by Equation (9.55)

deviation due to gyro noise of 0.00125 $\frac{deg}{\sqrt{hr}}$ (approximately 0.364×10^{-6} $\frac{radians}{\sqrt{s}}$). Although the noise values for both sensors are typical for navigation-grade, the error growth due to the gyro noise far exceeds that of the accelerometer.

In the next chapter, we will confirm via simulation that error due to gyro noise is the greater of the two even for long time periods. For periods longer than 5 min up to the length of a Schuler period (e.g., 90 min), an analysis that accounts for earth radius and gravity feedback is required. An example given by [5] approximates the position error variance due to gyro noise:

$$\sigma^2_{pos_{gyr}} = 6g^2 \left(\frac{1}{4\omega_s^4} t - \frac{1}{3\omega_s^5} \sin \omega_s t + \frac{1}{24\omega_s^5} \sin 2\omega_s t \right) \sigma^2_{gyr} \tag{9.56}$$

where ω_s is the Schuler frequency. For the same gyro noise as considered earlier (0.00125 $\frac{\text{deg}}{\sqrt{\text{hour}}}$), the position error standard deviation as given by (9.56) is depicted in Figure 9.16. As the plot shows, the error growth pauses at the end of the given time period of 90 min. Longer time periods will be examined via simulation in the next chapter. It is important to note these noise equations are "single channel" in that they do not account for the effects of cross-coupling between axes (which, again, will be explored in the next chapter). More detailed closed-form analyses are given in [6,7]. However, for nav-grade sensors, medium- and long-term system performance tends to be dominated by sensor biases, initialization errors, and maneuver-dependent errors.

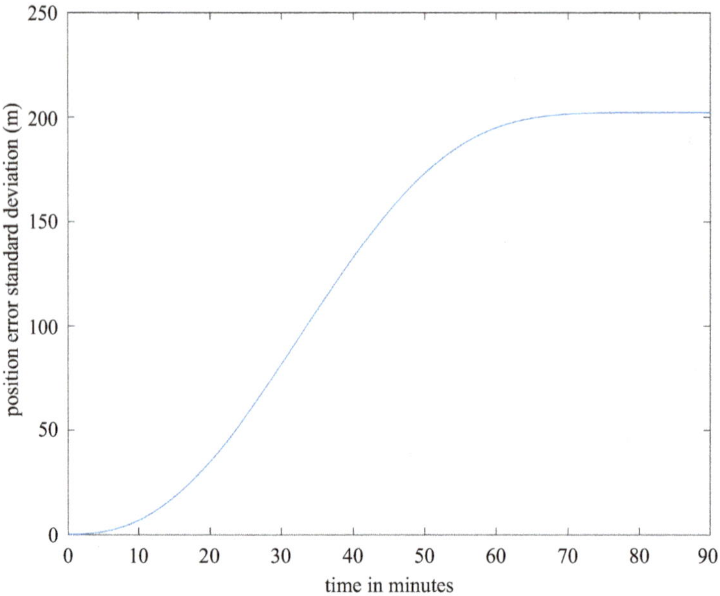

Figure 9.16 Standard deviation of position error due to gyro noise of 0.00125 $\frac{deg}{\sqrt{hr}}$ *as given by (9.56)*

Finally, it must be emphasized that the most significant impact of sensor noise is on initialization. Although the impact on position and velocity performance is small relative to other error sources, sensor noise essentially dictates a lower bound on the time required to achieve a given level of initialization performance. As we will explore in a later chapter, accelerometer measurements are used for coarse "leveling" (i.e., determination of initial pitch and roll) and gyro measurements are used for "gyrocompassing" (i.e., determination of initial yaw angle). A given level of sensor noise will dictate a certain amount of "averaging" (i.e., filtering) to achieve desired accuracy in the initial determination of pitch/roll/yaw.

References

[1] Kayton M, Fried W. *Avionics Navigation Systems*. 2nd ed. New York, NY: John Wiley & Sons; 1997.

[2] Ausman JS. Baro-Inertial Loop for the USAF Standard RLG INU. *NAVIGATION: Journal of the Institute of Navigation*. 1991;38(2):205–220.

[3] Ausman JS. A Kalman filter mechanization for the Baro-inertial vertical channel. In: *Proceedings of the 47th Annual Meeting of The Institute of Navigation*; 1991.

[4] Groves P. *Principles of GNSS, Inertial, and Multisensor Integrated Navigation Systems*. 2nd ed. Boston, MA: Artech House; 2013.

[5] Pue A. *Integration of GPS with Inertial Navigation Systems [Short Course Notes]*. NavtechGPS; 2007.

[6] Savage PG. *Strapdown Analytics*. Maple Plain, MN: Strapdown Associates; 2000.

[7] Blum C, Dambeck J. Analytical assessment of the propagation of colored sensor noise in strapdown inertial navigation. *MDPI Sensors*. 2020;20(23), 1–26.

Chapter 10
Inertial error simulation

10.1 Introduction

In the previous chapter, we analyzed the Schuler error loop in the horizontal channel and derived closed form expressions for position and velocity errors given accelerometer and gyro biases. The closed form expressions are actually first-order approximations of the real error behavior. In this chapter, we will use simulations to illustrate the cross-coupling phenomena between the north–south and east–west channels and will go on to analyze the dominance of certain error sources as a function of mission duration.

A so-called "whole value" simulator generates simulated accel and gyro measurements and processes them using the same algorithms as would be used in an actual INS. In a computer simulation, however, the input errors (e.g., accel/gyros biases) are controlled and known. The analyst can simulate a single error source on a single sensor and evaluate its impact on the entire system.

In all of the simulation results to follow, the INS is modeled as being in a vehicle that is stationary, level, north-pointing and is at a latitude of 45 degrees North.

10.2 North accelerometer bias

In this example, a 100 μg bias is simulated on the body x-axis accelerometer. Since the vehicle is level and is pointed northward, this accelerometer is sensitive in the north direction. The position, velocity, and Euler angle errors over a 4-hour period are plotted in Figures 10.1 and 10.2.

As would be expected, the errors are initially dominant in the north–south direction. As time goes on, however, errors start to build up in the east–west direction. The cross-coupling results from incorrect application of earth-rate (which results from an erroneous computation of latitude) in the n-frame update. When the INS thinks it is at a higher latitude than it actually is, for example, it will compute Earth rate components that are too large in the vertical direction and too small in the north direction. In a gimbaled system, this causes the platform to rotate around the vertical axis away from north and also rotate around the platform x-axis away from level. The "north" accelerometer is thus pointed away from north and thus has an east–west component. In addition, the tilt of the platform around the x-axis causes gravity to be coupled into the east–west axis as well. Thus, a north accelerometer error leads to east–west position/velocity/attitude errors.

Although not obvious at first glance, a closer inspection of Figure 10.1 reveals that the peak-to-peak values of the north errors are decreasing over time while the east-west errors are increasing. If the simulation is extended, the magnitudes (i.e., envelope) of

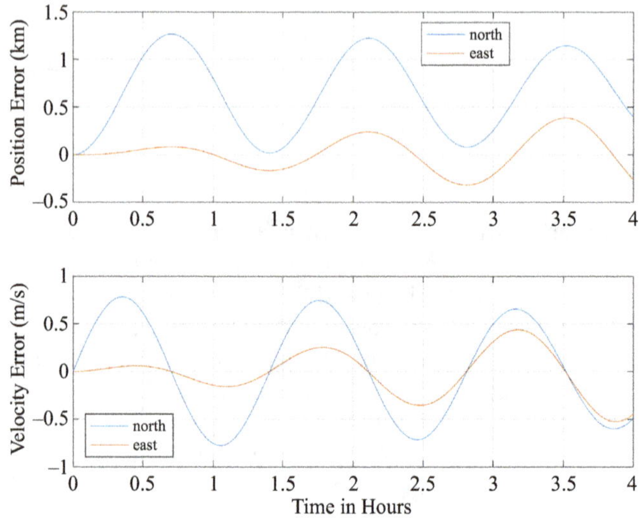

Figure 10.1 Position/velocity errors resulting from a 100 μg north accelerometer bias

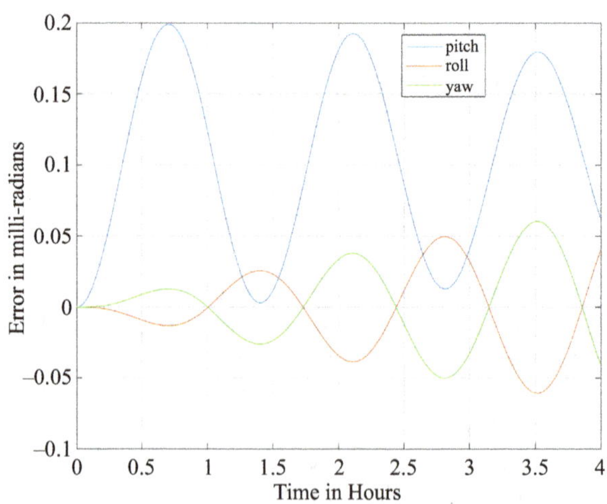

Figure 10.2 Euler angle errors resulting from a 100 μg north accelerometer bias

the east–west and north–south errors will slowly oscillate with a frequency equal to the local vertical component of Earth rate: $\omega_{Foucalt} = |\underline{\omega}_{ie}| \sin L$ (where L is latitude). This is known as the Foucault oscillation [1]. It is illustrated in Figures 10.3 and 10.4. In terms of communication theory, the Foucault oscillation can be thought of as the

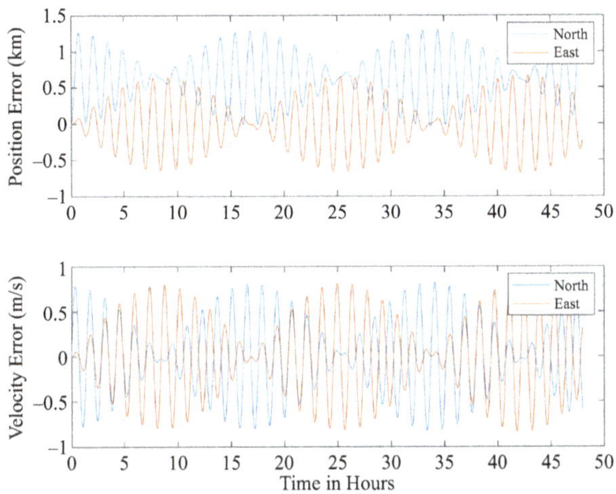

Figure 10.3 100 μg north accelerometer bias simulation extended to 48 hours. Position and velocity errors depicted. Since the vehicle is being simulated at a latitude of 45 degrees, the long-term Foucault oscillation has a period of approximately 34 hours

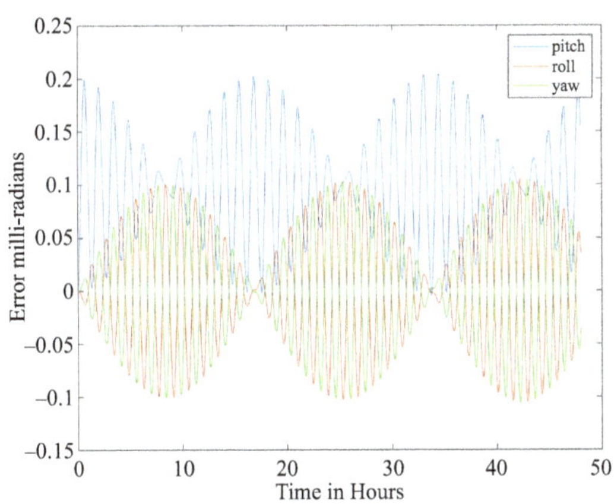

Figure 10.4 100 μg north accelerometer bias simulation extended to 48 hours. Euler angle errors depicted. Since the vehicle is being simulated at a latitude of 45 degrees, the long term Foucault oscillation has a period of approximately 34 hours

modulation in an amplitude modulated (AM) signal that is modulating the Schuler frequency "carrier." At first glance, it is tempting to assume the period of the Foucault oscillation in, for example, Figure 10.3 is about 17 hours. However, this is only the first half of the period. Computation of the vertical component of Earth rate given the aforementioned formula yields a frequency of approximately 5.1563×10^{-5} radians/s or a period of approximately 33.85 hours.

10.3 East gyro bias

In this example, a 0.01 degree/hour bias is simulated on the body *y*-axis gyroscope. Since the vehicle is level and pointed north, the gyro is sensitive in the east direction. The position, velocity, and Euler angle errors over a 4-hour period are depicted in Figures 10.5 and 10.6.

In this case, the Schuler periodicity is still present in the position error but the dominant effect is a longer-term growth trend that is approximately linear over the first few hours. Although somewhat counterintuitive, the horizontal gyro bias induces a longer-term error growth trend in the *yaw* angle, whereas the bounded Schuler oscillations are present in the roll and pitch angles. The linear growth in the north position error (in this example) does not continue indefinitely but, rather, is the beginning of a long period oscillation. As can be seen in Figures 10.7 and 10.8, the long-term oscillation has a 24-hour period and thus is distinct from the Foucault oscillation observed in the accelerometer bias simulations.

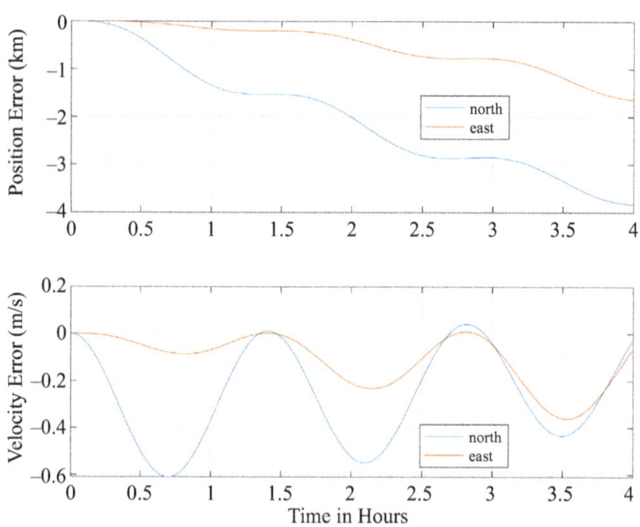

Figure 10.5 Position/velocity errors resulting from a 0.01 deg/hour east gyroscope bias

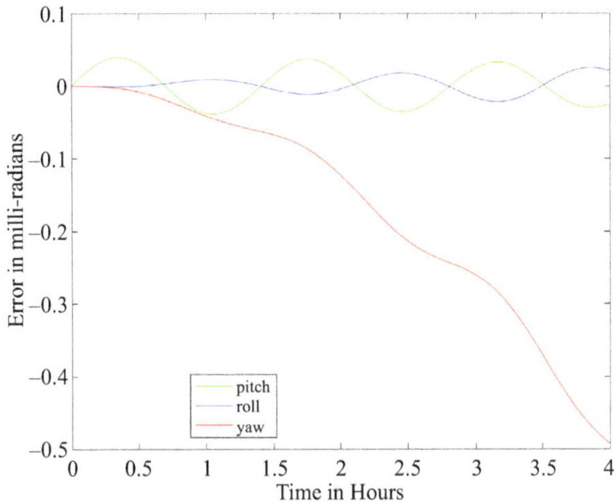

Figure 10.6 Euler angle errors resulting from a 0.01 deg/hour east gyroscope bias

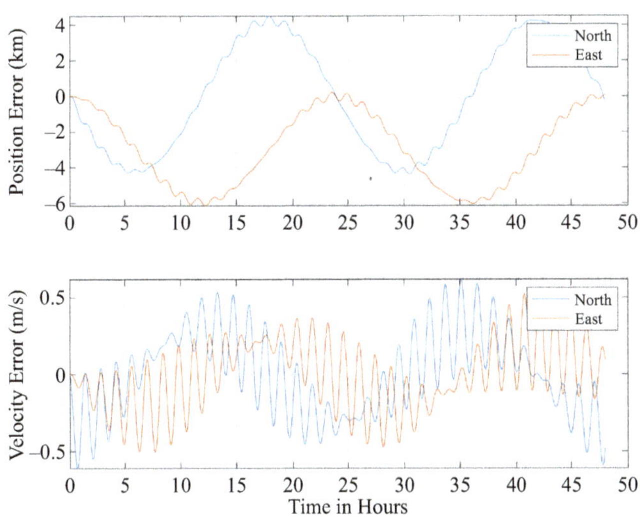

Figure 10.7 Extending the 0.01 deg/hour east gyroscope bias simulation to 48 hours. Position and velocity errors are depicted. A diurnal (24 hours) oscillation is observed

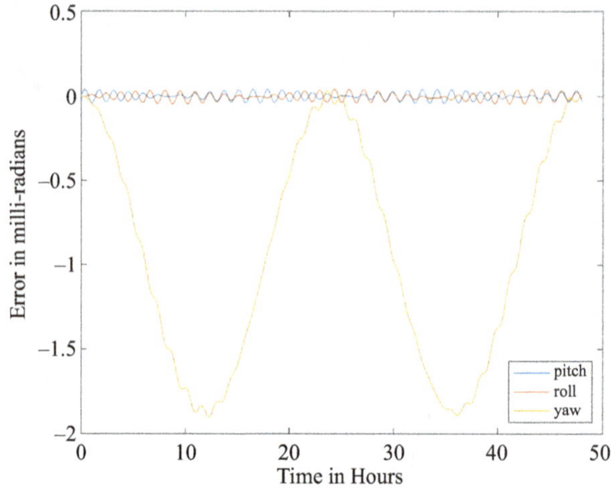

Figure 10.8 Extending the 0.01 deg/hour east gyroscope bias simulation to 48 hours. Euler angle errors are depicted. A diurnal (24 hours) oscillation is observed

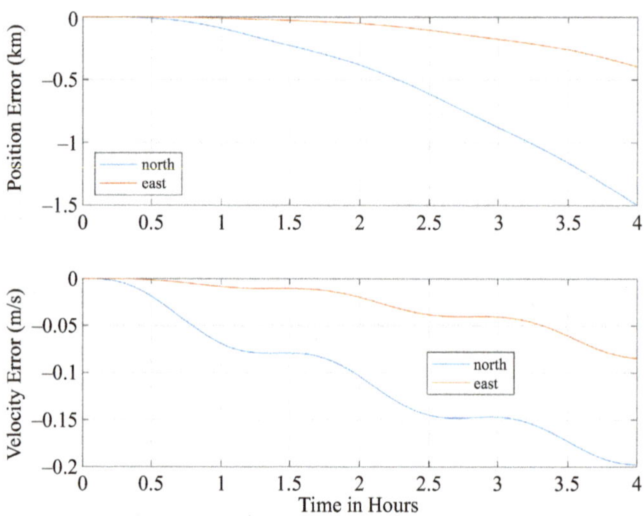

Figure 10.9 Position/velocity errors resulting from a 0.01 deg/hour vertical gyroscope bias

10.4 Vertical gyro bias

In this example, a 0.01 degree/hour bias is simulated on the body z-axis gyroscope. Since the vehicle is level, the gyro's sensitive axis is vertical. The position, velocity, and Euler angle errors over a 4-hour period are depicted in Figures 10.9 and 10.10.

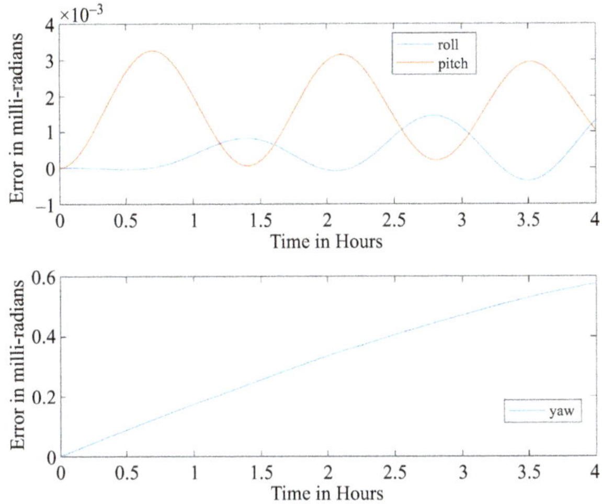

Figure 10.10 Euler angle errors resulting from a 0.01 deg/hour vertical gyroscope bias

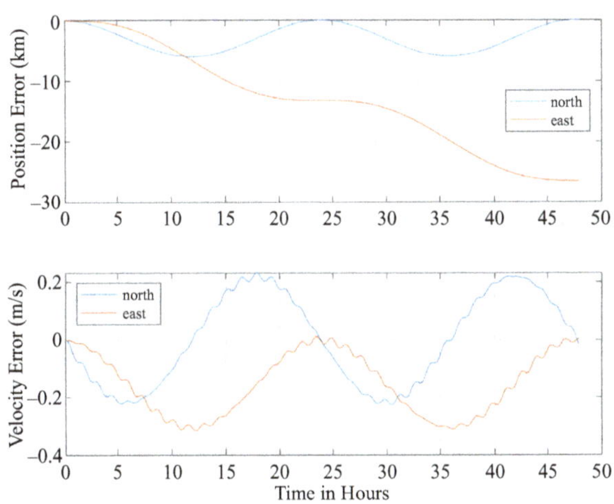

Figure 10.11 Extending the 0.01 deg/hour vertical gyroscope bias simulation to 48 hours. Position and velocity errors are depicted. A diurnal (24 hours) oscillation is observed

In this case, the linear error growth trend is observed on the velocity components and the position error growth is then quadratic. Nevertheless, the position error growth is still slower for the vertical gyro bias than was the case for the horizontal gyro bias. If we extend the duration of the simulation to 48 hours (Figures 10.11 and 10.12), we see a diurnal oscillation as was the case for the horizontal bias. However, we also see that the east position error has a linear trend over the entire 48 hours.

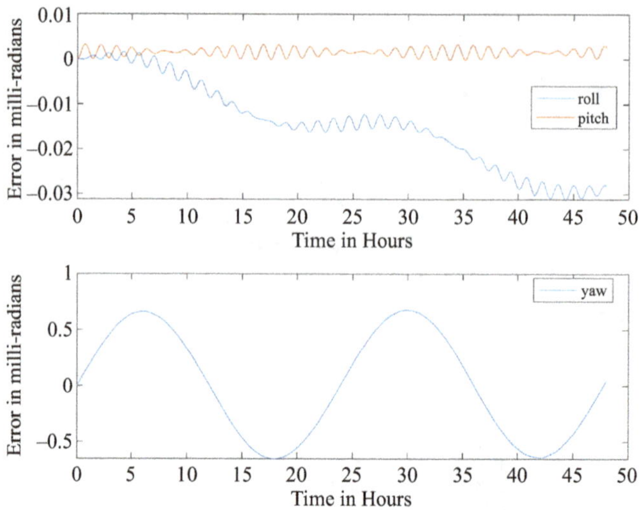

Figure 10.12 Extending the 0.01 deg/hour vertical gyroscope bias simulation to 48 hours. Euler angle errors are depicted. A diurnal (24 hours) oscillation is observed

10.5 Azimuth initialization error

In this example, a 1 milli-radian error in the initial azimuth (yaw) determination is simulated. The position, velocity, and Euler angle errors over a 4-hour period are depicted in Figures 10.13 and 10.14. It is noted that the position, velocity, and attitude error characteristics for an initial azimuth error are very similar to those for a horizontal gyro bias.

If we extend the duration of the simulation to 48 hours (Figures 10.15 and 10.16), the similarities with the horizontal gyro bias case are still apparent.

10.6 Velocity initialization error

In this example, a 0.1 m/s error in the initial north velocity determination is simulated. The position, velocity, and Euler angle errors over a 4-hour period are depicted in Figures 10.17 and 10.18. Virtually, pure Schuler oscillations are present and it is noted that all of the errors in this case are zero mean. As with the accelerometer bias case, Foucault oscillations are observed in the extended simulations shown in Figures 10.19 and 10.20. The magnitudes of the errors scale with the magnitude of the initial velocity error. An initial velocity error of 1 m/s would thus result in a peak position error of approximately 0.8 km.

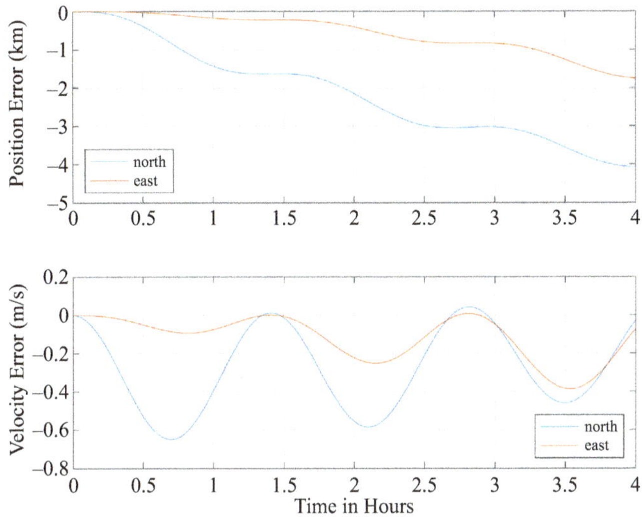

Figure 10.13 Position/velocity errors resulting from an initial azimuth error of 1 milli-radian

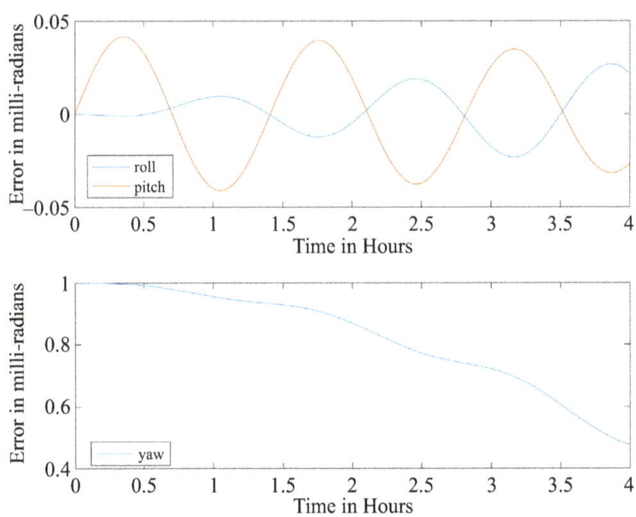

Figure 10.14 Euler angle errors resulting from an initial azimuth error of 1 milli-radian

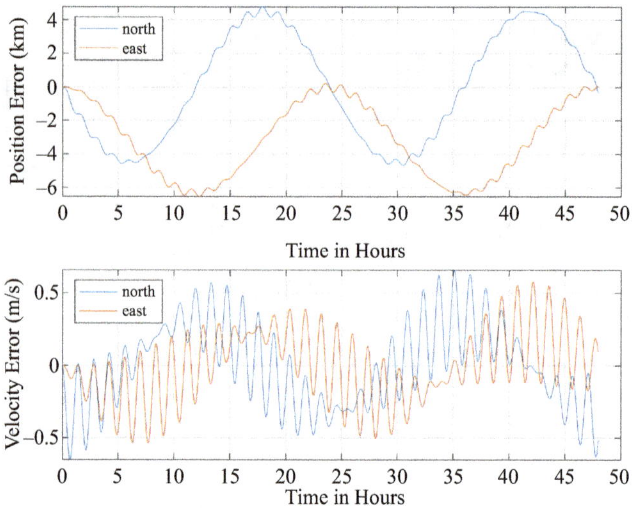

Figure 10.15 Extending the 1 milli-radian initial azimuth error to 48 hours. Position and velocity errors depicted

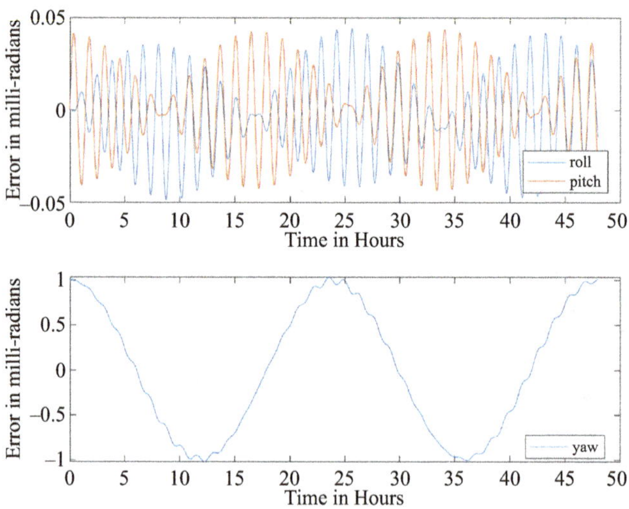

Figure 10.16 Extending the 1 milli-radian initial azimuth error to 48 hours. Position and velocity errors depicted

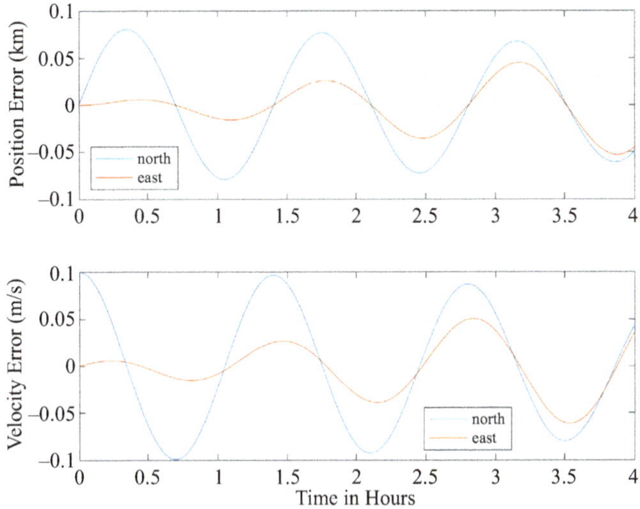

Figure 10.17 Position/velocity errors resulting from an initial north velocity error of 0.1 m/s

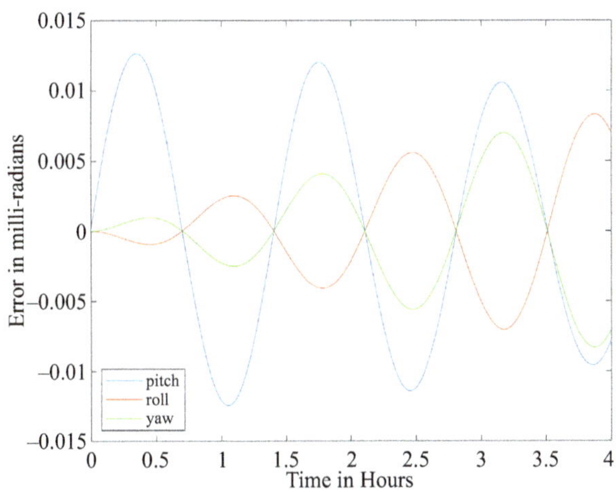

Figure 10.18 Euler angle errors resulting from an initial north velocity error of 0.1 m/s

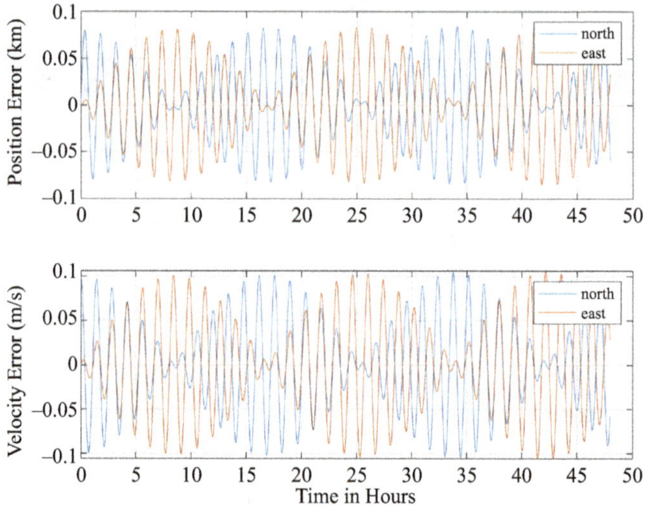

Figure 10.19 *Extending the initial north velocity error (0.1 m/s) simulation to 48 hours. Position/velocity errors depicted*

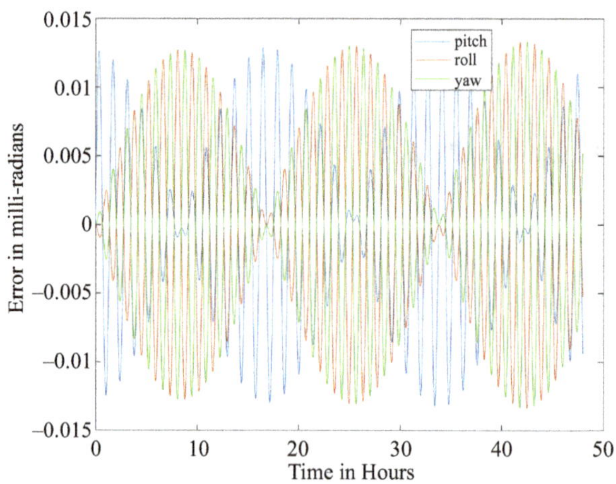

Figure 10.20 *Extending the initial north velocity error (0.1 m/s) simulation to 48 hours. Euler angle errors depicted*

10.7 Combination of errors

We now consider the combination of all the previously simulated single error sources: 100 μg north accelerometer bias; 0.01 deg/hour east gyro bias; 0.01 deg/hour vertical gyro bias; 1 milli-radian initial azimuth error; 0.1 m/s initial north velocity error. All six error sources are simulated simultaneously and the north position error will be plotted along with the previously shown results from each component. As we will show, the dominant error sources are a function of the length of the scenario. We start with a 15-min simulation. Figure 10.21 clearly shows the dominant error source is the accelerometer bias. The second largest contributor is the initial north velocity error followed by two errors that are nearly identical in terms of their impact: east gyro bias and initial azimuth error. Finally, the impact of the vertical gyro bias is negligibly small.

Next we extend the simulation to a period of 3 hours. As shown in Figure 10.22, by the end, the horizontal gyro bias and initial azimuth error are the dominant error sources. Notice the impact of the vertical gyro bias is slowly growing and the initial velocity error, which previously was significant (over a short interval) is now essentially negligible. When the simulation is extended to 12 hours (Figure 10.23), the dominant error source at the end is clearly the vertical gyro bias. Finally, when we extend the simulation to 30 hours (Figure 10.24), all three major frequency components may be observed. The "high" frequency Schuler oscillation is most noticeable in the accelerometer bias result but can also be observed in the other

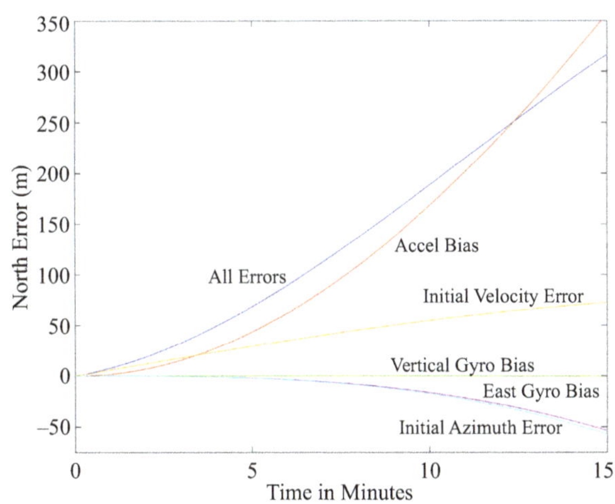

Figure 10.21 North position error resulting from various error sources depicted in the figure. Over this 15 min period, the accelerometer bias is clearly the dominant error source

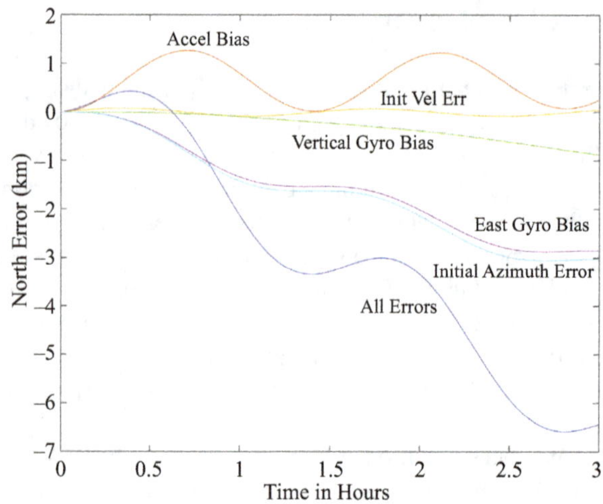

Figure 10.22 North position error resulting from various error sources depicted in the figure. At the end of this 3 hour period, the horizontal gyro bias and initial azimuth error are the dominant error sources

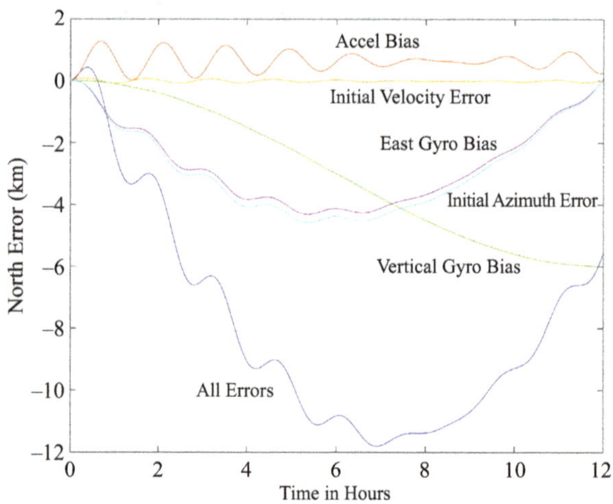

Figure 10.23 North position error resulting from various error sources depicted in the figure. By the end of 12 hours the dominant error source is the vertical gyro bias

error sources as well (except the vertical gyro bias). The dominant frequency over this long simulation is the diurnal oscillation observed in both gyro bias results and the initial azimuth error. Since these three are the dominant error sources, it is not surprising that that a diurnal oscillation is also observed in the combination of all error sources.

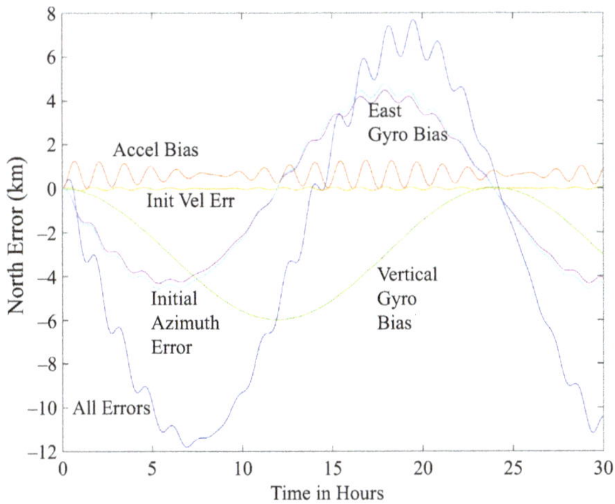

Figure 10.24 North position error resulting from various error sources depicted in the figure. Over this 30 hour period the three major frequencies (Schuler, Diurnal, Foucault) are readily apparent

Finally, the long period Foucault oscillation is also apparent in the accelerometer bias results although not as clearly as in the earlier figures (e.g., Figure 10.3).

We can thus conclude that for very short durations (i.e., less than 15 minutes), accelerometer errors (and equivalently, gravity modeling errors) will be the dominant error source. Over medium durations (e.g., 1–4 hours), the horizontal gyro errors and initial azimuth error dominate. For long durations (greater than 10 hours), the vertical gyro error is the dominant factor with horizontal gyro error and initial azimuth determination both following closely behind.

10.7.1 What about noise?

At the end of the previous chapter, we looked at a few equations that provided position error standard deviation due to sensor noise. These equations were limited to short time periods (two were applicable for periods less than 5 minutes; one for about an hour and a half) and all were "single axis" (only examining the effects of noise on one sensor and no consideration of cross-coupling). In the simulation environment, however, we can remove these limitations. Given a certain sensor noise standard deviation, we can simulate the noise on each sensor output and can then process them as with all the previously performed simulations. Since the sensor errors are stochastic, however, a single run of the simulation is not sufficient to characterize the performance. With the whole value simulations utilized in this chapter, we must perform so-called "Monte Carlo" simulations. That is, the simulation must be performed for a large number of trials with independent simulated noise error on each run. The results of all the trials may then be processed to estimate the standard deviation of the resulting system errors (e.g., position, velocity and/or attitude).

Let us first consider how noise error on a single sensor will cross-couple into the orthogonal horizontal axis (given the vertical channel instability, there is little point in examining it). Figure 10.25 depicts the position error standard deviation resulting from north accelerometer noise of 10 $\mu g/\sqrt{Hz}$. Both the general square-root of time error growth (on the north axis) and the cross-coupling are readily apparent. However, examination of the *y*-axis reveals the error is negligibly small in comparison with the errors previously considered in this chapter. Figure 10.26 depicts the position error standard deviation resulting from east gyro noise of 0.002 $deg/\sqrt{hour}$.

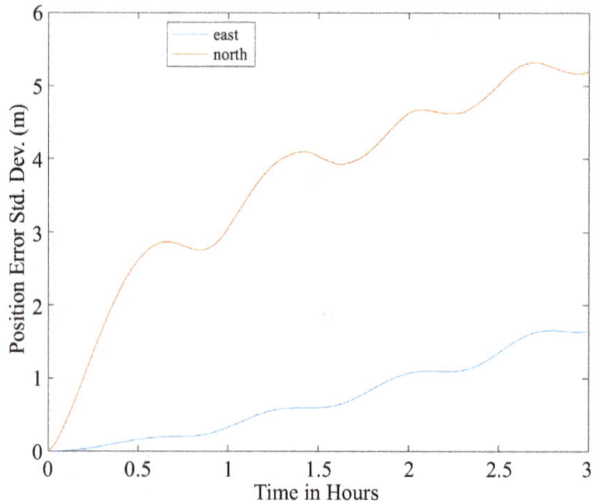

Figure 10.25 Position error standard deviation due to 10 $\mu g/\sqrt{Hz}$ on the north accelerometer only

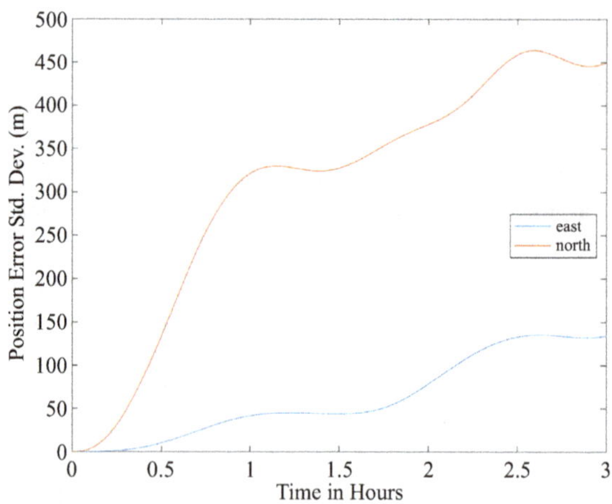

Figure 10.26 Position error standard deviation due to noise of 0.002 $deg/\sqrt{hour}$ on the east gyro only

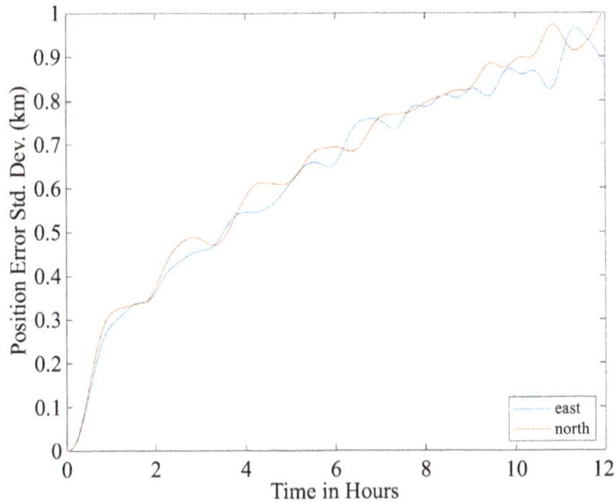

Figure 10.27 Position error standard deviation due to noise of 0.002 deg/$\sqrt{hour}$ on all three gyros and 10 μg/$\sqrt{Hz}$ on all three accelerometers

For typical nav-grade sensor noise values, it is clear that gyro noise is the dominant factor. Nevertheless, even for the gyro, the error growth is still relatively small in comparison to the previously considered error sources. To see that this is indeed the case even for more realistic scenarios, Figure 10.27 depicts the position error standard deviation resulting from 10 μg/$\sqrt{\text{Hz}}$ noise on all three accelerometers and 0.002 deg/$\sqrt{\text{hour}}$ noise on all three gyros. In comparison with the errors depicted in Figure 10.23, it is clear that nav-grade sensor noise plays a very small role in overall system performance.

10.8 Long-duration flight

Up to this point, we have considered the impact of various error sources and have examined their characteristics. We conclude this chapter with a sample simulated long-duration flight. Specifically, a flight was simulated from Houston, Texas in the United States to Auckland, New Zealand. A great-circle route was simulated as shown in Figure 10.28. The simulated path is 6,467 nautical miles in length and with a simulated velocity of 500 knots the duration is slightly less than 13 hours.

Although in reality inertial sensor errors will exhibit drift and variation over 13 hours, for the sake of the simulation, the bias errors are kept fixed. Horizontal accel biases were set to 100 μg, gyro biases were set to 0.01 deg/hour, and gyro noise was set to 0.01 deg/root-hour. The resulting errors are depicted in Figures 10.29, 10.30, and 10.31.

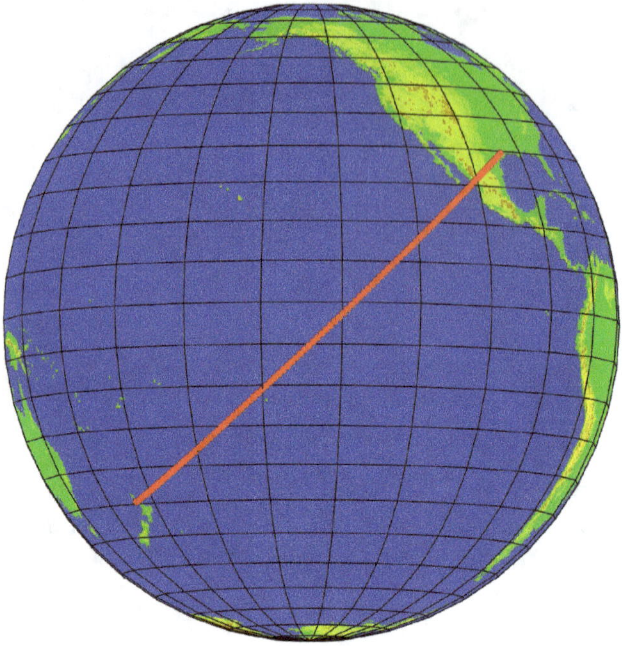

Figure 10.28 Great circle flight path from Houston, Texas, USA to Auckland, New Zealand

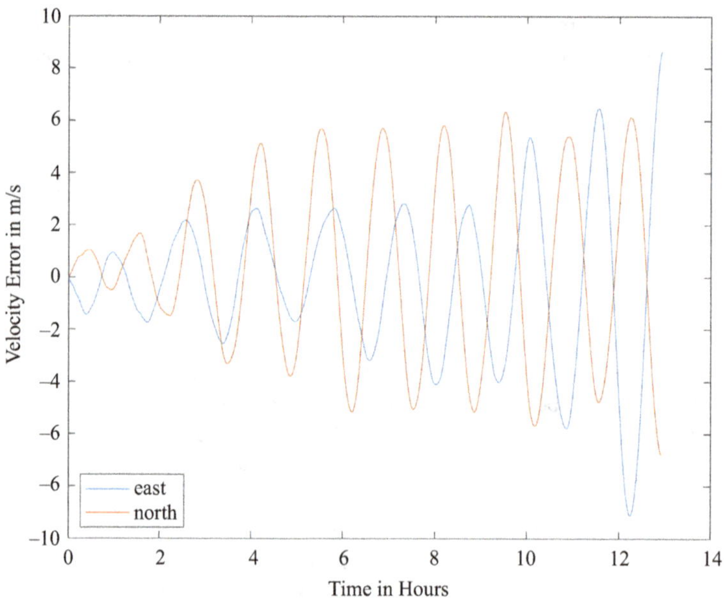

Figure 10.29 Horizontal velocity errors for the simulated Houston to Auckland flight

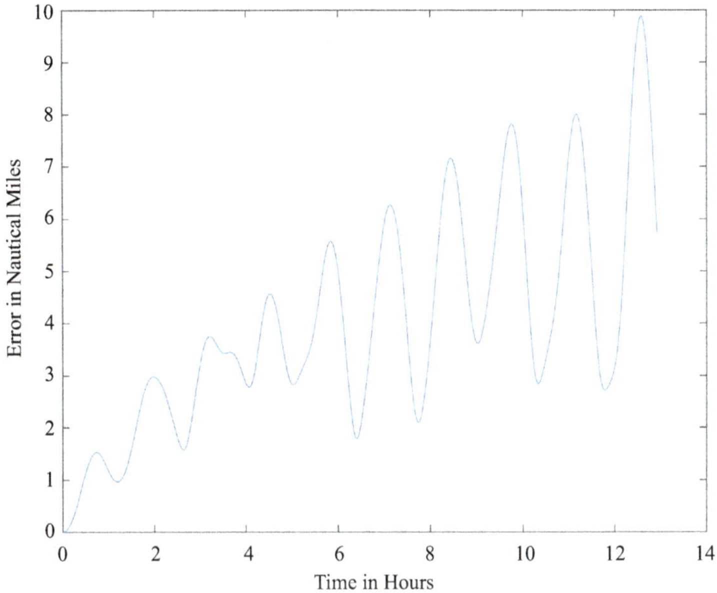

Figure 10.30 *Horizontal position error for the simulated Houston to Auckland flight*

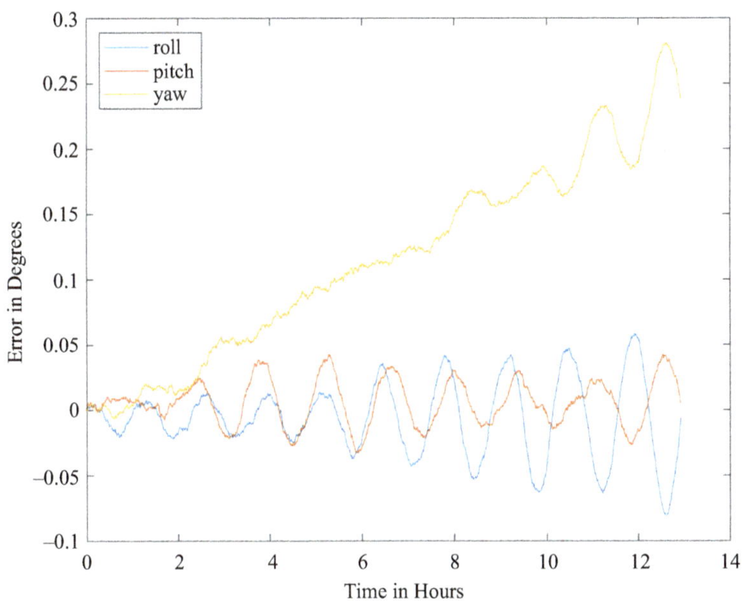

Figure 10.31 *Euler angle errors for the simulated Houston to Auckland flight*

To see the impact of the vertical gyro error, the simulation is repeated but with a vertical gyro bias of −0.01 deg/hour (all biases were positive in the previous case). All other error sources remain the same. The results are depicted in Figures 10.32, 10.33, and 10.34. Whereas the peak horizontal position error is less than 10 nautical

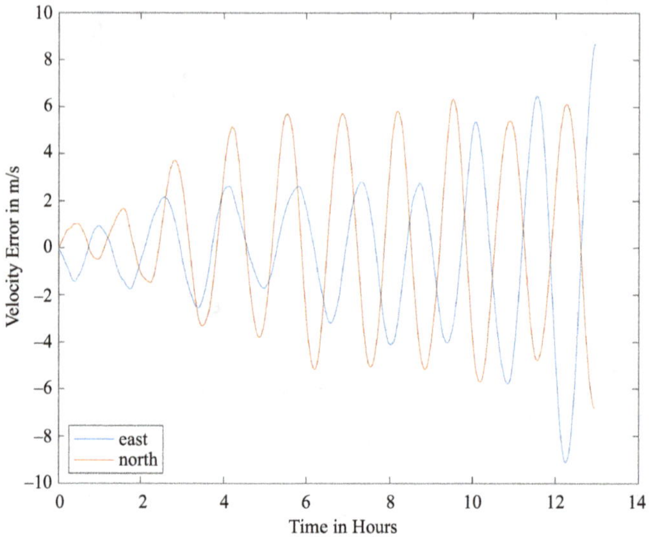

Figure 10.32 Horizontal velocity errors for the simulated Houston to Auckland flight with the vertical gyro bias negated

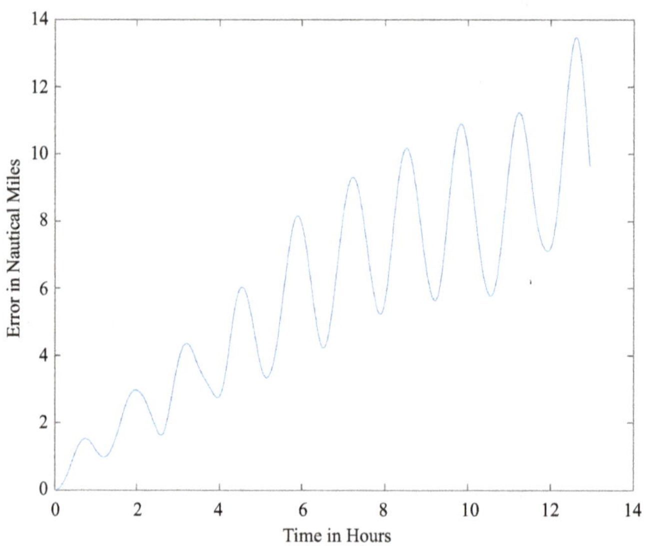

Figure 10.33 Horizontal position error for the simulated Houston to Auckland flight with the vertical gyro bias negated

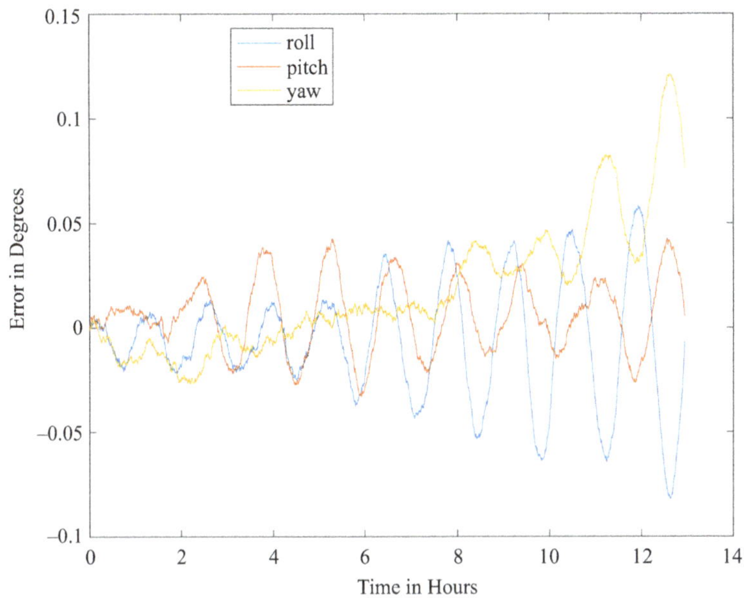

Figure 10.34 Euler angle errors for the simulated Houston to Auckland flight with the vertical gyro bias negated

miles in the first case, when the vertical gyro bias is negated the peak error is well over 13 nautical miles. Conversely, in the first case the maximum yaw error was approximately 0.28 degree but in the negated case is slightly over 0.12 degree.

Reference

[1] Titterton DH, Weston JL. *Strapdown Inertial Navigation Technology*. 2nd ed. Herts, UK: The Institution of Electrical Engineers; 2004.

Chapter 11

Inertial initialization—Part A

11.1 Introduction

This final chapter regarding inertial (only) processing addresses the procedures that are actually done *first* in the operation of an inertial system: initialization of position, velocity, and attitude. Initialization of attitude is sometimes referred to as "leveling and alignment." We had to wait until the end to learn about initialization since it depends on many of the concepts that we have developed up to this point including how we deal with gravity, earth rate, and various coordinate frames. We will introduce the major concepts in this chapter and will focus on so-called "coarse leveling" and "coarse alignment" [1]. Closer to the end of the book, we will come back to this topic to address fine leveling and alignment since they depend upon aiding concepts (typically utilizing a Kalman filter) that we will development in subsequent chapters.

11.2 Initialization of position

Initialization of position is a relatively simple matter of entering the position of the vehicle as determined by some external means. The simplest case is that of a stationary vehicle. The location is generally determined either by the use of some predetermined survey or through the use of a radionavigation system.

I remember flying through Europe on a business trip and making a connection in Frankfurt, Germany. There were not enough gates at the terminal so my plane had to park out on the ramp. We had to get out of the airplane and get onto a bus to go to the terminal. I remember when I was getting out of the airplane that I saw a sign on a post in front of what was essentially the aircraft's parking space. It had the latitude, longitude, and altitude of that aircraft parking spot. The purpose of that sign was to provide pilots with the information they needed to initialize position in the inertial navigation system.

If initialization is needed in flight (in some vehicles/applications, system resets can occur), then obviously there are no surveyed locations. An external position determination system is required. Frequently, this is done with satellite navigation systems such as the United States Global Positioning System (GPS). Historically, legacy ground-based radio navigation aids were used as well as flying over visual checkpoints where there was a known location. Depending upon the desired accuracy of the position initialization, the lever arm between the radionavigation antenna and the INS must be accommodated. Also, for in-flight initialization, the transport delay between the time-of-validity of the radionavigation-determined position and the position initialization of the INS must be taken into account.

11.3 Initialization of velocity

If the vehicle is stationary, then earth-referenced velocity is approximately zero (note that wind buffeting and personnel movement on a vehicle can induce short-term velocity "noise"). Again, if the vehicle is in motion, then velocity must be determined by some external source. That can be any global navigation satellite system (GNSS) or other radio navigation aid.

11.4 Initialization of attitude: leveling

The term "leveling" was coined at a time when all inertial navigation systems were gimbaled platforms. The platform containing the two horizontal accelerometers had to be physically leveled. In a strapdown system, the equivalent procedure involves determining the pitch and roll angles of the vehicle.

Consider a gimbaled platform with two orthogonal accelerometers mounted in the plane of the platform. If the vehicle is stationary, the only specific force is the reaction to gravity. Since that force lies, to first order, in the vertical direction, the platform is leveled by physically rotating it until the two in-plane accelerometers output zero. Now in reality all sensors output non-zero values at all times due to noise. Thus, even when there is no true input present, the accelerometers will output noise (the same is true for gyros). Thus, when we talk about rotating the platform until the two horizontal accelerometers output zero, we are really looking for an *average* of zero. For nav-grade accelerometers, it is necessary only to average over a few seconds.

In a strapdown system, leveling involves the estimation of pitch and roll (or the components of the body-to-nav DCM or quaternion that depend on those values). Leveling is actually accomplished in two steps. The first is coarse leveling in which a few seconds of accelerometer measurements are used to get a rough estimate of pitch and roll components. Fine leveling follows this and typically involves a Kalman filter to refine the estimates. As mentioned earlier, we will discuss fine leveling later in the book.

11.4.1 Coarse leveling

We are assuming the inertial is a strapdown type (i.e., body-mounted accelerometers) and we are further assuming that those accelerometers are mounted coincident with the body frame. Since we are assuming the vehicle is stationary, the vector of accelerometer measurements is simply the reaction to gravity expressed in the body frame. However, it can also be viewed as the reaction to gravity expressed in the nav-frame multiplied by the nav-to-body DCM:

$$\underline{A}^b = \begin{bmatrix} A_x^b \\ A_y^b \\ A_z^b \end{bmatrix} = -\underline{g}^b = C_n^b(-\underline{g}^n) = \begin{bmatrix} \text{DC} & \text{DC} & -\sin\theta \\ \text{DC} & \text{DC} & \sin\phi\cos\theta \\ \text{DC} & \text{DC} & \cos\phi\cos\theta \end{bmatrix} \begin{bmatrix} 0 \\ 0 \\ -g \end{bmatrix} \tag{11.1}$$

where the DC ("do not care") terms are irrelevant since they will be multiplied by zeros.

We know the reaction to gravity in the navigation frame is something that we can specify without any measurements (at least to the accuracy of our gravity model). We know that in the navigation frame, the horizontal components of the reaction to gravity are zero, and we only have a non-zero value in the local vertical (for most applications we can ignore the gravity deflection of the vertical). The reaction-to-gravity vector in the navigation frame is thus known. The reaction-to-gravity vector in the body frame *is* measured with the accelerometers. The nav-to-body direction cosine matrix relates the two. This relationship allows us to get a rough initialization of roll and pitch. Again, we do not care about six of the nine nav-to-body direction cosine matrix elements since those six get multiplied by zeroes anyway. We can thus simplify the relationship as follows:

$$\begin{bmatrix} A_x^b \\ A_y^b \\ A_z^b \end{bmatrix} = \begin{bmatrix} g\sin\theta \\ -g\sin\phi\cos\theta \\ -g\cos\phi\cos\theta \end{bmatrix} \tag{11.2}$$

Therefore, we can estimate roll by:

$$\phi = \text{arctan2}(-A_y^b, -A_z^b) \tag{11.3}$$

Notice that both arguments *must* be negated to cancel the effects of the minus signs in (11.2).

Proof:

$$\frac{A_y^b}{A_z^b} = \frac{-g\sin\phi\cos\theta}{-g\cos\phi\cos\theta} = \frac{\sin\phi}{\cos\phi} = \tan\phi \tag{11.4}$$

A four-quadrant arctangent must be used since roll ranges from 0 to ± 180 degrees. Note, however, that the formula fails if pitch is ± 90 degrees. Pitch is obtained as follows:

$$\theta = \tan^{-1}\left(\frac{A_x^b}{\sqrt{(A_y^b)^2 + (A_z^b)^2}}\right) \tag{11.5}$$

Proof:

$$\frac{A_x^b}{\sqrt{(A_y^b)^2 + (A_z^b)^2}} = \frac{g\sin\theta}{\sqrt{(-g\sin\phi\cos\theta)^2 + (-g\cos\phi\cos\theta)^2}}$$

$$= \frac{g\sin\theta}{\sqrt{(-g\cos\theta)^2(\sin^2\phi + \cos^2\phi)}}$$

$$= \frac{g\sin\theta}{\sqrt{(-g\cos\theta)^2(1)}} = \frac{g\sin\theta}{g\cos\theta} = \tan\theta$$

Note that only positive values of $\cos\theta$ will remain after the above processing and thus the expression is only valid for $\frac{-\pi}{2} \leq \theta \leq \frac{\pi}{2}$. Thus, we should use a two-quadrant arctangent function. This is perfectly acceptable, though, since pitch is defined to lie in the above range.

Recall that these expressions for roll and pitch are valid only for a stationary vehicle. Furthermore, the inputs must consist of accelerometer measurements that have been averaged over a sufficiently long interval of time (e.g., a few seconds for nav-grade sensors).

11.5 Initialization of attitude: alignment/gyrocompassing

Now we turn our attention to the concept of alignment. This can also be viewed as azimuth determination (i.e., where is north?). Once we have determined the level plane, then we need to determine where is north.

The way we do that is by taking advantage of the fact that if the vehicle is stationary, then the gyros only measure Earth rate. We then take advantage of the fact that Earth rate only exists in the north and vertical directions (which we can distinguish once we have leveled the platform). In a strapdown system, this is all done in software.

In a gimbaled system, once the platform is level, north can be determined by physically rotating the platform until the east gyro is outputting an average value of zero. If the horizontal component of Earth rate is measured in only one gyro, then that gyro is pointing at north. Systems built in this way were known as "north-finding" systems. In more recent times with readily available computational power, it is easier to determine azimuth by taking an arctangent of the ratio of the horizontal gyro outputs (this will be covered in more detail in the next section). This type of processing is known as "gyrocompassing."

It turns out that this process is more difficult than was the case for the accelerometers. The reason has to do with the magnitude of the input. Gravity is a large signal, $9.81\ \frac{\text{m}}{\text{s}^2}$. It is a large specific force to measure and thus it is relatively easy with nav-grade sensors to estimate pitch and roll with just a few seconds of accelerometer data. Unfortunately that is not the case for gyros measuring Earth rate.

Earth rate is a very tiny signal: 15 degrees-per-hour or 7.29×10^{-5} radians-per-second. The small magnitude can be better appreciated by considering the opposite extreme. An aerobatic aircraft performing a snap roll will rotate 360 degrees within just a couple of seconds. This is an angular rate of hundreds of degrees per second. Compare that to Earth rate which is only 15 degrees per hour. When you couple a relatively small signal with relatively large sensor noise, the result is that gyrocompassing is not a quick process. It generally takes several minutes to resolve azimuth to the necessary small fraction of a degree.

If the vehicle is on the equator, there is no vertical component of Earth rate. Earth rate lies entirely in the north direction. That is helpful because it is the horizontal component of Earth rate that we need for azimuth determination. However, for locations away from the equator, the magnitude of the horizontal component of Earth rate decreases. At the extreme (the North or the South Pole), the entirety of Earth rate lies in the vertical direction. Thus, there is no horizontal component of Earth rate when we are at a pole. Therefore, not only would the east gyro be outputting zero, the north gyro would be outputting zero too because there is no horizontal component of Earth rate.

The point is that at very high latitudes, it becomes very difficult to do gyrocompassing because the horizontal component of Earth rate is so small. It is small to begin with, but then it becomes very, very small the closer you get to the poles. That is why nav-grade inertial units will specify a certain alignment time up to some maximum absolute value of latitude. The inertial manufacturers will tell their customers that if the unit is initialized at a higher latitude, the unit will still operate but there is no guarantee on how quickly it will be able to determine initial azimuth.

11.5.1 Coarse gyrocompassing

Let us take a look at how we can estimate azimuth on a static platform. The idea is that we are going to be averaging gyro measurements just as we were averaging accelerometer measurements before. In fact, what we have to do to start off with in this case is the coarse leveling. With coarse initial estimates of pitch and roll, we can compute a body-to-local-level DCM:

$$C_b^n(\phi,\theta)\Big|_{\psi=0} = C_b^{LL} = \begin{bmatrix} \cos\theta & \sin\phi\sin\theta & \cos\phi\sin\theta \\ 0 & \cos\phi & -\sin\phi \\ -\sin\theta & \sin\phi\cos\theta & \cos\phi\cos\theta \end{bmatrix} \quad (11.6)$$

where "LL" stands for "local-level." Note this body-to-LL DCM is obtained from the theoretical expression for the body-to-nav DCM by setting yaw equal to zero. We do not yet know yaw (azimuth), but coarse estimates of pitch and roll allow us to convert the angular rate measurements from the body-frame to a frame that is approximately locally level.

Since the vehicle is assumed to be stationary, the vector of gyro measurements is simply the earth rate vector expressed in the body-frame (we will need to average these measurements to deal with the noise). We can then convert the measured (and

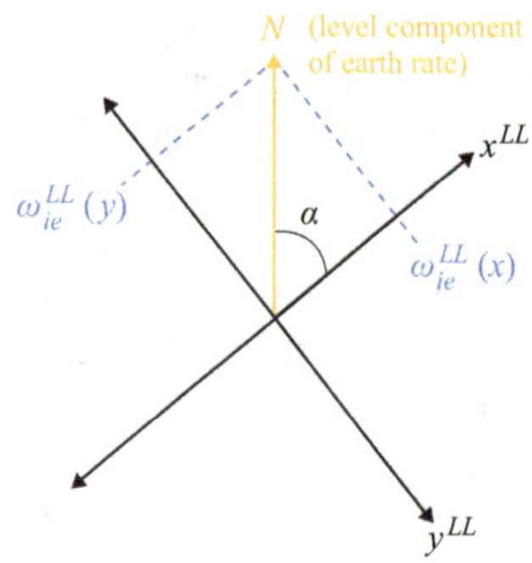

Figure 11.1 The horizontal component of earth rate lies in the north direction. Thus, the ratio of the horizontal components sensed in the local-level frame can be used to determine the rotation angle of the local-level frame from north. The angle in the diagram is denoted as the wander angle (α). This is a design choice. The designer can let the initial azimuthal offset be represented by a non-zero wander angle (thus implying initial yaw = 0) or the azimuth offset can be applied to yaw (thus implying initial wander angle = 0)

averaged) earth-rate vector from the body-frame to the local-level frame with the above DCM:

$$\underline{\omega}_{ie}^{LL} = C_b^{LL}\underline{\omega}_{ie}^{b} \approx C_b^{LL}\underline{\omega}_{meas}^{b} \tag{11.7}$$

where $\underline{\omega}_{meas}^{b}$ is the measured angular rate vector. Since the horizontal component of Earth rate lies solely in the north direction, the level components of the resolved Earth rate vector are simply projections of the north component of Earth rate onto the horizontal axes of the local-level frame. This is illustrated in Figure 11.1. Notice in the figure that the north component of earth rate projects onto the positive local-level x-axis but onto the negative local-level y-axis. Thus, a negative sign is needed in the equation for the azimuth angle:

$$\alpha = \text{arctan2}(-\omega_{ie}^{LL}(y), \omega_{ie}^{LL}(x)) \tag{11.8}$$

11.6 A comment regarding recursive estimation

We need to remind ourselves that the preceding algorithms work with a few seconds of data to estimate initial level and a bit more data to estimate initial alignment. This works as long as the platform is stationary, and we still have to remember that the result is just a *rough* initial value that then needs to be turned over to a Kalman filter. Why is that the case? If we tried to utilize this batch processing over long periods of time for fine alignment, a challenge, certainly for legacy systems, is that storing large qualities of measurements is difficult in an embedded system. Furthermore, even if the data could be stored, the batch processing cannot be extended to a moving base alignment situation. To do the initialization in flight, one cannot simply average data since the true values are not constant.

Early generation systems performed refinement of the initial estimate with a so-called "fixed gain" filter. In modern times, it is performed with a Kalman filter. The details will be discussed in a later chapter but for now suffice it to say the filter uses external information to achieve enhanced accuracy. Once coarse leveling and alignment have been achieved, the inertial system can begin to compute position, velocity, and attitude. System errors can then be determined by comparing the computed position/velocity/attitude with externally known quantities. For example, so-called zero-velocity updates (ZUPTs) exploit the fact that a stationary vehicle is not moving. Non-zero velocity values computed by the inertial system thus constitute observed errors that the filter can process to improve system accuracy.

11.7 Conclusion

Although position and velocity must obviously be initialized, the challenge in inertial system initialization is leveling and alignment. For stationary vehicles, coarse leveling can be performed in just a few seconds. Coarse alignment can take longer especially at higher latitudes. Fine leveling and alignment can take several minutes (again, longer for higher latitudes) and is done almost exclusively with some form of Kalman filter.

Reference

[1] Kayton M, Fried W. *Avionics Navigation Systems*. 2nd ed. New York, NY: John Wiley & Sons; 1997.

Chapter 12

Introduction to integration and estimation theory

12.1 Why integrate?

It is a reasonable question. Why would we want to go through the time and effort to integrate two different navigation systems? The simple answer is that we integrate multiple systems when no one system by itself will satisfy all the necessary requirements. More generally, we integrate multiple systems in order to construct a super-system that in some way outperforms any of the individual systems.

We are going to be spending a lot of time developing the framework behind the Kalman filter and ultimately driving toward the goal of developing an architecture that will enable us to integrate multiple navigation systems into a single integrated unit. We will start off by looking at navigation system integration and consider some key issues along with goals and requirements. We will investigate subsystem characteristics and will look at some challenges to integration.

Navigation engineers must be cognizant of the fact that their product is a component within a larger system. This larger system is typically a vehicle (specifically, a land, air, sea, or space vehicle). As I have told many students throughout my years of teaching: an inertial navigation system will not transport me from New York to Chicago. It will tell me where I am. It will tell me how fast I am going. It will tell me the current attitude of my vehicle and how fast it is turning. However, it will not answer the most basic navigation question that all drivers, pilots, and captains must answer in order to get where they want to go: Which way to steer?

The position, velocity, and attitude information determined by the navigation system, in virtually every case, is handed off to another system. Frequently it is either a flight management system or a flight control system or some kind of display for the pilot (or driver or captain...). It is also possible that it may be pointing a sensor or a weapon or providing the position and velocity needed to process synthetic aperture radar images. In any case, the navigation system needs to provide its information in a form and with a quality that satisfies the needs of the downstream user or system.

12.2 A high level look at an aircraft flight control system

Let us consider, as an example, an aircraft navigation system. The system is comprised of position/velocity/attitude sensors, a database, a computer, and a flight control system. The need for the positioning sensor is obvious. A database is needed to store the key components of the desired flight plan (e.g., waypoints along the route) and a computer is needed to determine the difference between the current position and the desired route. Finally, a flight control system is needed to maneuver the airplane in order to maintain it on the desired route. All of these subsystems together comprise the overall aircraft navigation system. As we will discuss further, successful operation of this system requires some specific performance characteristics in the position/velocity/attitude sensors. Although it is a bit of a misnomer, we will stay consistent with standard parlance by referring to the position/velocity/attitude determination system(s) as "navigation system(s)" (e.g., the inertial navigation system). Although systems such as an INS or GNSS are actually position/velocity determination systems (and are only a component of an overall vehicle navigation system) they are nevertheless almost universally referred to simply as navigation systems.

Figure 12.1 depicts the various errors in an aircraft navigation system (the same principles apply to other types of vehicles as well). Total system error (TSE) is the difference between the actual aircraft position and the desired position (e.g., the desired position on the desired flight path). It is composed of navigation system (or sensor) error (NSE) and flight control system error (also known as flight technical error or FTE). NSE is the difference between the actual aircraft position and the (navigation) sensor-derived position. NSE is thus the error in the positioning system. FTE is the difference between the desired position and the (navigation) sensor-derived position (since, of course, the flight control system does not know the actual position of the aircraft). The figure thus depicts three "paths": where the aircraft actually is, where the aircraft thinks it is; and where the aircraft wants to be. Notice that both NSE and FTE must be small in order to satisfy the overall goal of keeping the aircraft on or near the desired path. A perfect navigation (sensor) system with a lousy flight

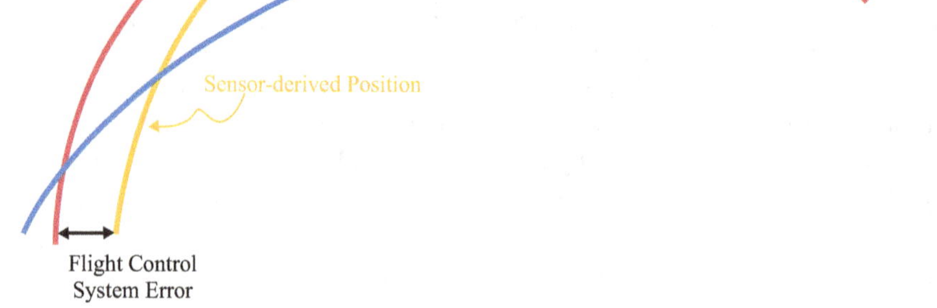

Figure 12.1 Aircraft navigation system error components: total system error is the difference between the actual aircraft position and the desired position. It is composed of navigation system error and flight control system error

control system will not be acceptable. Similarly, a perfect control system being fed lousy position/velocity data will not work either.

12.3 Navigation system requirements

Our focus is on the navigation sensor or sensors. There are a number of requirements that must be satisfied. Chief among these are:

1. Accuracy
2. Integrity
3. Availability
4. Data rate
5. Data latency

Accuracy is the most obvious requirement. The accuracy must adhere to an error budget that has been derived for the particular vehicle and for the operational scenario (e.g., the accuracy requirement for domestic en-route navigation is different than for precision approach and landing). Integrity has to do with the degree of trust that can be placed on the navigation system output. For example, say that 95% of the time the navigation sensor is determining position with an accuracy of better than 2 m and the requirement is to fly from New York to Los Angeles. An accuracy of 2 m is perfectly fine. However, what happens if once every 10,000 hours the error is 100 miles and it builds up gradually so that you do not notice it? The 95% accuracy is still 2 m since this large excursion happens such a small percentage of time. Would you want to fly on an airplane with that kind of a sensor? No, you certainly would not since you do not want to be on the airplane when the sensor has a 100-mile error. You might end up slamming into a mountain or another airplane. The system has acceptable accuracy but not integrity. For the rare occasions when the navigation system far exceeds its error budget, some kind of monitor is needed. Either an internal or external monitor is needed to provide an alert or an alarm when the required accuracy is not being provided.

Availability is defined as the percentage of time that the system is in service. There are two types of availability: short term and long term. Short-term availability (frequently referred to as continuity) is defined as the probability that the system will provide service throughout a given mission. Long-term availability is defined as the probability that the system is functioning properly at the beginning of the mission.

For example, say you want to take a flight from New York to Paris. Is the navigation system working properly at the beginning of the flight? If so, it is considered "available." Now let us say that in the middle of the Atlantic Ocean, the integrity monitor determines that something is wrong and the positioning system is not reliable. The system has passed the integrity test but it has not provided the needed continuity. Note that a lack of availability impacts convenience and revenue but not safety. If the navigation system is not working properly to begin with, then the flight simply is canceled. However, a continuity failure is a safety issue. What is the pilot going to do in the middle of the Atlantic Ocean without a functioning navigation system? Radio-navigation system providers operate their systems to meet certain availability/continuity requirements on the so-called "signal-in-space." On the user side, availability/continuity requirements are typically addressed through reliability metrics (e.g., mean time between failure or MTBF). Some operations ("phases of flight") cannot be adequately supported by a single, stand-alone, navigation system and thus integration of multiple systems is needed.

Sensor data rate and latency requirements are needed to satisfy the flight control system. For example, if the positioning sensor only outputs a data point once every half hour it is not going to be of much use for the flight control system. However, if the sensor provides 100 Hz data (perfectly adequate for most aircraft flight control systems), but the data arrives at the control system 5 sec after it was computed, that is also unacceptable. A 5-sec latency will drive a flight control system unstable. The navigation system must provide not only a sufficiently high data rate but also a sufficiently low data latency to keep the control system stable.

Not surprisingly, the goal of an integrated navigation system is to combine the data from multiple sensors in such a way that the requirements are satisfied. What we will see is that various navigation subsystems or sensors meet some of the requirements but not all. The goal of an integrated system is to get a combination of sensors such that the combination meets all of the requirements.

12.4 Case study: GNSS and INS

We will now consider two commonly integrated navigation systems: global navigation satellite systems (GNSS) and INS. Neither system meets all of the aforementioned requirements. When integrated properly, however, the combination can. In this book we will focus on accuracy and will achieve acceptable data rate and latency as well. Integrity and availability issues will not be addressed here but the literature is replete with the work of others in this regard.

The characteristics of GNSS are:

- Excellent long-term accuracy
- Noisy in the short term
- Low data rates
- Long data latency
- Challenges in determination of attitude

GNSS exhibits excellent long-term accuracy. The error is on the order of 5 m now and it will be the same an hour from now, 12 hours from now and 2 days from now. It has excellent stability. The error does not grow with time. However, over the short term, it is somewhat noisy and receivers have low data rates (1 Hz is very typical). Higher data rates are possible but the measurement noise increases significantly in proportion to the data rate. Along with a relatively low data rate, GNSS receivers also typically have relatively long data latencies (on the order of hundreds of milliseconds) that are unacceptable for flight control systems. Finally, additional complexity is required to use GNSS to determine attitude. The techniques for attitude determination with GNSS require multiple antenna/receiver pairs and thus add complexity to a given installation. Even then, the data rate and data latency issues remain.

We can contrast this with inertial navigation:

- Poor long-term accuracy
- Excellent short-term accuracy

- High data rates
- Short data latencies
- High-quality attitude determination is a by-product of the processing

We know that inertial navigation has poor long-term accuracy. The inertial system error grows with time. Conversely, over the short term the inertial data is relatively noise-free. Although the raw sensor outputs are noisy, they are integrated twice in order to determine position. As a result, the bulk of the noise is smoothed out. We also know that inertial systems can provide high data rates (hundreds of samples per second) along with very short latencies (for flight control purposes it is nearly instantaneous). Finally, we also know that high-quality attitude data is a by-product of the inertial navigation processing.

Thus, it is very clear that GNSS and INS complement each other extremely well. Each system has strengths where the other has weaknesses. The challenge in integration is to maximize the strengths of both systems while minimizing the weaknesses of each. Obviously we want to get the long-term stability out of the GNSS, but we want the low noise, high data rates, and low data latency out of the inertial navigation system.

There are some additional issues. For example, there is the need to deal with asynchronous data streams. The GNSS data are not synchronous with the inertial data. We would also like to have robust operation even when less than four satellites are available. We will see later that it is possible to combine the GNSS and inertial in such a way that useful information can be extracted from the GNSS measurements even when a GNSS-only position solution is not possible. Another issue is the lever-arm between the inertial and the GNSS antenna. The two systems are determining position and velocity with respect to two different points on the vehicle and thus the lever-arm must be taken into account when processing the measurements.

Let us start this integration journey by learning about estimation: the theory behind the algorithms that extract useful information from noisy data.

12.5 Introduction to estimation theory: optimality

How do we take a noisy set of data and derive an estimate of a set of underlying true parameters? First we must consider the concept of optimality.

Our goal in an estimation process is to do the best job possible of extracting the underlying "truth" from a noisy set of measurements. We desire to do this in an optimal fashion. That sounds nice. It sounds like what you want to do. It sounds like it is best in some sense, but what does it mean? What is optimal? In general, an optimal solution, in the context we are talking about, is one that minimizes a certain cost function. You have to define the cost function, and then the optimum solution will minimize that cost function.

Cost functions can take on different forms depending on the situation. For example, the optimum cruise control system in a car might minimize the number of overshoots as it is striving to maintain a certain velocity. Or, it might minimize the amount of overshoot. In the case of an *optimal estimator*, we want to *minimize* the *error of the estimate*. That sounds obvious, but let us parse out this sentence structure a bit and figure out what these terms mean.

An *estimator* is an algorithm whereas the result (i.e., output) of the algorithm is an *estimate*. Let us take the simple example of the sample mean estimator (i.e., the sample average). The estimator is the algorithm given by:

$$\hat{\mu} = \frac{1}{N}\sum_{i=1}^{N} z_i \tag{12.1}$$

However, the estimate is the result obtained when we evaluate Equation (12.1) with a specific set of numbers. For example:

$$\hat{\mu} = \frac{2+3+1+3}{4} = \frac{9}{4} = 2.25$$

Let us generalize this and specify Θ to be the true value of some parameter of interest and $\hat{\Theta}$ to be the estimate of that parameter. For optimal estimation, we are seeking to minimize the difference between the truth and the estimate. We define our estimation error as:

$$e = \hat{\Theta} - \Theta \tag{12.2}$$

Accuracy is normally specified in terms of average performance (i.e., over a given period of time, 95% of the error is less than some value). However, the sample mean of the estimation error is not a good cost function (i.e., performance metric) due to the fact that negative errors and positive errors cancel out. For example, two notional error plots, both having average values of zero, are depicted in Figure 12.2. However, clearly the top plot is better than the bottom plot. We could solve this problem by weighting the positive and negative errors equally by taking the absolute value:

$$|e| = |\hat{\Theta} - \Theta|$$

The difficulty with the absolute value function, however, is that it has a discontinuous slope as shown in Figure 12.3. This makes the absolute value function mathematically awkward and hard to apply normal calculus and minimization algorithms. The better solution is to form the square of the error:

$$e^2 = (\hat{\Theta} - \Theta)^2 \tag{12.3}$$

Negative and positive errors are thus both weighted equally but with a function that has no discontinuities. We can then form the average of the squared error in order to characterize performance over some interval of interest. The metric is known as mean squared error (MSE):

$$\text{MSE} = E[e^2] = E[(\hat{\Theta} - \Theta)^2] \tag{12.4}$$

where "E" is the statistical expectation operator. In the general case where the error (before squaring) has a bias in it as well as some variation around that bias, it can be shown that mean squared error is equal to the variance plus the square of the bias.

There are a variety of minimum mean squared error estimators and many are designed to have zero bias and minimum variance. Most of us are familiar with the

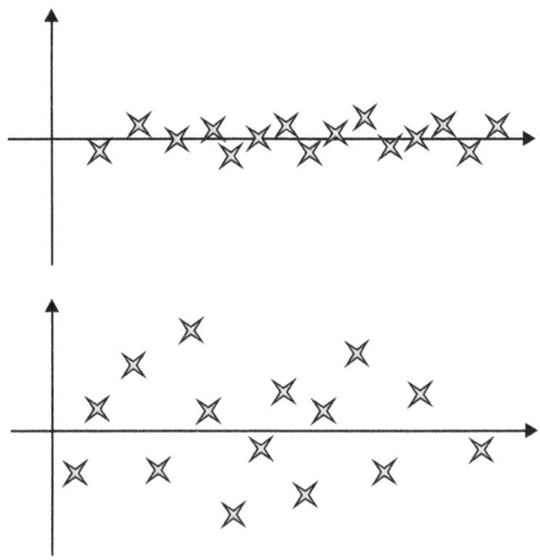

Figure 12.2 Two notional error plots. Both are zero mean but clearly the top plot is better than the bottom

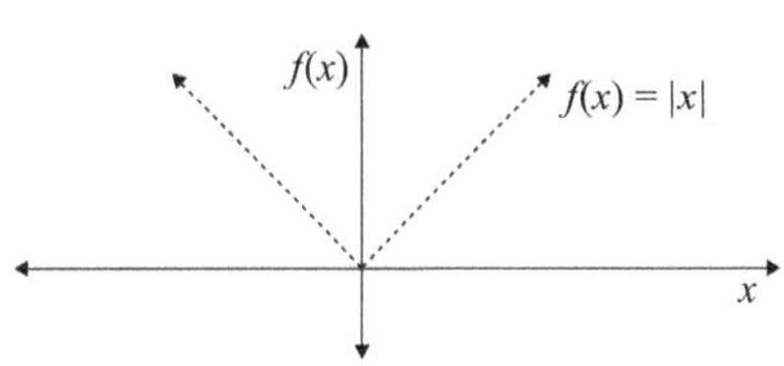

Figure 12.3 The absolute value function is mathematically awkward due to the fact that the slope is discontinuous

mean and variance of a random variable, but what is the mean and variance of an estimator? What is the mean and variance of an algorithm?

Reference to the mean and variance of an estimator is actually a reference to the mean and variance of the estimation error. Consider the example of the sample mean algorithm given in Equation (12.1). It can be shown that the sample mean is an optimal estimation algorithm. Specifically, it has zero average estimation error and it minimizes the variance of the estimation error. It is referred to as an "unbiased" estimator since its average estimation error is zero.

Let us explore this more with some examples. Assume we have a large set of Gaussian random numbers with a true mean of 18 and a variance of a 9 (i.e., a standard deviation of 3). Note the specific values for the mean and variance are completely arbitrary but are easier to visualize than using generic variables. A plot of 1,000 samples from this set is given in Figure 12.4. If we take the first one hundred numbers and compute the sample mean, we will get an estimate that should be somewhat close to the true value of 18. We will, however, get more insight into the performance of the

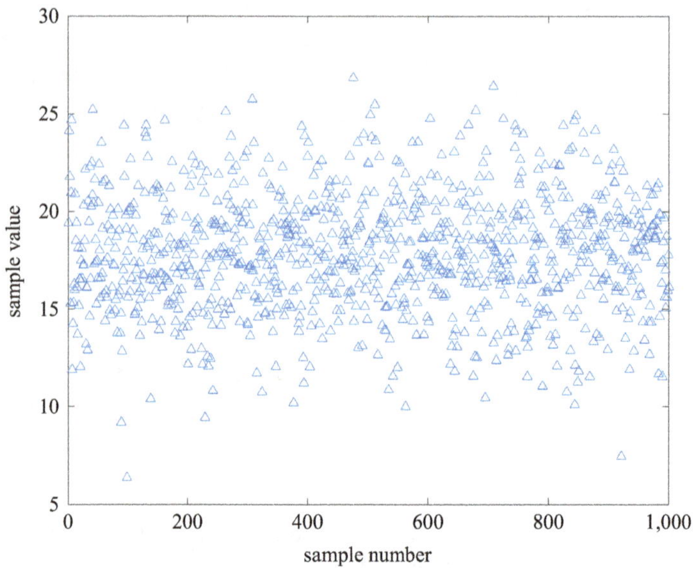

Figure 12.4 One thousand samples from a set of Gaussian random numbers with a mean of 18 and a variance of 9

estimator if we compute many, many estimates. Let us take one thousand sets of 100 numbers and compute the sample mean of each set. Figure 12.5 shows the results.

As expected, the estimates are close to the true value. We can get a better picture of the distribution of these estimates if we compute a histogram (Figure 12.6). The histogram appears to show that the estimates are roughly Gaussian distributed. This is not a coincidence. It can be shown [1] that when the sample-mean estimator operates on Gaussian random numbers, *the resulting estimates themselves* are Gaussian distributed with a mean and variance given by:

$$E[\hat{\mu}] = \mu \qquad VAR[\hat{\mu}] = \frac{\sigma^2}{N} \tag{12.5}$$

where $\hat{\mu}$ is the sample-mean estimate; μ is the true value of the mean of the random numbers; σ^2 is the true variance of the random numbers; and N is the number of samples used in the estimate. As mentioned earlier, the sample-mean is an unbiased estimator since its average value is equal to the truth. As might be expected, the variation in the estimates is directly proportional to the variation in the underlying random numbers and also is inversely proportional to the number of samples used in the estimate.

Let us verify this by repeating the experiment with 1,000-sample averages. The estimates and histogram are given in Figures 12.7 and 12.8. As we would expect when averaging over a larger number of samples, the results are better. Equation (12.5) quantifies exactly how much better the estimator will do for a given number of samples.

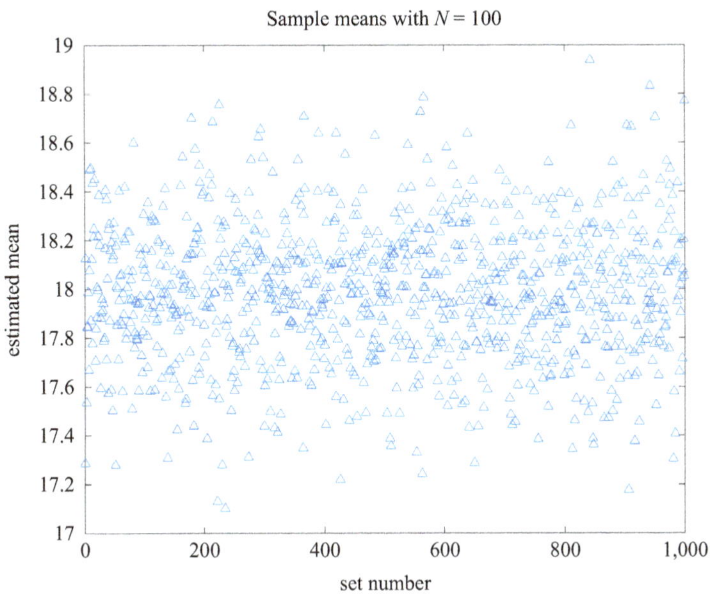

Figure 12.5 One thousand 100-sample-mean estimates of Gaussian random numbers with a true mean of 18 and a standard deviation of 3

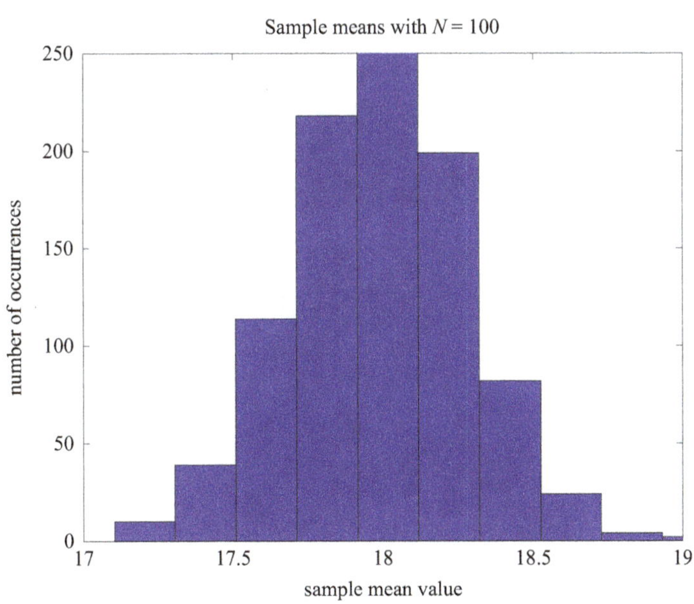

Figure 12.6 Histogram of the 100-sample-mean estimates plotted in Figure 12.5

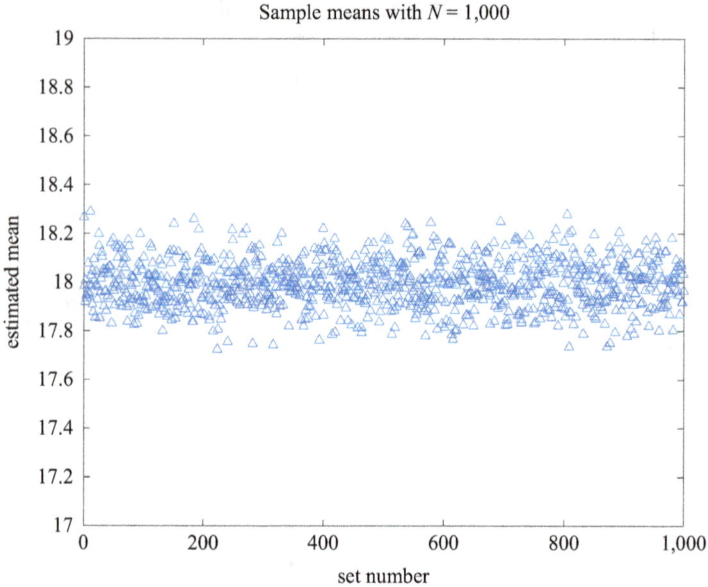

Figure 12.7 One thousand 1,000-sample-mean estimates of Gaussian random numbers with a true mean of 18 and a standard deviation of 3

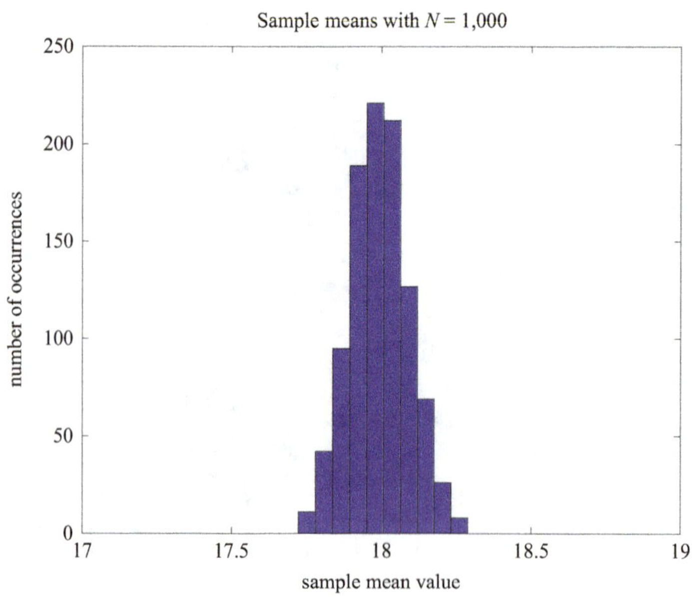

Figure 12.8 Histogram of the 1,000-sample-mean estimates plotted in Figure 12.7

12.6 Optimal estimation with data only: least-squares

The sample-mean algorithm that we previously investigated is an example of a so-called "least squares" estimator. Specifically, it minimizes the square of the estimation error. It turns out that in the absence of any additional information (i.e., you have a set of samples or measurements and nothing else), a least-squares estimate is the best you can do. Thus, when you have a set of data (and nothing else), the optimal estimator is a least squares estimator. In general, the least squares estimator is used in the context of the so-called "linear model":

$$\underline{y} = H\underline{\beta} \tag{12.6}$$

where $\underline{\beta}$ is a vector of unknown parameters that we are trying to estimate; $\underline{y}$ is the vector of measurements (data); and H is the so-called data matrix that relates the measurements to the parameters. It can be shown [2] that the least-squares estimator for the unknown parameter vector is given by:

$$\hat{\underline{\beta}} = (H^T H)^{-1} H^T \underline{y} \tag{12.7}$$

Equation (12.7) is known as the ordinary least squares (OLS) estimator. The OLS stands in contrast to the so-called weighted least squares (WLS) estimator. A WLS estimator weights each data point in inverse proportion to known measurement error (i.e., if certain measurements are known to be better than others, they are given more weight). We are not considering WLS since we are assuming that we do not know anything about the quality of the data. We have the measurements, and nothing else.

Although it can be shown that the sample mean (Equation (12.1)) is a special case of Equation (12.7), let us consider a problem that is just slightly more complex: determining a best-fitting straight line for a set of measurement data. Recall the equation for a straight line:

$$y = mx + b \tag{12.8}$$

For a given set of measurements (y) collected at some independent variable of interest (x), the problem of fitting a straight line requires the estimation of two parameters: the y-intercept (b) and the slope (m). Let us start with an error-free case (i.e., perfect measurements):

$$x = \begin{bmatrix} 1 & 2 & 3 & 4 \end{bmatrix}$$
$$y = \begin{bmatrix} 3 & 5 & 7 & 9 \end{bmatrix}$$

With just one glance it is obvious that the slope is 2 and the intercept is 1, but let us put this in the form of the linear model (Equation (12.6)):

$$\underline{y} = H\underline{\beta}$$

$$\begin{bmatrix} 3 \\ 5 \\ 7 \\ 9 \end{bmatrix} = \begin{bmatrix} 1 & 1 \\ 1 & 2 \\ 1 & 3 \\ 1 & 4 \end{bmatrix} \begin{bmatrix} \beta_1 \\ \beta_2 \end{bmatrix} = \begin{bmatrix} 1 & 1 \\ 1 & 2 \\ 1 & 3 \\ 1 & 4 \end{bmatrix} \begin{bmatrix} b \\ m \end{bmatrix} = \begin{bmatrix} 1 & 1 \\ 1 & 2 \\ 1 & 3 \\ 1 & 4 \end{bmatrix} \begin{bmatrix} 1 \\ 2 \end{bmatrix} \tag{12.9}$$

On the extreme left is the measurement vector $\underline{y}$. β_1 is the unknown y-intercept and β_2 is the unknown slope (keep in mind that, in a real situation, we would not know what the true values are). H is the data matrix that relates the dependent variables (i.e., the measurements) to the independent variables. It turns out that the columns of the H matrix are actually the values of x (the independent variable) raised to various powers. Recall the values of x are 1, 2, 3, 4. The first column of the H matrix thus consists of the values of x raised to the zeroth power. Thus, they are all 1. The second column consists of the values of x raised to the first power. Thus, the second column is just the values of x: 1, 2, 3, 4. Note that if we were fitting a quadratic instead of a straight line, H would have three columns and the third column would consist of the values of x raised to the second power. This can be extended arbitrarily to an nth order polynomial.

So we have considered the error-free case. In reality, measurement data are noisy. For example:

$$y = \begin{bmatrix} 3.05 & 4.73 & 7.04 & 9.06 \end{bmatrix} \tag{12.10}$$

and the values of x are the same as in the error-free case. The linear model needs to be expanded to account for this error vector:

$$\underline{y} = H\underline{\beta} + \underline{\epsilon}$$

$$\begin{bmatrix} 3.05 \\ 4.73 \\ 7.04 \\ 9.06 \end{bmatrix} = \begin{bmatrix} 1 & 1 \\ 1 & 2 \\ 1 & 3 \\ 1 & 4 \end{bmatrix} \begin{bmatrix} \beta_1 \\ \beta_2 \end{bmatrix} + \begin{bmatrix} \epsilon_1 \\ \epsilon_2 \\ \epsilon_3 \\ \epsilon_4 \end{bmatrix} = \begin{bmatrix} 1 & 1 \\ 1 & 2 \\ 1 & 3 \\ 1 & 4 \end{bmatrix} \begin{bmatrix} 1 \\ 2 \end{bmatrix} + \begin{bmatrix} 0.05 \\ -0.27 \\ 0.04 \\ 0.06 \end{bmatrix} \tag{12.11}$$

Note that the data matrix (H) is the same and we are still trying to estimate the same underlying unknowns. Now, of course, in reality we do not know what the measurement error is. We are trying to compute an estimate that minimizes it and we only have the measurements available to us. Using the noisy measurements (Equation (12.10)), the least squares estimate of the slope and y-intercept (Equation (12.7)) yields:

$$\underline{\hat{\beta}} = \begin{bmatrix} 0.885 \\ 2.034 \end{bmatrix}$$

Thus, we have achieved reasonable estimates of the true y-intercept ($b = 1$) and the true slope ($m = 2$). Although it was obvious given the contrived example, we made one important assumption when we processed the data. We assumed that a straight-line model was correct. Given a set of real data, it might not be obvious whether, for example, a straight-line or quadratic model should be used. As we will explore more later, the assumptions we make (whether good or bad) about a set of data will have a significant impact on our estimation.

References

[1] Freund JE, Walpole RE. *Mathematical Statistics.* 4th ed. Englewood Cliffs, NJ: Prentice-Hall, Inc.; 1987.

[2] Gelb A, editor. *Applied Optimal Estimation*. Cambridge, MA: The M.I.T. Press; 1974.

Chapter 13
Estimation theory and introduction to the Kalman filter

The fundamental concept in recursive estimation theory is:

Estimates are weighted combinations of predictions and measurements

We can write this in equation form as:

$$Estimate = WF_{pred} \times Prediction + WF_{meas} \times Measurement \tag{13.1}$$

where "WF" are the weighting factors.

This equation is deceptively simple in appearance. There are four terms and at least three of them will prove to be challenging (the fourth is the measurement term which is self-evident in our context). We have the challenge of figuring out how to form an intelligent prediction of what we think the parameter-to-be-estimated should be. Then, having made that prediction, how do we form intelligent weighting factors for the prediction and measurement in order to form an estimate?

First, we should address an important point. Why should we simply accept Equation (13.1) in the first place? Although not a proof, the equation can be justified with a simple example. Assume a sonar system is used to determine the velocity of a submarine. Let us also assume the sonar measurements are quite noisy. The sonar may indicate, due to noise, that the velocity has changed a relatively large amount. However, since we know that the acceleration of a large submersible vessel is quite small, it is unreasonable just to accept the noisy sonar-based velocity measurement. If we have a realistic model of the dynamics of the vessel, we can form a prediction of its current velocity based on the model and knowledge (perhaps estimated) of the vessel's previous velocity. We then have both a prediction and a measurement of the velocity. The weighting factors must then be derived based on the relative accuracies of the prediction and the measurement.

We will explore this in more detail later but let us start off with a very simple example: the least squares estimator of the mean, i.e., the sample average. Recall the formula for the sample average:

$$\hat{\mu} = \frac{1}{N}\sum_{i=1}^{N} z_i \tag{13.2}$$

To cast this as a recursion, we expand (13.2) for $N = 1$, 2, and 3. We will generalize the terms and use "x" to indicate the parameter-to-be-estimated, $\hat{x}$ as the estimate of that parameter, and "z" to indicate the measurement. For $N = 1$ then:

$$\hat{x}_1 = \frac{1}{1} z_1 \tag{13.3}$$

As we already knew, if we only have a single measurement, the estimate of the mean is just the measurement itself. The result for $N = 2$ is almost as simple:

$$\hat{x}_2 = \frac{1}{2}(z_1 + z_2) \tag{13.4}$$

However, we can rearrange (13.4) such that it is a function of the result obtained for $N = 1$:

$$\hat{x}_2 = \frac{2-1}{2} z_1 + \frac{1}{2} z_2 = \frac{2-1}{2} \hat{x}_1 + \frac{1}{2} z_2 \tag{13.5}$$

The reason for the strange-looking coefficient on the "1" term will become apparent in a moment. Continuing on to $N = 3$:

$$\begin{aligned} \hat{x}_3 &= \frac{1}{3}(z_1 + z_2 + z_3) \\ &= \frac{1}{3}\left[2 \cdot \frac{1}{2}(z_1 + z_2)\right] + \frac{1}{3} z_3 \\ &= \frac{1}{3}[2 \cdot \hat{x}_2] + \frac{1}{3} z_3 \\ &= \frac{3-1}{3} \hat{x}_2 + \frac{1}{3} z_3 \end{aligned} \tag{13.6}$$

We can then write the expression for arbitrary index "k":

$$\hat{x}_k = \frac{k-1}{k} \hat{x}_{k-1} + \frac{1}{k} z_k \tag{13.7}$$

The key result here is that the current estimate is a function of the previous estimate and the current measurement. Equation (13.7) has the form of Equation (13.1) with the prediction being given by the previous estimate. Why is the previous estimate a prediction? In the absence of any other a priori knowledge, we can form no better prediction than what is given by the previous estimate. Later we will consider cases where we do have other information available to us that will permit the formation of better predictions.

Note the weighting factors in (13.7) are not constant. For $k = 1$, the weighting factor on the "prediction" (i.e., the previous estimate) is zero and the weighting factor on the measurement is unity:

$$\hat{x}_1 = \frac{1-1}{1} \hat{x}_0 + \frac{1}{1} z_1 = 0 \cdot \hat{x}_0 + 1 \cdot z_1$$

This makes sense since there is no previous estimate and thus there is no prediction. 100% of the weight needs to be placed on the measurement. Now carry it to the other extreme. As k approaches infinity, the weighting factor on the prediction is unity* and the weighting factor on the current measurement is zero:

$$\hat{x}_{\infty} = \frac{\infty - 1}{\infty}\hat{x}_{\infty-1} + \frac{1}{\infty}z_{\infty} = 1 \cdot \hat{x}_{\infty-1} + 0 \cdot z_{\infty}$$

As k gets larger and larger in Equation (13.7), the importance of each new data point decreases. We are placing increasing weight on the previous estimate (a.k.a., our best current prediction). Although recursive estimation of the mean may not have great practical value, the simple example helps lay the groundwork for non-trivial cases that we will explore later.

13.1 Case study: bucketful of resistors

As we will see later, the Kalman filter is a powerful, yet complex, estimation algorithm. Under the right conditions, it provides an optimum way to form predictions as well as the weighting factors. Its ability to estimate dozens of parameters simultaneously will prove to be, as we shall see, highly useful. However, starting an explanation of the full Kalman filter with its usual matrix form is, in the opinion of this author (and teacher), counterproductive. We seek to understand the principles underlying the filter's structure and not get hopelessly lost in the matrix manipulations. To do this, we will start with a simple scalar problem (i.e., estimation of a single parameter). After fully describing the scalar Kalman filter, we will be able to attack the multivariate filter from a position of strength.

The particular scalar problem that we will consider is known as the Bucketful of Resistors problem and was originated back in 1967 by Roger M. Du Plessis at North American Rockwell Electronics Group [1]. Assume you have a bucketful of resistors. You pick one out of the bucket but you want to know the precise value of its resistance. Assume, for the moment, that the resistor has no markings so you have no idea what its resistance is. How will you determine the resistance? Your first instinct is probably to grab an ohmmeter or a multi-meter. Let us say you work at a company that makes ohmmeters and there are hundreds of them in inventory. You pick one off the shelf and measure the resistance. The measured value is 101.5 ohms. How do you know if that value is correct? The meter has some error in it and, for the sake of the example, let us say you have no idea how big the errors might be. Perhaps you then take another ohmmeter off the shelf and use that to measure the resistor also. The measurement from the second ohmmeter is 99.8 ohms. Now you are starting to get a better feel for the actual resistance. If you take N ohmmeters off the shelf and make measurements, what is the optimum estimation algorithm you can use? Since we are assuming we have no information other than the measurements, the best estimate you can compute is simply the average.

Now, what if we did have more information? What advantage could be gained from additional information that we might reasonably have in this hypothetical scenario? We will assume now that we have the following information: the resistor is

*Of course, infinity over infinity is undefined. However, simple application of L'Hopital's Rule shows the result is equal to 1.

labeled (if you are an electrical engineer you might remember the resistor color code) and all of the resistors in the bucket are labeled the same. Assume the resistor markings indicate the nominal value of the resistors in the bucket is 100 ohms. Assume there is a gold band on the resistors thus indicating 1% tolerance. Finally, assume the standard deviation of the error on each of the ohmmeters is 3 ohms (unrealistically large, I know, but it helps with the estimation theory concepts that will be explored shortly).

We are also going to assume that the 1% tolerance represents a standard deviation and that the actual values of the resistors are Gaussian distributed. The resistors in the bucket thus have a mean value of 100 ohms and a standard deviation of 1% (1 ohm). Thus, approximately 67% of the resistors in the bucket are between 99 and 101 ohms; 95% of the resistors in the bucket are between 98 and 102 ohms, etc. [again, this is not realistic since manufacturing processes will ensure more of a uniform distribution rather than a Gaussian distribution]. Similarly, we will assume the ohmmeter measurement noise is Gaussian distributed. Finally, we will assume the errors in the ohmmeters are statistically independent of each other.

Recall the least squares estimator of the mean (Equations (13.2) or (13.7)) assumes that only the measurement data is available. All other information is effectively thrown away. Intuitively, it just seems reasonable that we ought to be able to do a better job in the estimation process if we can take the additional information into account. We are given the nominal value of the resistor. We are given the tolerance so we know the distribution of the actual values of the resistors in the bucket. We also know the statistics of the ohmmeter measurement error as well. We should have a gut feeling that there must be a better way to do estimation if all of that information can be taken into account.

13.2 The resistor bucket meets the Kalman filter

So, can we do better? Is our gut-check correct? Of course, the answer is a firm and solid yes. An optimum estimator for this problem is the Kalman filter. The Kalman estimator has the following form:

$$\hat{x}_k^+ = \hat{x}_k^- + K_k(z_k - H_k\hat{x}_k^-) \tag{13.8}$$

where $\hat{x}_k^+$ is the estimated value of x at time index k;

$\hat{x}_k^-$ is the predicted value of x for time index k;

K_k is a weighting factor (called the Kalman gain);

z_k is the measurement at time index k;

H_k is the so-called data (or observation) matrix (reduced to a scalar for the current example).

As we will learn later, the data matrix relates the measurement domain to the solution domain. In the simple example of the resistor estimation, however, it is just a scalar and is simply equal to 1 (the simplicity arises from the fact that we are measuring and estimating the same quantity, resistance in this case). We note in passing the term "gain" (as in Kalman gain) is a holdover from fields such as controls engineering where, for example, the loop gain is the product of all the multiplicative factors in the feedback loop.

Although Equation (13.8) is the "standard form" of the Kalman estimator, it needs to be manipulated to show that it actually is a form of Equation (13.1). Since the data matrix is equal to 1 for the current case:

$$\begin{aligned}\hat{x}_k^+ &= \hat{x}_k^- + K_k(z_k - \hat{x}_k^-) \\ &= \hat{x}_k^- + K_k z_k - K_k \hat{x}_k^- \\ &= (1 - K_k)\hat{x}_k^- + K_k z_k\end{aligned} \tag{13.9}$$

Thus, the weighting factor on the prediction is $(1 - K_k)$ and the weighting factor on the measurement is K_k. We will see shortly that for scalar problems, such as our resistor example, the Kalman gain ranges from 0 to 1.

If we reexamine the recursive least squares algorithm (Equation (13.7)), we see the weighting factors are purely functions of the measurement index. For the Kalman filter to perform better than least squares, the Kalman gain must embody the additional information that is known. There is nothing else in Equation (13.9) that can be manipulated.

Since the derivation of the Kalman gain is not particularly helpful when first learning how to use it, it is stated without proof:

$$K_k = \frac{P_k^- H_k^T}{H_k P_k^- H_k^T + R_k} \tag{13.10}$$

where P_k^- is the prediction error covariance matrix;
H_k^T is the transpose of H_k;
R_k is the measurement error covariance matrix.

We will explore these matrices in more detail later (and derivations of the Kalman filter equations may be found in a variety of references such as [2,3]). For the resistor example, though, they are simply scalars and the transpose operation can be ignored. Thus, P_k^- is simply the prediction error *variance* and R_k is the measurement error *variance*. We will explain later how they are computed. For now, it is sufficient to understand that these variances are just the usual statistical quantity (e.g., mean and variance of a random variable). The prediction error variance is thus proportional to the uncertainty of the prediction and the measurement error variance is proportional to the uncertainty of the measurement.

Since H_k is equal to 1 for the resistor problem, Equation (13.10) can be reduced to:

$$K_k = \frac{P_k^-}{P_k^- + R_k} \tag{13.11}$$

In order to show that Equation (13.11) actually makes sense, let us consider some limit cases. In the limit as the prediction gets better and better, the prediction error variance approaches zero. The Kalman gain is thus:

$$\lim_{P_k^- \to 0} K_k = \lim_{P_k^- \to 0} \frac{P_k^-}{P_k^- + R_k} = \frac{0}{0 + R_k} = 0 \tag{13.12}$$

Recalling Equation (13.9), a Kalman gain of zero places 100% weight on the prediction and zero weight on the measurement. This makes sense if it is known that the prediction is perfect (which is implied by $P_k^- = 0$). Now let us consider the opposite extreme. What happens if the prediction is horrible?

$$\lim_{P_k^- \to \infty} K_k = \lim_{P_k^- \to \infty} \frac{P_k^-}{P_k^- + R_k} = 1 \tag{13.13}$$

Note that L'Hopital's rule was needed to evaluate (13.13). Again, Equation (13.9) shows that a Kalman gain of 1 places zero weight on the prediction and 100% weight on the measurement. Again, this makes sense if it is known that the prediction is completely useless.

To assess the impact of the measurement error variance, consider a case where the prediction error variance has some finite, but greater than zero, value (recall that variances are always positive). If the measurement is excellent:

$$\lim_{R_k \to 0} K_k = \lim_{R_k \to 0} \frac{P_k^-}{P_k^- + R_k} = \frac{P_k^-}{P_k^- + 0} = 1 \tag{13.14}$$

Again, Equation (13.9) shows that this places 100% weight on the measurement as should be done if it is known that the measurement is perfect. Conversely, if the measurements are horrible:

$$\lim_{R_k \to \infty} K_k = \lim_{R_k \to \infty} \frac{P_k^-}{P_k^- + R_k} = \frac{P_k^-}{P_k^- + \infty} = 0 \tag{13.15}$$

Which, again, is the desired result since Equation (13.9) places 100% weight on the prediction if the Kalman gain is zero. In actual operation, of course, both the predictions and the measurements are going to lie someplace in between perfect and useless. The Kalman gain strikes just the right balance weighting the prediction and the measurement given a known amount of uncertainty in each.

13.3 Back to the bucket

Now that we have dipped our toes in the Kalman Filter pool, let us apply it to the resistor estimation example. Based on the a priori information we have, we can do three things. We can form an initial prediction, we can determine the variance of the prediction error and we can determine the variance of the measurement error:

$$\begin{aligned} \hat{x}_1^- &= 100 \text{ ohms} \\ P_1^- &= (1 \text{ ohm})^2 = 1 \text{ ohm}^2 \\ R &= (3 \text{ ohms})^2 = 9 \text{ ohms}^2 \end{aligned}$$

Since the resistors in the bucket have an average value of 100 ohms, that is the best initial prediction we can form. For any given resistor that we pull out of the bucket, there is a 50–50 chance that its actual value will be above or below 100 ohms. Thus, the average value of all the resistors in the bucket is the best prediction we can form before any measurements have been taken. Since we also know/assume that the resistors in the bucket are Gaussian distributed with a standard deviation of 1 ohm, the variance of the error in the initial prediction is the square of that standard deviation.

Finally, the variance of the ohmmeter error is the square of the standard deviation of the error.

Notice the measurement error variance (R) has no subscript since it is a constant in this example. Conversely, the prediction error variance has a subscript since it will change over time as new predictions are formed. As measurements are taken, new estimates are formed using Equation (13.8) or (13.9). We will need an equation to quantify the uncertainty of the estimate. The so-called estimation error covariance matrix is given (again, without proof) by:

$$P_k^+ = (I - K_k H_k) P_k^- \tag{13.16}$$

For our resistor example, all the terms in Equation (13.16) are scalars and the identity matrix (I) is obviously just equal to 1. Recalling that the data matrix is also equal to 1 for the resistor example, (13.16) reduces to:

$$P_k^+ = (1 - K_k) P_k^- \tag{13.17}$$

Since the Kalman gain has a value between 0 and 1, we see the estimation error variance will be less than the prediction error variance. This makes sense since the estimate is optimally including the information provided by the measurement that has been incorporated into the estimate. Let us manipulate (13.17) to be able to check some limit cases. If we substitute (13.11) into (13.17) and do a little algebra we get:

$$P_k^+ = \left(1 - \frac{P_k^-}{P_k^- + R}\right) P_k^- = \frac{P_k^- R}{P_k^- + R}$$

Now let us look at the limit cases. If the measurement error variance is much larger than the prediction error variance:

$$R >> P_k^- \Rightarrow P_k^+ \approx \frac{P_k^- R}{R} = P_k^-$$

In this case, the estimation error variance is approximately equal to the prediction error variance. This makes sense when the measurement error variance is large since, as we showed earlier, the Kalman estimate will rely primarily on the prediction in this situation. The other limit case also provides the expected result:

$$P_k^- >> R \Rightarrow P_k^+ \approx \frac{P_k^- R}{P_k^-} = R$$

If the prediction error variance is large, the Kalman estimate will rely primarily on the measurement and the estimation error variance is equal to the measurement error variance.

Once we have formed the estimate (and associated estimation error variance) for the current measurement, we then need to form the prediction for the next moment in time (or, more generally, for the next measurement however it occurs: temporally, spatially or otherwise). For the resistor problem, the prediction is simple. We are assuming that the true value of the resistor is *constant* and thus does not change over time. Later we will explore the more general case where there are dynamics in the parameters of interest. For now, though, we are just going to assume the simple static case where the true value is constant. As a result, the best prediction we can form is simply the estimate that we just computed:

$$\hat{x}_{k+1}^{-} = \hat{x}_k^{+} \tag{13.18}$$

Now we need to compute the uncertainty in this prediction (i.e., the prediction error variance). Again, in this simple example, the prediction error variance is just the estimation error variance since the prediction is equal to the previous estimate:

$$P_{k+1}^{-} = P_k^{+} \tag{13.19}$$

Later we will see that in the general case where the parameters-of-interest have dynamics, the prediction error covariance expression will be more complicated.

13.4 Summary of the Kalman filter for the bucketful of resistors example

Step 0: Determine the initial prediction, initial prediction error variance, measurement error variance, and initialize the index:

$$\begin{aligned} \hat{x}_1^{-} &= 100 \text{ ohms} \\ P_1^{-} &= (1 \text{ ohm})^2 = 1 \text{ ohm}^2 \\ R &= (3 \text{ ohms})^2 = 9 \text{ ohms}^2 \\ k &= 1 \end{aligned}$$

Step 1: Compute the Kalman gain (Equation (13.11)):

$$K_k = \frac{P_k^{-}}{P_k^{-} + R_k}$$

Step 2: Take a measurement (z_k) and compute the Kalman estimate (Equation (13.9)):

$$\hat{x}_k^{+} = (1 - K_k)\hat{x}_k^{-} + K_k z_k$$

Step 3: Compute the estimation error variance (Equation (13.17)):

$$P_k^{+} = (1 - K_k)P_k^{-}$$

Step 4: Predict ahead to the next moment in time (Equation (13.18)):

$$\hat{x}_{k+1}^{-} = \hat{x}_k^{+}$$

Step 5: Compute the prediction error variance (Equation (13.19)):

$$P_{k+1}^{-} = P_k^{+}$$

Step 6: Increment the index ($k = k + 1$) and go to Step 1

13.5 Theoretical performance comparison

We have shown how to apply the Kalman filter to the bucketful of resistors problem. We did so with the knowledge that the filter could do a better job than recursive least squares by taking advantage of additional a priori information. We can show this theoretically by a statistical comparison of the estimation error of the two. Recall, from the previous chapter, the estimation error variance for the recursive least squares estimator is:

$$VAR[\hat{r}_{LS}] = \frac{\sigma^2_{ohmmeter}}{k} = \frac{(3)^2}{k} = \frac{9}{k} \tag{13.20}$$

where $\hat{r}_{LS}$ is the least-squares estimate of the resistance and $\sigma^2_{ohmmeter}$ is the variance of the ohmmeter measurement error. Thus, the standard deviation of the estimation error is:

$$\sigma_{LS} = \sqrt{\frac{9}{k}} = \frac{3}{\sqrt{k}} \tag{13.21}$$

Recall the computation of the estimation error variance is an integral part of the Kalman filter. It was Step 3 (Equation (13.17)) and the standard deviation is simply the square-root of the variance. If you study the steps of the Kalman filter, you will notice that the estimation error variance can be computed without the need to take measurements and explicitly compute the estimate. After initialization (Step 0), then Steps 1, 3, 5 and 6 can be performed recursively. The Kalman filter and least squares estimation error standard deviation for the resistor problem are depicted in Figure 13.1.

At the first reading, the least squares estimation error standard deviation is 3 ohms. This is consistent with Equation (13.21). The Kalman filter, however, starts off with an estimation error standard deviation slightly less than 1 ohm. Since the ohmmeter measurement error variance (9 ohms^2) is much larger than the initial prediction error variance (1 ohm^2), we expect the Kalman filter to rely more on the prediction than on the measurement. Nevertheless, the filter is able to extract some information from the first measurement by weighting it appropriately. As the numbers of readings increases, the accuracies of the two estimators appear to be converging. This is not surprising since the extra information utilized by the Kalman filter is exploited primarily in the initial prediction. The value of the enhanced initial prediction diminishes as the total number of measurements increases.

At this point, we should acknowledge that we have stacked the deck in the favor of the Kalman filter. The accuracy of modern ohmmeters are typically much less than 0.1 ohm when measuring a nominally 100 ohm resistor. Another important point is that the Kalman filter expects and assumes that each measurement presented to it is statistically independent of all of the previous measurements. Put another way, the filter assumes the measurement errors are uncorrelated. In a modern ohmmeter, there are a variety of error sources and some are correlated over time. The point here is not to provide a highly realistic situation. This contrived problem does, however, give us

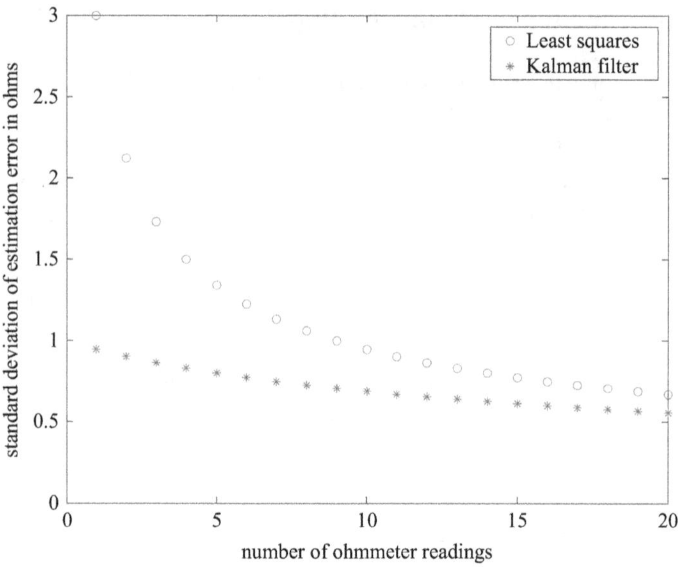

Figure 13.1 Theoretical performance comparison of least squares and the Kalman filter for the bucketful of resistors problem

insight into how the Kalman filter exploits information about the system (resistor in this case) and sensor (ohmmeter in this case) that the least-squares estimator ignores.

13.6 Monte Carlo simulations of the recursive least squares estimator

It is instructive to perform simulations both to verify the theoretical results and to gain additional insight into the behavior of the estimators. Figure 13.2 illustrates the results of a simulation of the least squares estimator. A random resistor has been taken out of the bucket. In this particular example, the true value is approximately 99.5 ohms. The simulated measurements are scattered around the truth in proportion to the measurement error standard deviation. Note that at the first reading, the measurement and the estimate are identical. As the number of readings increases, the estimates roughly converge to the true value. This is a so-called "single run" of the simulation since only a single set of 20 readings was simulated.

What would the estimation error plots look like if we simulated these 20 readings over and over and over again? For each new single run, we simulate (with a random number generator) picking a new resistor out of the bucket and we simulate (again, with a random number generator) a new set of measurements on that resistor. The simulated measurements for each single run are passed through the recursive least squares estimator. The results for 200 runs are shown in Figure 13.3.

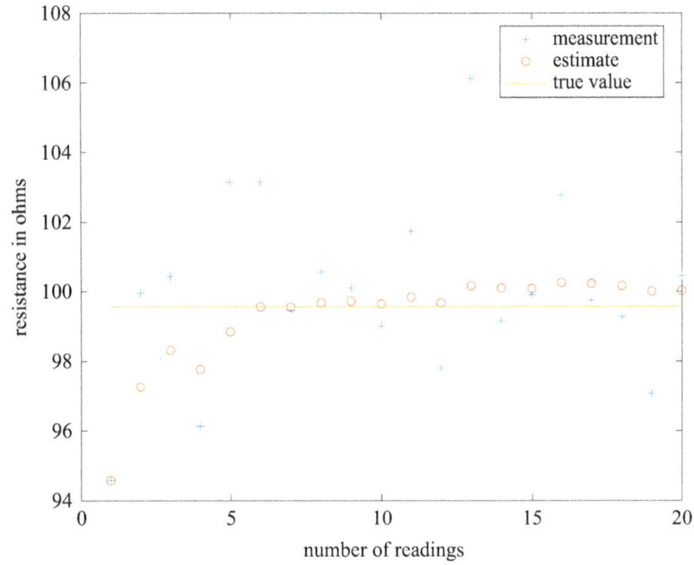

Figure 13.2 Results for a simulated single run of the recursive least squares estimator

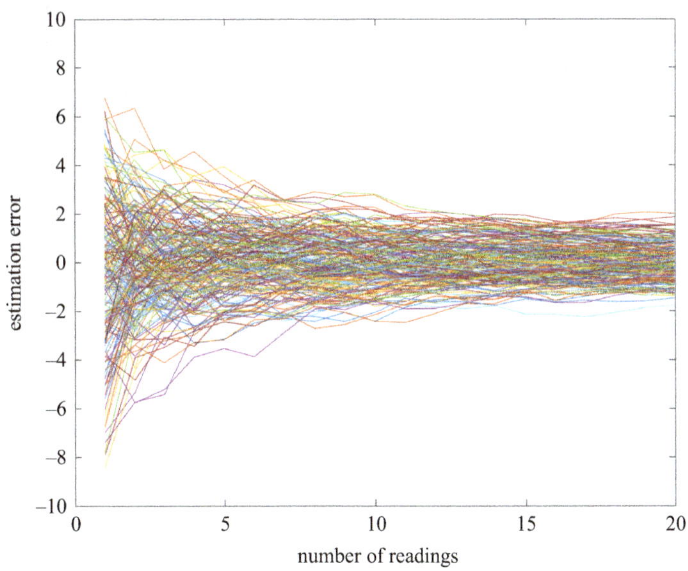

Figure 13.3 Estimation error for 200 simulated runs of the recursive least squares estimator

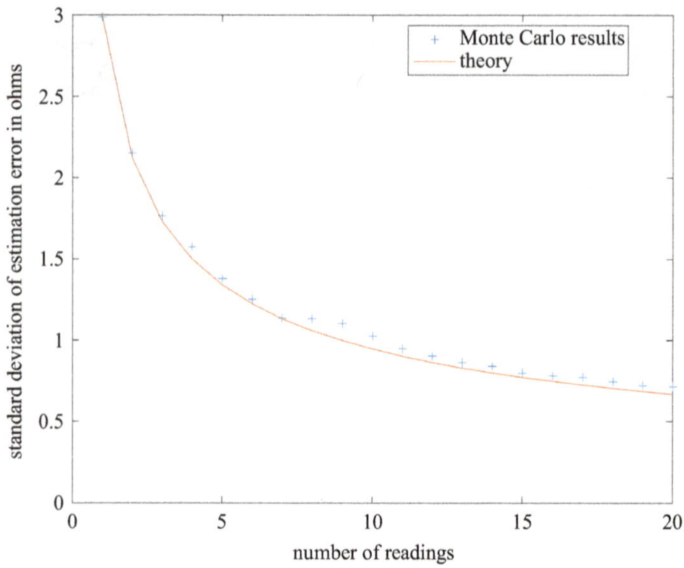

Figure 13.4 Comparison of theory and simulation results for the recursive least squares estimator

How well do the simulation results compare with the theory (Figure 13.1)? In other words, are the simulation results consistent with the theory? To do this assessment, we need to compute, at each reading, the standard deviation of the estimation error in Figure 13.3. Visually, we can see that the standard deviation of the estimation error at reading number 1 is larger than it is for, say, reading number 6. We can compute the standard deviation of the estimation error (in Figure 13.3) for each reading. At each reading we are thus computing a standard deviation across the *ensemble* of runs. A comparison of this with the theory is given in Figure 13.4. Clearly the simulation results match the theory quite well even with just 200 Monte Carlo runs. If we increased the number of runs to, say, 10,000, the theory and simulation results would be visually indistinguishable.

13.7 Monte Carlo simulations of the Kalman filter

If we use the Kalman filter to process the identical simulated measurement data presented in Figure 13.2, the results are as shown in Figure 13.5. The single run clearly shows the Kalman is performing better than least squares. This is verified when 200 Monte Carlo trials are performed (Figure 13.6). Again, the simulation results match well with the theory (Figure 13.7).

13.7.1 The danger of mismodeling

It is important to realize that at this point we have shown that the Kalman filter can outperform least squares if all the assumptions are valid. We had an idealized

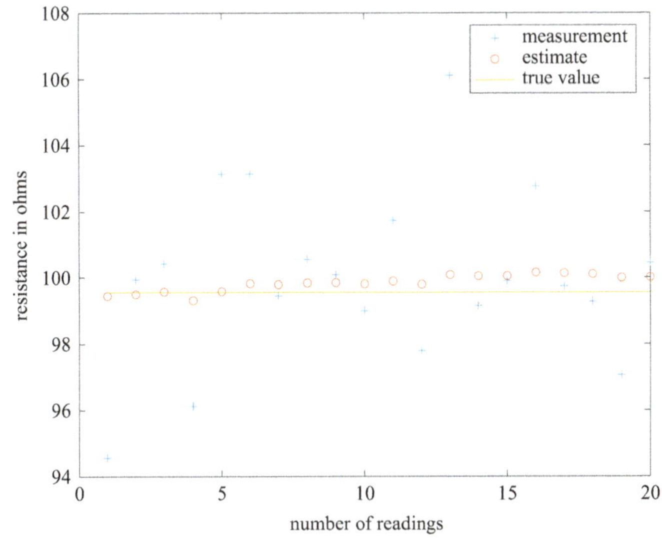

Figure 13.5 Single-run results for the Kalman filter. The simulated measurement data is the same as that used in Figure 13.2

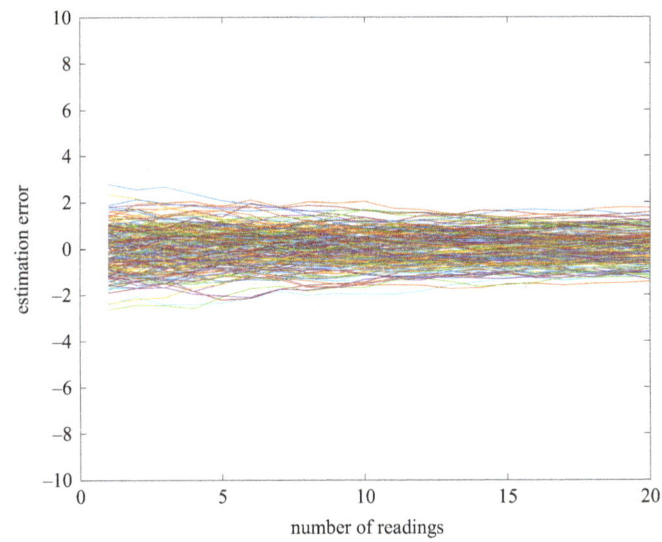

Figure 13.6 Estimation error for 200 simulated runs of the Kalman filter

case where all statistical quantities (e.g., measurement error, distribution of the actual values of the resistors in the bucket) were purely Gaussian and were known exactly. In the real world, the statistical quantities are not known precisely. In some cases, data can be collected and analyzed to estimate the actual distributions but these approximations have varying degrees of validity depending upon the exact situation. It is critically

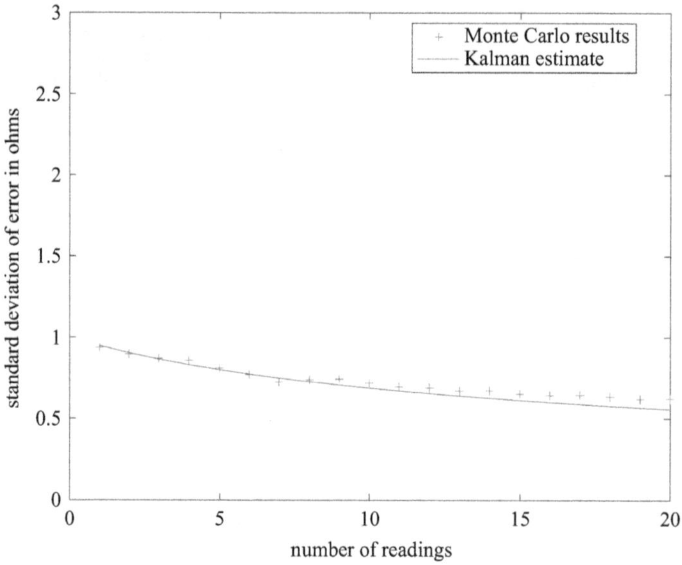

Figure 13.7 Comparison of theory and simulation results for the Kalman filter

important to realize this a priori information built into the Kalman filter has a dramatic impact on performance.

To illustrate this point, we will rerun the previous simulation but we will simulate a bucketful of 10% tolerance resistors instead of 1% tolerance resistors. Thus, 67% of the resistors will have true values ranging from 90 to 110 ohms instead of 99 to 101. However, we will *lie* to the filter by still setting the initial prediction error variance to 1 ohm^2 just as we did before. "Lie" is a harsh word. Maybe the bucket was mislabeled as containing 1% resistors. Incorrect information given to the filter is referred to as *mismodeling*. As we will explore further in later chapters, the "extra" information that we give to the filter (versus only measurements) is based on implicit models of the system (that we are estimating) and the sensor (from which we obtain measurements). In this first example, we are mismodeling the system (i.e., the resistors) by telling the filter that the distribution of the resistors in the bucket is narrower than it actually is. The results of a set of 200 Monte Carlo runs are shown in Figures 13.8 and 13.9.

The poor performance of the filter as depicted in Figure 13.8 makes sense given the fact that we told the filter that the initial prediction was very good when in actuality it was not. As a result, the filter placed too much weight on its prediction and not enough weight on the measurements. The results shown in Figure 13.9 are more than just a comparison of theory and simulation. The simulation results show how well the filter is actually performing when it is operating with bad assumptions. The "theory," conversely, shows how well the filter *thinks* it is performing. Computation of estimation error variance is one of the steps in the filter algorithm but its validity is completely dependent upon the assumptions designed into the filter. If we lie to the filter, the filter's own computation of estimation error variance will be garbage.

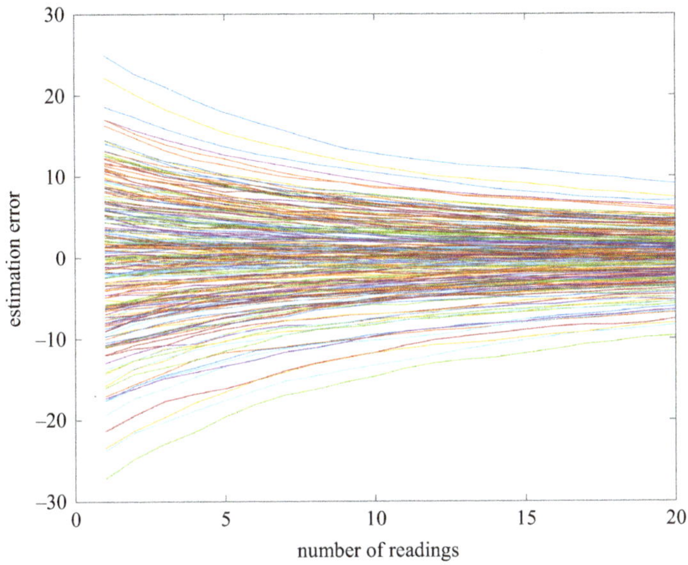

Figure 13.8 Estimation error for 200 simulated runs of the Kalman filter with the initial prediction error variance set too low

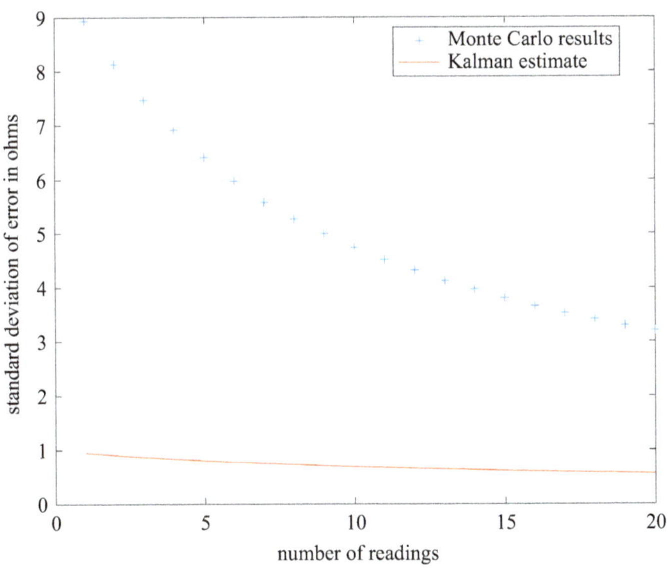

Figure 13.9 Comparison of theory and simulation results for the Kalman filter with the initial prediction error variance set too low

Figures 13.10 and 13.11 show the results for another example of mismodeling. In this case, the measurement error variance is actually 36 ohms^2 but is still being modeled in the filter as being 9 ohms^2. Thus, the actual measurement error standard deviation (6 ohms) is double what is being modeled (3 ohms). As a result, the filter places too much weight on the measurements and not enough weight on the

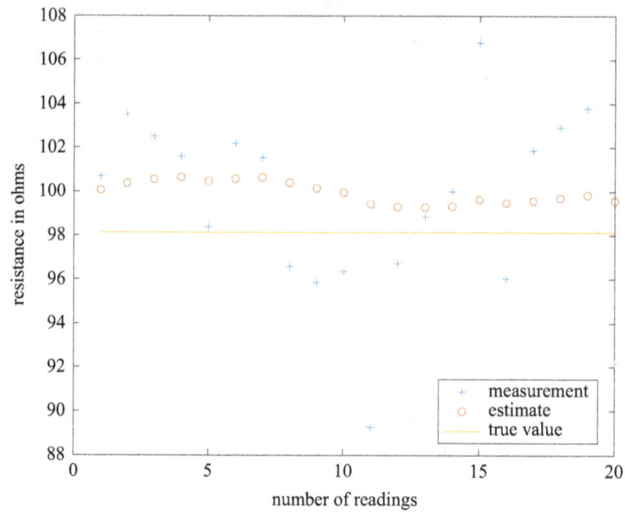

Figure 13.10 Estimation error for 200 simulated runs of the Kalman filter with the measurement error variance set too low

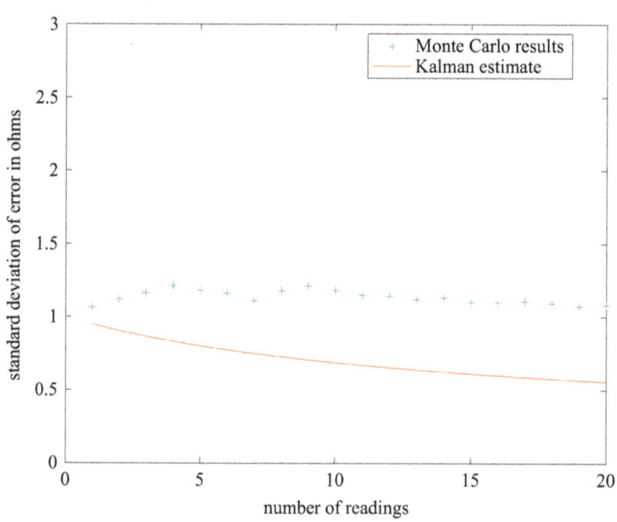

Figure 13.11 Comparison of theory and simulation results for the Kalman filter with the measurement error variance set too low

predictions. The convergence time is longer and the filter's own estimation error variance values are too small.

We have shown that the Kalman filter can take advantage of the additional information that least squares ignores. However, we must always keep the assumptions in mind. Those assumptions are not always valid. In actual practice, we would like to be able to evaluate the filter performance by processing real data with known errors. That requires having some kind of truth reference. Sometimes such "truth" data are available but not always. Nevertheless, the designer must keep in mind that assumptions are inherent in a Kalman filter and their degree of validity will impact performance.

References

[1] Du Plessis RM. *Poor Man's Explanation of Kalman Filtering or How I Learned to Stop Worrying and Learned to Love Matrix Inversion*. Anaheim, CA: North American Rockwell Electronics Group; 1967.

[2] Gelb A, editor. *Applied Optimal Estimation*. Cambridge, MA: The M.I.T. Press; 1974.

[3] Brown RG, Hwang PYC. *Introduction to Random Signals and Applied Kalman Filtering: with MATLAB® exercises*. 4th ed. Hoboken, NJ: John Wiley & Sons, Inc.; 2012.

Chapter 14

The multivariate Kalman filter

Previously we explored the Kalman filter within the framework of a simple problem. In the bucketful of resistors problem, we were estimating a single static parameter. We pulled one resistor out of the bucket and attempted to estimate its resistance given some successive measurements. Obviously we need to go beyond that for most practical applications. We need to be able to handle cases where the parameter of interest is actually changing over time and we want to be able to form an estimate of its value over time as it is changing. Furthermore, we want to be able to handle cases where we are estimating multiple parameters simultaneously and those parameters are related to each other. It is not a simple matter of five parameters and thus the need for five single-parameter Kalman filters running at the same time. If we are trying to estimate multiple parameters that have some kind of relationship to each other, then we want to be able to exploit those relationships in the course of our estimation and thus we need to expand our Kalman filter accordingly.

We are thus transitioning from univariate parameter estimation to multivariate parameter estimation and from static systems to dynamic systems. Along the way we will explore state variables, state vectors, and state transition matrices. We will learn about the fundamental equations upon which the Kalman filter is based: the so-called "system equation" and the "measurement equation." We will need to learn about covariance matrices and then eventually we will develop the full Kalman recursion. We will then take our bucketful of resistors example and modify it slightly in order to ease into these new topics.

Up to this point, we have only considered the estimation of a single parameter (the univariate case) with a value that did not change over time (a static system). We need to generalize the Kalman filter to be able to handle dynamic multivariate cases. The single parameters are going to become vectors of parameters and the variances are going to become covariance matrices. We will model system dynamics within the filter (the state transition matrix) and will model complex relationships between the states and the measured quantities (the H matrix will no longer be equal to 1).

14.1 A simple radar example

Let us start with state variables, vectors, and transitions. State variables are the formal name we call the parameters used to describe the current condition or state of the system. The state vector is simply a vector of the state variables. The state transition matrix describes the dynamics of the systems by quantifying how the state vector "propagates" (i.e., changes) over time.

Let us consider a simple radar example so that we can get our hands and our minds wrapped around these concepts. Consider a radar that forms both range and range-rate measurements to a particular aircraft. We are going to make the example very simple by assuming the aircraft is flying with nominally constant velocity and is flying either directly toward or away from the radar. As a result, the state vector at discrete-time index k is:

$$\begin{bmatrix} d_k \\ r_k \end{bmatrix} = \begin{bmatrix} d \\ r \end{bmatrix}_k \tag{14.1}$$

where d_k is the distance (range) and r_k is the range-rate (velocity). Since the time index is the same for both state variables, it can be moved outside of the square brackets. Assuming the aircraft velocity is something other than zero, the distance will clearly change over time. We know from our basic physics that, in continuous-time, distance is the integral of velocity and velocity is the integral of acceleration:

$$\begin{aligned} d(t) &= d(0) + \int_0^t r(\lambda)d\lambda \\ r(t) &= r(0) + \int_0^t a(\lambda)d\lambda \end{aligned} \tag{14.2}$$

where "a" is the acceleration. The Kalman filter is a linear estimator and we must represent the system dynamics with a linear approximation. As mentioned earlier, we are assuming a nominally constant-velocity flight path and thus Equations (14.2) can be approximated by:

$$\begin{aligned} d(t) &\approx d(0) + r \cdot t \\ r(t) &\approx r(0) \end{aligned} \tag{14.3}$$

Now in the back of your head you may be thinking, "What on earth does this have to do with practical situations in navigation where the vehicles (e.g., aircraft) clearly have nontrivial acceleration?" Your instincts are correct and we will see later that the Kalman filter's linear assumption poses some interesting challenges. However, we will also see that there are some clever techniques to address these challenges as well. For now just hang on while we explore the simple examples. In discrete-time, Equation (14.3) can be rewritten in matrix form as:

$$\begin{bmatrix} d \\ r \end{bmatrix}_{k+1} = \begin{bmatrix} 1 & T \\ 0 & 1 \end{bmatrix} \begin{bmatrix} d \\ r \end{bmatrix}_k \tag{14.4}$$

where "T" is the sampling period or sampling interval (i.e., the time-difference between time index k and $k+1$). The 2×2 matrix is known as the state transition matrix. It is the matrix that describes how the state vector transitions ("changes" or "propagates") from one time instant to the next.

We can modify Equation (14.4) to be slightly more realistic by taking some turbulence into account. Some random perturbations from the constant velocity flight path can be modeled by adding what we call system noise (also known as process noise or plant noise):

$$\begin{bmatrix} d \\ r \end{bmatrix}_{k+1} = \begin{bmatrix} 1 & T \\ 0 & 1 \end{bmatrix} \begin{bmatrix} d \\ r \end{bmatrix}_k + \begin{bmatrix} w_1 \\ w_2 \end{bmatrix}_k \tag{14.5}$$

In the context of this particular example, we are assuming that the aircraft has some flight control system that is trying to maintain it on a constant velocity flight path. There are some perturbations due to wind gusts and disturbances, but the control system is actively returning the aircraft back to its nominal constant velocity flight path and the aircraft is not deviating significantly from it.

14.2 The Kalman filter "System Equation" and "Measurement Equation"

Equation (14.5) is an example of the so-called Kalman filter "system equation." The general form is given by:

$$\underline{x}_{k+1} = \Phi_k \underline{x}_k + \underline{w}_k \tag{14.6}$$

where $\underline{x}_k$ is the state vector, $\underline{w}_k$ is the system noise vector, and Φ_k is the state transition matrix (all at time k). It is very important to understand that the system equation is the Kalman filter's *model* of how the system behaves. It is a theoretical construct that, in practice is, at best, an approximation of how the real system behaves.

The companion to the system equation is the measurement equation. Like the system equation, the measurement equation is a linear model. The measurement equation is the Kalman filter's *model* of the relationship between the measurements and the states. It is ***not*** an equation for how the measurements are actually formed.

Since the measurements in the radar example are in the same dimensions as the state variables (we are measuring range and range rate and we are trying to estimate range and range rate), the measurement equation is pretty trivial in this case:

$$\begin{bmatrix} d_{meas} \\ r_{meas} \end{bmatrix}_k = \begin{bmatrix} 1 & 0 \\ 0 & 1 \end{bmatrix} \begin{bmatrix} d \\ r \end{bmatrix}_k + \begin{bmatrix} \nu_1 \\ \nu_2 \end{bmatrix}_k \tag{14.7}$$

where d_{meas} and r_{meas} are the measured range and the range-rate, d and r are the true range and range-rate, and ν_1 and ν_2 are the range and the range-rate measurement noise.

For a slightly less trivial case, if the radar *only* provided range measurements but we still wanted to estimate both range and range-rate, the system equation would be identical (Equation (14.5)) but the measurement equation would be:

$$\begin{bmatrix} d_{meas} \end{bmatrix}_k = \begin{bmatrix} 1 & 0 \end{bmatrix} \begin{bmatrix} d \\ r \end{bmatrix}_k + \begin{bmatrix} \nu_1 \end{bmatrix}_k \tag{14.8}$$

Both Equations (14.7) and (14.8) are special cases of the general measurement equation:

$$\underline{z}_k = H_k \underline{x}_k + \underline{\nu}_k \tag{14.9}$$

where z_k is the measurement (or observation) vector, ν_k is the measurement noise vector, and H_k is the data (or observation) matrix (again, all at time k).

You may be surprised to learn that neither the system equation nor the measurement equation is used verbatim in the Kalman filter algorithm. They are used, instead, in the *design* phase of the filter and they implicitly embody the system and measurement models that the filter uses to provide improved performance over least-squares. The state transition and data matrices are used in the filter but, obviously, in real life we cannot know the exact, real-time, values of either the system or measurement noise. What we do instead is attempt to characterize the noise *statistically*. In simulations, we are free to generate noise values that can be statistically characterized perfectly but in real life we will always be faced with the need to approximate.

What statistical characterization is used? Mean and variance. More specifically, we model the noise terms as zero-mean independent Gaussian random variables (i.e., white noise). Thus, the variance is the sole statistical parameter needed to characterize the probability distribution. Later we will learn how the filter can deal with correlated noise processes but even then the system and measurement noise vectors are composed of white noise terms. Most folks who are new to the Kalman filter are familiar with the variance of a (scalar) random variable but frequently are not familiar with the concept of the variance of a random *vector*. Let us explore that now.

14.3 Covariance matrices

The Kalman system and measurement equations are the models of the key relationships needed to design the filter. However, we do not assume that we can know either the system or measurement noise exactly in real time. We do assume that we can know, or at least approximate, the statistics of these noise values. The Kalman filter assumes that all noise values are Gaussian distributed and a Gaussian random variable is completely characterized by its mean and variance. Assuming the mean is zero, only the variance is needed. Since, in general, we will have multiple states and/or multiple measurements (at a given moment in time), we cannot use variance alone to describe the statistics since variance applies to a single parameter. We need something called a covariance matrix that contains both the variances of the individual parameters along with the covariance between the parameters.

To ease into the covariance matrix, let us start with the definition of the variance:

$$VAR(X) = \sigma_x^2 = E([X - E(X)]^2) \tag{14.10}$$

where "X" is a random variable and "E" is the expectation operator. The term inside the square brackets is the difference between the random variable and its mean value. This difference is squared (so that the positive and negative values are treated equally) and the outer expectation operation determines the mean value of the squared differences. If a given random process has a large variance, then the average value of the squared differences will be large.

For reasons that will become obvious shortly, let us expand (14.10) as follows:

$$VAR(X) = \sigma_x^2 = E([X - E(X)][X - E(X)]) \tag{14.11}$$

Referring to (14.11), the equation for the covariance between two different random variables is straightforward:

$$COV(X, Y) = \sigma_{xy}^2 = E([X - E(X)][Y - E(Y)]) \tag{14.12}$$

Equation (14.12) shows us that variance is a special case of covariance (specifically, the covariance of a random variable with itself).

At this point, you may be thinking, "I understand what variance means. Variance is an indication of the spread of the data around its average value. But what is a covariance?" The covariance gives us an indication of the relationship between two random variables. Do they both tend to follow the same trends? Are they both usually positive at the same time or usually negative at the same time, or are they opposite? Are they both large at the same time? Et cetera.

Figure 14.1 illustrates some examples. In the upper left plot, the covariance is a "larger" positive value. If X increases in value, Y is highly likely to increase in value (and vice versa). In the upper right plot, the covariance is still positive but is smaller. The trend is the same as in the upper left plot but Y has more variation. Given, say, an increase in the value of X, there is not as high a likelihood that Y will increase (in some cases Y will actually decrease in value). The lower left plot is similar to the upper left except the trend is reversed. An increase in X corresponds to a decrease

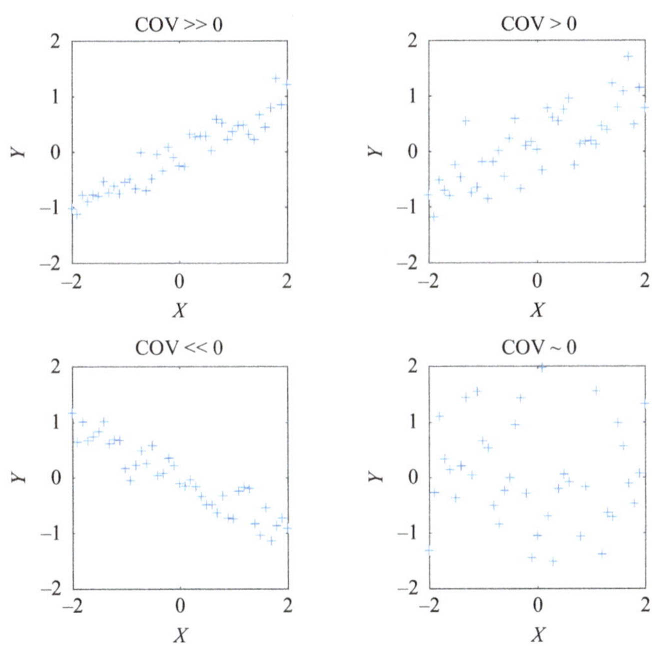

Figure 14.1 Plots of random variables X and Y to demonstrate covariance. Upper left: larger positive covariance; upper right: smaller positive covariance; lower left: larger negative covariance; lower right: covariance nearly zero

in Y. The covariance is thus negative. Finally, in the lower right plot, there is no discernible trend (also known as correlation) and thus the covariance is close to zero. A change in X does not imply any particular change in Y.

We can now generalize these results. If we have a vector of random variables:

$$\underline{X} = \begin{bmatrix} X_1 & X_2 & \cdots & X_N \end{bmatrix}^T \tag{14.13}$$

Then the general covariance is given by:

$$C = \begin{bmatrix} E([X_1 - E(X_1)]^2) & E([X_1 - E(X_1)][X_2 - E(X_2)]) & \cdots \\ E([X_2 - E(X_2)][X_1 - E(X_1)]) & \cdots & \cdots \\ \cdots & \cdots & \cdots \\ E([X_N - E(X_N)][X_1 - E(X_1)]) & \cdots & E([X_N - E(X_N)]^2) \end{bmatrix} \tag{14.14}$$

Equation (14.14) can be written more succinctly as:

$$C = \begin{bmatrix} VAR(X_1) & COV(X_1,X_2) & \cdots & COV(X_1,X_N) \\ COV(X_2,X_1) & \cdots & \cdots & \cdots \\ \cdots & \cdots & \cdots & \cdots \\ COV(X_N,X_1) & \cdots & \cdots & VAR(X_N)) \end{bmatrix} \tag{14.15}$$

You can see that the main diagonal is just the variances of the elements and the off diagonal elements consist of the covariances between the variables within the vector. For example the (1,2) element is the covariance of X_1 with X_2. The covariance of the (2,1) element is the covariance of X_2 with X_1. Now if you say, "Wait a minute, there isn't any difference between the two," you are absolutely right. The matrix is symmetric so the covariance at (2,1) is equal to the covariance at (1,2) and that symmetry exists throughout the entire matrix. In fact, you can characterize the entire matrix with just the upper right triangular portion or the lower left triangular portion.

If the random variables in (14.13) are unbiased (i.e., have means of zero), then Equation (14.15) can be written even more succinctly:

$$C = E\left[\underline{X}\,\underline{X}^T\right] = E\left\{ \begin{bmatrix} X_1 \\ X_2 \\ \cdots \\ X_N \end{bmatrix} \begin{bmatrix} X_1 & X_2 & \cdots & X_N \end{bmatrix} \right\} \tag{14.16}$$

Note that Equation (14.14) reduces to (14.16) when the random variables in X have means of zero. The term $[\underline{X}\,\underline{X}^T]$ is referred to as an 'outer product.'*

14.4 System and measurement noise covariance matrices

We now are able to characterize, statistically, the system and measurement noise vectors (Equations (14.6) and (14.9)). From Equation (14.16), we can write:

$$Q_k = E\left[\underline{w}_k\, \underline{w}_k^T\right] \tag{14.17}$$

*As opposed to the 'inner product' formed by: $\underline{X}^T\underline{X}$. Recall the inner product is the generalization of the "dot product" to an N-element vector.

$$R_k = E\left[\underline{\nu}_k\ \underline{\nu}_k^T\right] \tag{14.18}$$

where Q_k is referred to as the system noise covariance matrix and R_k is referred to as the measurement noise covariance matrix. In actual practice, the off-diagonal noise covariance values are frequently negligible. As a result, Q and R are implemented as diagonal matrices with the main diagonal elements filled with the variances of the state variables (for Q) and measurements (for R).

14.5 Estimation and prediction error covariance matrices

At this point, we have defined almost everything we need in order to describe the full Kalman filter. There are just a few more items. First, the estimation error is defined as before:

$$\underline{e}_k = \hat{\underline{x}}_k^+ - \underline{x}_k \tag{14.19}$$

where the hat (or carrot) symbol (with superscript +) indicates an estimate and no hat indicates truth.[†] The estimation error covariance matrix is then given by:

$$P_k^+ = E\left[\underline{e}_k\ \underline{e}_k^T\right] \tag{14.20}$$

Similarly, the prediction error is defined as:

$$\underline{e}_k^- = \hat{\underline{x}}_k^- - \underline{x}_k \tag{14.21}$$

where the superscript "-" indicates a predicted (as opposed to an estimated) quantity. The prediction error covariance matrix is then given by:

$$P_k^- = E\left[\underline{e}_k^-\ (\underline{e}_k^-)^T\right] \tag{14.22}$$

14.6 The discrete-time Kalman filter

We can now describe the complete discrete-time Kalman filter.

Step 0: Set the time index, k, to 1. Choose the initial prediction for the state vector $(\hat{\underline{x}}_1^-)$ and determine the initial value of the prediction error covariance matrix (P_1^-).

Step 1: Compute the Kalman gain:

$$K_k = P_k^- H_k^T (H_k P_k^- H_k^T + R_k)^{-1} \tag{14.23}$$

You may take a look at (14.23) and say, "Wait a minute, that's not exactly what we had for the Kalman gain previously." Actually, it is, but the $(HPH^T + R)$ term was depicted in the denominator because we were dealing with scalars in the bucketful of resistors example. Since we are now considering full matrices, we cannot perform division in the scalar sense and thus we have to perform the matrix inverse of the quantity in parentheses. It is the same equation that we had before but in a matrix-friendly format. It should also be noted that this is the *optimal* gain [1].

[†]Note the absence of a superscript is an implied plus sign.

Step 2: Take the current measurement and form the current estimate:

$$\hat{\underline{x}}_k^+ = \hat{\underline{x}}_k^- + K_k(\underline{z}_k - H_k\hat{\underline{x}}_k^-) \tag{14.24}$$

This is the main "output" of the filter and is frequently referred to as the Kalman "update" equation. The idea is that we have updated our prediction of the state vector using the information from the current measurement vector.

Step 3: Form the estimation error covariance matrix:

$$P_k^+ = (I - K_kH_k)P_k^- \tag{14.25}$$

Note that Equation (14.25) assumes that the optimum Kalman gain has been used in the estimate. There are situations (e.g., sensitivity analyses) where non-optimal gains may be used but the reader is referred to [1] for details.

Step 4: Project ahead to the next step: compute the state transition matrix and form the state prediction for the next moment in time:

$$\hat{\underline{x}}_{k+1}^- = \Phi_k\hat{\underline{x}}_k^+ \tag{14.26}$$

Notice that the prediction is a straightforward application of the system equation (14.6). First, the current estimate is substituted for the true state vector. Since the system noise is both zero-mean and completely unpredictable, it is ignored for the purposes of prediction (the best prediction one can make of a zero-mean Gaussian random variable is zero). Note the superscript "+" is sometimes not explicitly shown. The absence of a superscript minus sign implies a plus sign.

Step 5: Compute the prediction error covariance matrix [1,2]:

$$P_{k+1}^- = \Phi_kP_k^+\Phi_k^T + Q_k \tag{14.27}$$

We see that the prediction error covariance matrix is a function of the estimation error covariance, the state transition matrix, and the system noise covariance matrix. The first term in (14.27) takes the estimation error covariance matrix and projects it forward in accordance with the system dynamics described by the state transition matrix. In accordance with the Kalman filter system equation, the system noise covariance matrix is added in (14.27) to account for the system uncertainty not contained in the state transition matrix. Alternate forms of this equation may be found in the literature and, again, are used in situations such as sensitivity analyses.

Step 6: Go back to Step 1 for the next moment in time.

That is it. We have the complete Kalman filter. We started with the univariate case and expanded it to the multivariate case. We started with a static system and we expanded to a dynamic system. We were introduced to covariance matrices and how they are used in the Kalman filter. And we have described the complete filter. Now let us take a look at a couple of examples.

14.7 Case study 1: an aging resistor

Assume that our meter readings, instcad of being seconds apart, are years apart. The effects of aging or some harsh environment cause the true value of the resistor to decay over time. Now we need to account for the fact that we actually have a dynamic system: a system with values that are not constant. For the sake of this exercise, we will model the true value of the time-varying resistance as:

$$x_{k+1} = 0.99x_k + w_k \tag{14.28}$$

where x_k is the current true value of the resistance and w_k is the non-deterministic component of the change in the resistance value. Note that the true value decays by 1 per cent during each time period (e.g., a year) plus whatever the system noise term is. We will model the system noise as a zero-mean Gaussian with a standard deviation of 0.2 ohm. The resistor is nominally 100 ohms and has 1% precision. We will assume the standard deviation of the error of the ohmmeter is 2 ohms (thus a little better than in our previous example).

Designing a Kalman filter amounts to determining eight vector/matrices. The first three are derived from the Kalman system equation: state vector ($\underline{x}$); state transition matrix (Φ); system noise covariance matrix (Q). The second three are derived from the Kalman measurement equation: measurement vector ($\underline{z}$); data matrix (H); measurement noise covariance matrix (R). Finally, the initial prediction for the state vector ($\hat{\underline{x}}_1^-$) and the initial prediction error covariance matrix (P_1^-) must be specified.

For this time-varying resistor problem, the aforementioned vectors and matrices are all scalars but we will refer to them by their formal names for the sake of consistency. The state vector is thus a scalar (x_k) consisting of the true resistance of the resistor at a given point in time (k). We have made this problem particularly easy by simulating the true time varying resistance (Equation (14.28)) in the form of the Kalman system equation (Equation (14.6)). Thus, we see by inspection of these two equations that $\Phi = 0.99$. Also, since we stated that w_k had a standard deviation of 0.2 ohm, then we know that $Q = (0.2)^2 = 0.04$ since the system noise covariance matrix is, for this problem, just the variance of the scalar system noise.

The measurement vector (z) is just the current scalar value of the measured resistance as it was in our previous resistor examples. The data matrix (H) is also still equal to 1 since we are both measuring and estimating resistance in ohms. The measurement error covariance matrix (R) is just the variance of the ohmmeter error. Since we have stated the ohmmeter error standard deviation is 2 ohms, then $R = 2^2 = 4$.

Finally, we need to specify the initial prediction (x_1^-) and the prediction error covariance matrix (P_1^-). Since the aging process does not start until $k = 2$ (i.e., the resistor is "brand new" when we pick it out of the bucket), then as in our previous resistor examples, $x_1^- = 100$ and $P_1^- = 1$ (we are thus still assuming the bucket is filled with resistors labeled as 100 ohms and the variance of the actual resistor values is 1 ohm-squared).

Figures 14.2 and 14.3 show the results of a simulation of this time-varying resistor example. Figure 14.2 shows the results for a single run of the simulation. The filter is not able to track the true resistance perfectly (due to the unpredictable system

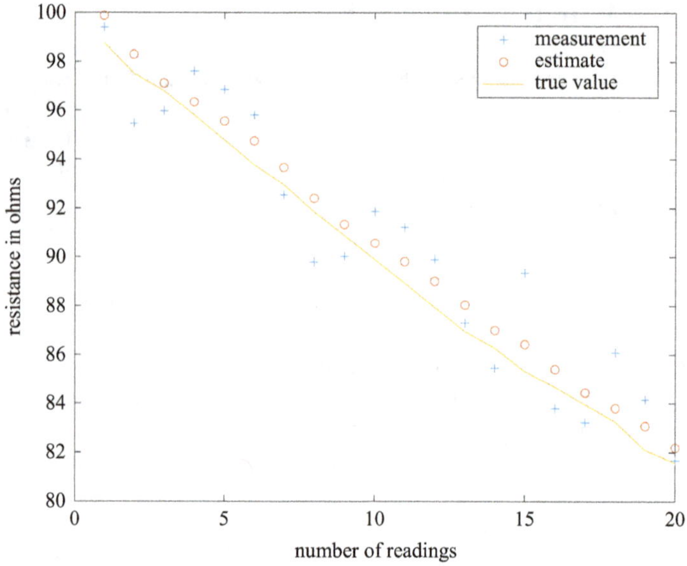

Figure 14.2 Simulated results for the time-varying resistor example

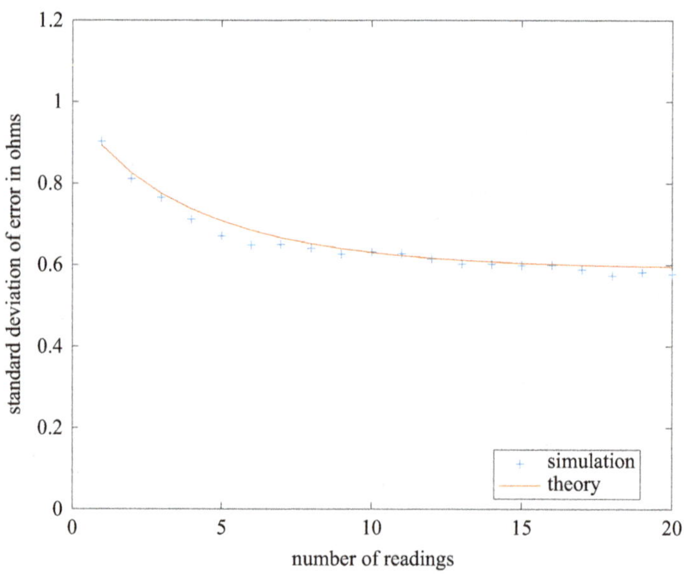

Figure 14.3 A plot of the square-root of the Kalman filter's estimation error covariance (theory) and the corresponding standard deviation of the estimation error obtained from 1,000 Monte Carlo runs of the simulation of the time-varying resistor example

noise) but is able to do much better than the individual noisy measurements alone. Figure 14.3 depicts both how well the Kalman filter thinks it is doing along with how well it is actually doing. The curve marked "theory" is the square root of the filter's estimation error covariance (actually, just variance in this case). The curve marked "simulation" was obtained by computing the standard deviation of the estimation error obtained from 1,000 Monte Carlo runs of the simulation. We see excellent agreement between theory and simulation which is to be expected since the Kalman filter system and measurement equations perfectly matched the actual behavior. However, it is important to note that the inherent system noise (as modeled by Q in the filter) imposes a "floor" on the filter's performance. The estimation error variance (or standard deviation) reaches a non-zero minimum value in steady-state. For this particular example, that floor is approximately 0.6 ohm. This is consistent with theory. As described in [3], for random walk processes in which $Q \ll R$, the estimation error covariance converges approximately to $P = \sqrt{(QR)}$.

It is important to note that most practical systems have non-zero system noise and thus must be modeled with the Q matrix in the filter. As a result, the steady-state error of the filter will always be non-zero. Furthermore, regardless of the actual system noise, it turns out that it is useful always to have some Q in the filter (even for systems with negligible uncertainty). At first, this seems counterintuitive. Why would I tell the filter that the system has some uncertainty when in reality it does not? There are two reasons. One is that it helps to maintain good numerical performance in the filter. This is less of a concern nowadays with modern double precision arithmetic but historically it was quite important. The second reason is that non-zero Q values ensure that the filter always "pays attention" to the measurements. If Q is set to zero, the prediction error covariance will converge to zero and that causes the Kalman gain to converge to zero. As we have studied previously, if the Kalman gain is zero, the filter completely ignores the measurements and simply sets the estimate equal to its own prediction. If there is no system noise and the assumptions in the filter (i.e., the system and measurement equations) are perfect, then this is the correct behavior. However, in practical applications, we are never able to model the system and measurements perfectly. There is always some inherent uncertainty and thus Q should always be set to a non-zero value. In subsequent chapters, we will learn more about the effect that Q has on filter performance.

14.8 Case study 2: simple radar

At the beginning of this chapter, we considered a simple radar example. Let us return to that and design a Kalman filter to estimate distance (range) and velocity (range-rate) of the target aircraft under the assumption that the radar only provides noisy distance measurements. Recall we assumed the aircraft is flying either directly toward or away from the radar and that it is flying at roughly constant velocity. For the sake of this simple example, we will assume the following:

- Initial range to the target is known to be approximately 1,000 m with an uncertainty that is Gaussian with a variance of 2,000 m^2
- Initial velocity is unknown

- Radar measurements are made at a rate of 1 Hz and the measurement noise is Gaussian with zero mean and a standard deviation of 10 m
- Target is moving at approximately constant velocity but turbulence is causing perturbations in the true range. The statistics of the perturbations are unknown but have been estimated to be zero mean Gaussian with a variance of 4 m^2 and perturbations in the true velocity have been estimated to be zero mean Gaussian with a variance of 8 $\left(\frac{m}{s}\right)^2$. For this example, we will further assume the range and velocity perturbations are uncorrelated.

Recall from the previous example that there are eight vectors/matrices that need to be determined in the process of designing the filter. We defined the Kalman system equation for this problem in Equation (14.5). Thus, the state vector consists of the range and velocity of the target:

$$\underline{x}_k = \begin{bmatrix} d \\ r \end{bmatrix}_k \tag{14.29}$$

Again, from Equation (14.5), we see the state transition matrix is given by:

$$\Phi_k = \begin{bmatrix} 1 & T \\ 0 & 1 \end{bmatrix} \tag{14.30}$$

In this example, the state transition matrix is a constant and thus the time subscript (k) is not actually needed. The theoretical equation for the system noise covariance matrix (Q) was given in Equation (14.17) but from the final bullet in the assumptions we see that:

$$Q_k = \begin{bmatrix} 4 & 0 \\ 0 & 8 \end{bmatrix} \tag{14.31}$$

where the units were specified in the list of assumptions. Again, the time index is superfluous here. The measurement equation for this example was given in (14.8). Thus, the measurement vector is simply the radar distance measurement:

$$\underline{z}_k = \begin{bmatrix} d_{meas} \end{bmatrix}_k \tag{14.32}$$

From inspection of (14.8), we see the data matrix is:

$$H = \begin{bmatrix} 1 & 0 \end{bmatrix} \tag{14.33}$$

The last item we determine from the measurement equation is the measurement noise covariance matrix. Since the radar is providing scalar measurements (range), then R is also a scalar. Specifically, it is the variance of the radar measurement noise. From our assumptions about the radar, we see that:

$$R = 10^2 = 100 \tag{14.34}$$

where the units are m^2. Finally, we need to determine the initial prediction of the state vector and the initial prediction error covariance matrix. Again, from the assumptions we note the initial range is 1,000 m but the initial velocity is unknown. In a real system, the designer might choose to use the first few range data points to determine a rough approximation of the velocity. For this example, however, in order to demonstrate the convergence of the filter, we will simply set the initial velocity prediction to zero:

$$\underline{x}_1^- = \begin{bmatrix} 1,000 \\ 0 \end{bmatrix} \tag{14.35}$$

We complete the filter design by determining the initial prediction error covariance matrix. From our assumptions we see that the variance of the initial range prediction is 2,000 m^2. The uncertainty of the initial velocity prediction, however, is unknown. This is not an uncommon situation in real life. Approximations/assumptions frequently need to be made and the designer is frequently faced with the task of "tuning" the filter (a nice word for iterating the design). If the initial prediction error covariance matrix values are too small, the filter will not give enough weight to the measurements and filter convergence can end up being prolonged. If the values are too high, the filter may behave erratically during convergence. This particular example is sufficiently simple that the filter performance does not change drastically with the choice of initial velocity prediction error variance. We will set it at 20 $\left(\frac{\text{m}}{\text{s}}\right)^2$. Although the initial prediction range uncertainty and velocity uncertainty are likely correlated, we will set the off-diagonal elements to zero. For this example, the impact on performance is minimal and, in actual practice, it is common for designers to make the initial matrix diagonal. For the current example then:

$$P_1^- = \begin{bmatrix} 2,000 & 0 \\ 0 & 20 \end{bmatrix} \tag{14.36}$$

A 200 sec trajectory is simulated with an average true velocity of 100 m/s. A plot of the true and estimated range (Figure 14.4) does not reveal the error due to the scale. The error in both the raw radar measurements and the filter's range estimates are depicted in Figure 14.5. Since the true range is inherently noisy in this example (recall $Q(1,1)$), the filter improves on the raw measurements but not by very much.

The true velocity and estimated velocity are depicted in Figure 14.6. Again, with the true velocity being so noisy, there is only so much the filter can do. However, its performance is better shown by contrasting the error in the estimated velocity with the "raw" velocity obtained by simply differencing the radar range measurements (Figure 14.7).

To begin to see the effects of "tuning" the filter, let us arbitrarily set the values of Q to be extremely small. Given the a priori known behavior of the target aircraft, Q was correctly specified back in Equation (14.31). What will happen if we set the non-zero elements of Q to small values? Such as:

$$Q_k = \begin{bmatrix} 0.01 & 0 \\ 0 & 0.01 \end{bmatrix} \tag{14.37}$$

First, we should immediately recognize that we are knowingly mis-modeling the system. We are telling the filter that both the true distance and the velocity are much less noisy than they actually are. As a result, the filter will place much less gain on the measurements and thus much more on its predictions. The predictions assume (recall the system equation for this example) a *constant* velocity. The results are shown in Figures 14.8, 14.9, and 14.10. The errors in the estimated range (Figure 14.8) seem

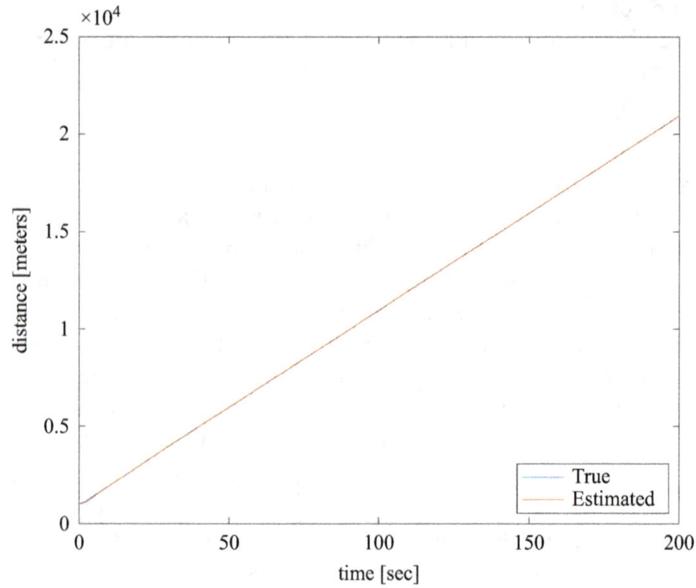

Figure 14.4 True and estimated range for the simple radar example

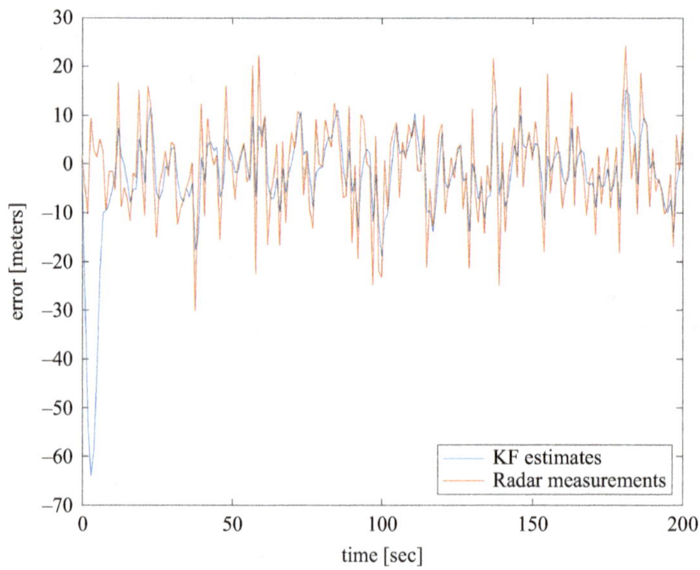

Figure 14.5 Range errors from the radar measurements and the filter estimates

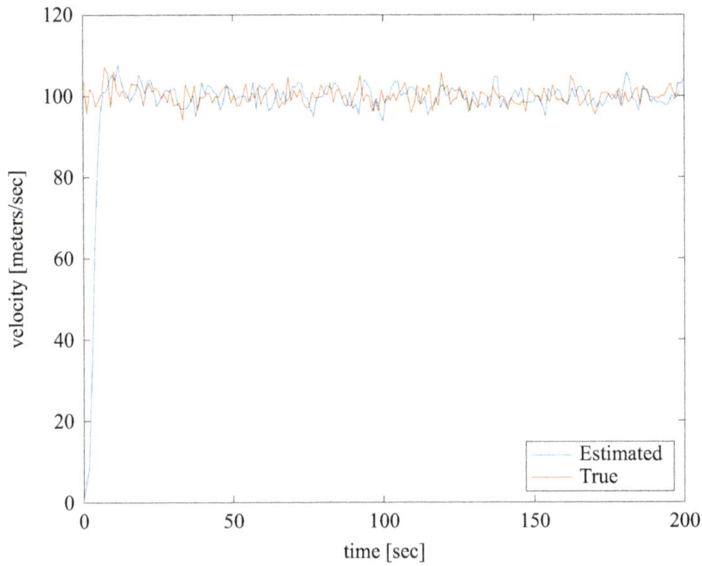

Figure 14.6 True and filter-estimated velocity

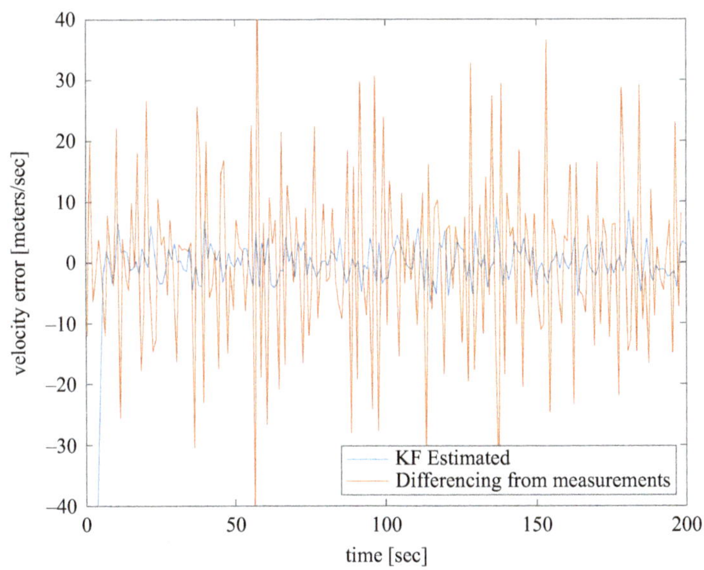

Figure 14.7 Errors in velocity obtained from the filter and from simple differencing of the radar measurements

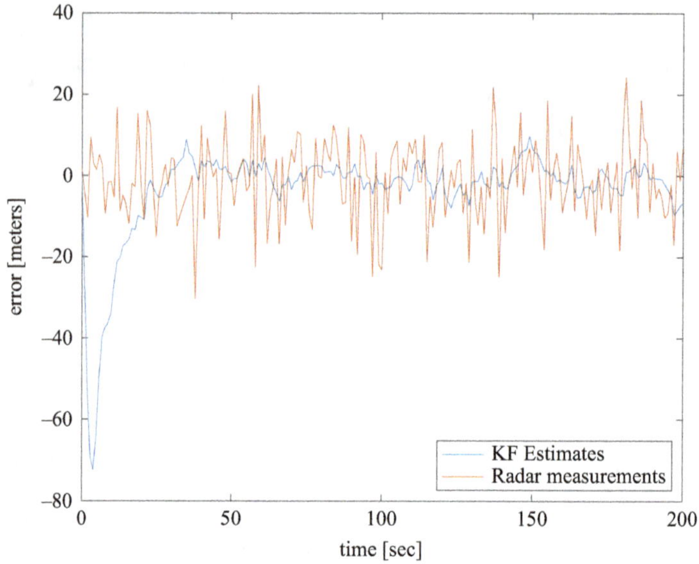

Figure 14.8 Range errors from the radar measurements and the filter estimates when specifying very small system noise covariance values

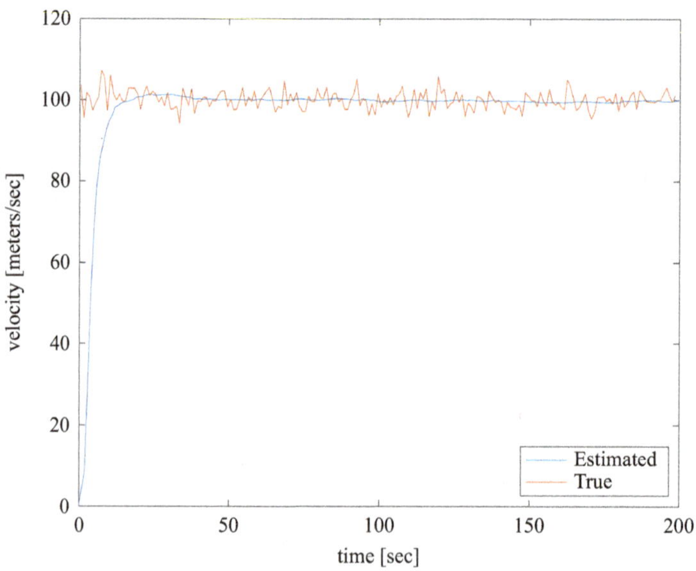

Figure 14.9 True and filter-estimated velocity (very small Q values)

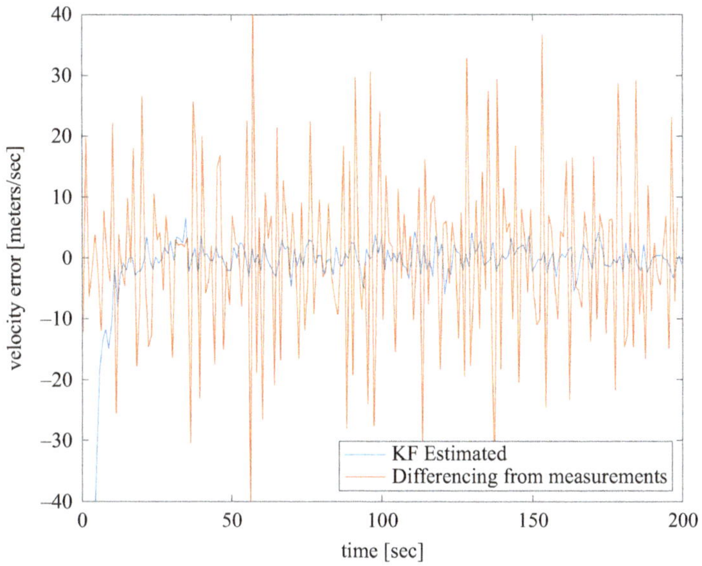

Figure 14.10 Errors in velocity obtained from the filter and from simple differencing of the radar measurements (very small Q values)

to be somewhat smaller than in the previous case (Figure 14.5) indicating that the previously specified position error covariance may have been too high. However, Figure 14.8 also clearly demonstrates the longer convergence time incurred due to the low Q value.

The true and estimated velocity values depicted in Figure 14.9 clearly show that the filter is only estimating the underlying deterministic component of the true velocity. Nevertheless, the error in the estimated velocity is somewhat smaller than in the previous filter. The example is, of course, simple and contrived but it begins to show us how filters are tuned in real life. In real life, the system equation is always, at best, an approximation. Thus, in real life, we are always mis-modeling to a certain extent. Adjustment of the values in Q is frequently done as part of filter performance improvement process.

References

[1] Gelb A, editor. *Applied Optimal Estimation*. Cambridge, MA: The M.I.T. Press; 1974.

[2] Brown RG, Hwang PYC. *Introduction to Random Signals and Applied Kalman Filtering: with MATLAB® exercises*. 4th ed. Hoboken, NJ: John Wiley & Sons, Inc.; 2012.

[3] Pue A. Integration of GPS with Inertial Navigation Systems [Short Course Notes]. NavtechGPS; 2007.

Chapter 15

Radionavigation Kalman filtering: case studies in two dimensions

So far we have spent a good deal of time estimating scalars and even the radar example involved a vehicle traveling in essentially a straight line. It is time to move into two dimensions and see where that takes us.

We will investigate horizontal positioning and, as an example, will consider the distance measuring equipment (DME) system. DME comprises ground stations (known as transponders) that respond to RF interrogations transmitted from a unit in an aircraft (known as an interrogator). The two-way transmissions enable the airborne unit to determine the distance (range) to the ground station. The measured ranges are an approximation to the horizontal distance from the ground station to the aircraft. Range measurements made to at least two ground stations may be used (along with the known ground station locations) to compute the horizontal position of the aircraft. Since there is no need for receiver clock bias estimation, the use of DME for horizontal positioning provides us with a simpler solution that is possible with GPS and allows us to ease into the topic. Throughout this chapter (and the next), we will consider positioning via radionavigation only.

15.1 Linearized Kalman filter

The first thing that we have to deal with is the fact that the relationship between ranging measurements and two-dimensional position is nonlinear and, as a result, the Kalman filter is going to have to be set up to deal with these kinds of measurements. We address this with either the "linearized Kalman filter" or the "extended Kalman filter" [1]. Both are methods to allow the Kalman filter, which inherently requires linear relationships, to accommodate nonlinear situations.

The first approach to deal with nonlinear measurement and/or system equations is simply to linearize them. If you have some approximate point about which you can perform a Taylor series expansion, and truncate it to the first couple of terms, you can linearize the relationships and use that approximation in the Kalman filter. In a sense, we have seen this already when we took a look at the earlier stand-alone radar example. We imposed a linear approximation by assuming that the flight was straight, level and constant velocity. We neglected the influence of the acceleration of the vehicle and thus were linearizing the system dynamics.

We will consider two scenarios. First, we will examine the case where a nominal vehicle trajectory is known a priori. The Kalman filter will then estimate the offset between the nominal and the actual vehicle position (or state, in general). This is known as an ordinary linearized Kalman filter. Later we will examine the case where there is no a priori knowledge of the vehicle trajectory and linearization will be done with respect to the state predicted, in real-time, by the filter. This is known as an extended Kalman filter.

15.2 Case study: DME–DME positioning with an a priori known nominal trajectory

Consider horizontal positioning with ranging measurements made from two DME ground transponders with known locations (Figure 15.1). The true range between the vehicle and the *i*th DME transponder is given by:

$$\rho_i = \sqrt{(x_u - x_i)^2 + (y_u - y_i)^2} \tag{15.1}$$

where "*u*" indicates the "user" or vehicle. ρ is the ideal measured range and the *i*th DME transponder coordinates (x_i, y_i) are known. In order to solve for the two components of user position (given range measurements to at least two transponders), we need a linearized approximation of Equation (15.1). We will use the linearized approximation first to form an ordinary least squares position solution and then will derive a Kalman filter estimator for position.

To derive the linear approximation, we expand (15.1) in a Taylor Series about a nominal user position. For the current example, the nominal position will be obtained from an a priori known nominal trajectory (with a corresponding nominal position at

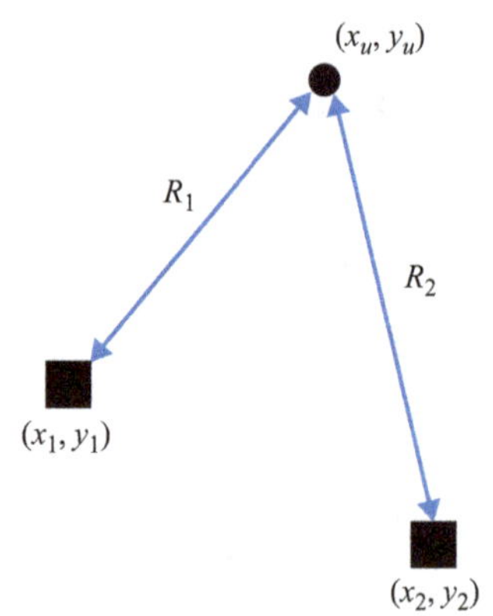

Figure 15.1 Horizontal positioning using two distance measuring equipment (DME) ground transponders. (x_1, y_1) *and* (x_2, y_2) *are the coordinates of the first and second DME transponders and* R_1 *and* R_2 *are the ranging measurements made by the user located at* (x_u, y_u)

any given epoch or moment in time). Retaining only the zeroth and first-order terms in the Taylor Series results in the following approximation:

$$\rho_i \approx \rho_{io} + \frac{x_o - x_i}{\rho_{io}}(x_u - x_o) + \frac{y_o - y_i}{\rho_{io}}(y_u - y_o) \tag{15.2}$$

where the nominal user position is given by (x_o, y_o) and the zeroth order term is the distance from the transponder to the nominal user position:

$$\rho_{io} = \sqrt{(x_o - x_i)^2 + (y_o - y_i)^2} \tag{15.3}$$

Define the difference between the nominal and true user positions as:

$$\begin{bmatrix} \Delta x \\ \Delta y \end{bmatrix} = \begin{bmatrix} x_u - x_o \\ y_u - y_o \end{bmatrix} \tag{15.4}$$

These "deltas" are the offsets of the true position from the nominal position and they are the unknowns that we will be solving for initially.

Substituting (15.4) into (15.2) yields:

$$\rho_i \approx \rho_{io} + \frac{x_o - x_i}{\rho_{io}}\Delta x + \frac{y_o - y_i}{\rho_{io}}\Delta y \tag{15.5}$$

A linear equation in the unknowns (x and y position offsets) is created by moving the zeroth order term (which is a constant) to the left side of the equation:

$$\Delta\rho_i = \frac{x_o - x_i}{\rho_{io}}\Delta x + \frac{y_o - y_i}{\rho_{io}}\Delta y \tag{15.6}$$

where $\Delta\rho_i = \rho_i - \rho_{io}$. Since there are two unknowns in (15.6), we need at least two measurements (i.e., range measurements to two different DME transponders). Two instances of (15.6) for two different transponders yields:

$$\begin{bmatrix} \Delta\rho_1 \\ \Delta\rho_2 \end{bmatrix} = \begin{bmatrix} \left(\frac{x_o - x_1}{\rho_{1o}}\right) & \left(\frac{y_o - y_1}{\rho_{1o}}\right) \\ \left(\frac{x_o - x_2}{\rho_{2o}}\right) & \left(\frac{y_o - y_2}{\rho_{2o}}\right) \end{bmatrix} \begin{bmatrix} \Delta x \\ \Delta y \end{bmatrix} \tag{15.7}$$

which can be written more succinctly as:

$$\underline{\Delta\rho} = H\underline{\Delta x} \tag{15.8}$$

where

$$\underline{\Delta x} = \begin{bmatrix} \Delta x \\ \Delta y \end{bmatrix} \tag{15.9}$$

In reality, there are measurement errors that need to be included in (15.8) to form the complete measurement equation:

$$\underline{\Delta\rho} = H\underline{\Delta x} + \underline{\nu} \tag{15.10}$$

where $\underline{\nu}$ is the vector of measurement errors. It is important to keep in mind that (15.10) is a theoretical model and it does not explain how the real measurements are actually formed. For example, distance measurements in DME are made by timing transmission and reply pulses and this process, of course, is not reflected in any way in the theoretical measurement equation. Although we will not go through the derivation here, the "solution" to (15.10) involves finding an estimate of $\underline{\Delta x}$ that best fits the noisy measurements in a least-squares sense. The result is the ordinary least squares "solution" (or more accurately, "estimate"):

$$\underline{\hat{\Delta x}} = (H^T H)^{-1} H^T \underline{\Delta \rho} \tag{15.11}$$

where the hat symbol signifies an estimate. We will shortly develop a Kalman filter to achieve enhanced performance over least-squares. Be careful, though, not to forget that solving for the position offsets is *not* the final step. Recalling Equation (15.4), we need to add the estimated offsets to the nominal position to get the total estimated position:

$$\begin{bmatrix} x_u \\ y_u \end{bmatrix} = \begin{bmatrix} x_o \\ y_o \end{bmatrix} + \begin{bmatrix} \hat{\Delta x} \\ \hat{\Delta y} \end{bmatrix} \tag{15.12}$$

Note that (15.7) can easily be expanded by simply adding rows to incorporate additional measurements. As a sidenote, we also point out that if the nominal position is sufficiently distant from the truth so as to make the linearization (15.2) a poor approximation, the entire process can be iterated using the result of (15.12) as the new nominal position. The process can be iterated until the magnitude of the solution (Equation (15.11)) is sufficiently small. For our current purposes, however, the nominal positions will be close enough to the truth that iteration will not be needed. (For the sake of completeness it should be noted that it is possible to specify the nominal position far enough from the truth that the aforementioned iterative process will fail. However, if the error in the nominal position is significantly less than the distance of the true position from the transponders, this will not happen.)

The linearized Kalman filter measurement equation is obtained by simply incorporating measurement noise into Equation (15.8):

$$\underline{z} = \underline{\Delta \rho} = H \underline{\Delta x} + \underline{\nu} \tag{15.13}$$

However, it is important to remember that the Kalman measurement equation is a *theoretical* model describing the relationship between the measurement vector and the state vector. The purpose of deriving the measurement equation is to be able to determine the data matrix (H) and the measurement noise covariance matrix (R). The measurement equation is *not* a formula for computing the measurement vector! This is a point that many people miss when first learning about Kalman filters. For the current example, the measurement vector is computed using the left side of (15.6). The ith element of the measurement vector is thus given by:

$$z_i = \rho_i - \rho_{io} = \Delta \rho_i \tag{15.14}$$

recalling that ρ_i is the *measured* range from the user to the ith DME ground transponder.

For the current example, we are assuming that we have an a priori known nominal trajectory. Specifically, we will follow an example originated by [1] in which the vehicle is assumed to travel in the positive direction "up" the y-axis as indicated in Figure 15.2.

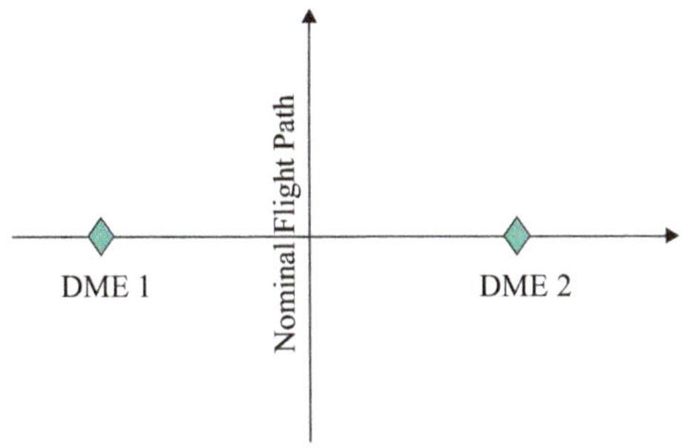

Figure 15.2 The nominal flight path lies along the y-axis. The two DME ground transponders are located on the x-axis. Although not shown in the figure, for the simulations shown herein, the DMEs are located at −10 km and +10 km, respectively, on the x-axis and the starting and ending points of the flight path are located at −10 km and +10 km, respectively, on the y-axis

The actual vehicle trajectory is simulated as follows with a nominal x-velocity of zero and a nominal y-velocity of 100 m/s:

$$\begin{bmatrix} x_u \\ y_u \end{bmatrix}_{k+1} = \begin{bmatrix} x_u \\ y_u \end{bmatrix}_k + \begin{bmatrix} 0 \\ 100 \end{bmatrix} \Delta t + \begin{bmatrix} w_x \\ w_y \end{bmatrix} \tag{15.15}$$

where position disturbances, w_x and w_y, are simulated as zero-mean Gaussian random variables each as having a variance of 400 m^2. The initial position is simulated as:

$$\begin{bmatrix} x_u \\ y_u \end{bmatrix}_1 = \begin{bmatrix} 0 \\ -10,000 \end{bmatrix} + \begin{bmatrix} n_x \\ n_y \end{bmatrix} \tag{15.16}$$

where n_x and n_y are also zero-mean Gaussian random variables each having a variance of 2,000 m^2. The a priori known nominal trajectory is obtained from (15.15) by ignoring the position disturbances:

$$\begin{bmatrix} x_o \\ y_o \end{bmatrix}_{k+1} = \begin{bmatrix} x_o \\ y_o \end{bmatrix}_k + \begin{bmatrix} 0 \\ 100 \end{bmatrix} \Delta t \tag{15.17}$$

With the a priori known nominal initial position similarly obtained from (15.16):

$$\begin{bmatrix} x_o \\ y_o \end{bmatrix}_1 = \begin{bmatrix} 0 \\ -10,000 \end{bmatrix} \tag{15.18}$$

Comparison of (15.15) and (15.17) shows that the difference between the true and nominal position components is a random walk process:

$$\begin{bmatrix} x_u \\ y_u \end{bmatrix}_k = \begin{bmatrix} x_o \\ y_o \end{bmatrix}_k + \sum_{i=1}^{k} \left\{ \begin{bmatrix} w_x \\ w_y \end{bmatrix}_i \right\} \tag{15.19}$$

Recall the Kalman filter system equation requires that we determine the state vector, the state transition matrix and the system noise vector. The state vector consists of the position offsets (Equation (15.4)) and the system equation is then given by:

$$\underline{\Delta x}_{k+1} = \begin{bmatrix} 1 & 0 \\ 0 & 1 \end{bmatrix} \underline{\Delta x}_k + \underline{w}_k \tag{15.20}$$

Although the 2 × 2 identity matrix may be considered superfluous at this point, we will need it shortly.

At this point we have all the necessary information to construct the filter. The initial state prediction consists of zeros since the offsets between the truth and nominal are unknown zero-mean Gaussian processes:

$$\underline{\Delta x}_1^- = \begin{bmatrix} 0 \\ 0 \end{bmatrix} \tag{15.21}$$

The initial prediction error covariance matrix is obtained from (15.16) where it was specified that the actual initial offset components had variances of 2,000 m^2:

$$P_1^- = \begin{bmatrix} 2,000 & 0 \\ 0 & 2,000 \end{bmatrix} \tag{15.22}$$

Note the zeros on the off-diagonal elements indicate that the Gaussian random variables n_x and n_y in (15.16) are independent/uncorrelated.

From Equation (15.20), the state transition matrix is simply:

$$\Phi = \begin{bmatrix} 1 & 0 \\ 0 & 1 \end{bmatrix} \tag{15.23}$$

The system noise covariance matrix is obtained from (15.15) where it was specified that the "noise" (or variations) of each of the position components had a variance of 400 m^2:

$$Q = \begin{bmatrix} 400 & 0 \\ 0 & 400 \end{bmatrix} \tag{15.24}$$

Finally, the DME range measurements are simulated as having independent white Gaussian noise errors with a variance of 225 m^2. Thus, the measurement error covariance is given by:

$$R = \begin{bmatrix} 225 & 0 \\ 0 & 225 \end{bmatrix} \tag{15.25}$$

In order to appreciate the performance of the Kalman filter, we will first show the performance of the ordinary least squares (OLS) estimator. The true simulated flight path is shown in Figure 15.3.

The errors of the OLS position estimates are shown in Figures 15.4 and 15.5. Given the fact that the DME (ranging) measurement errors have a standard deviation of 15 m, the x position error is quite reasonable. In the middle of the run, however, the y error becomes extremely large. What is happening here?

Welcome to the influence of geometry. The large errors in the y component are due to the poor geometry that exists when the user crosses the x-axis. At this point the

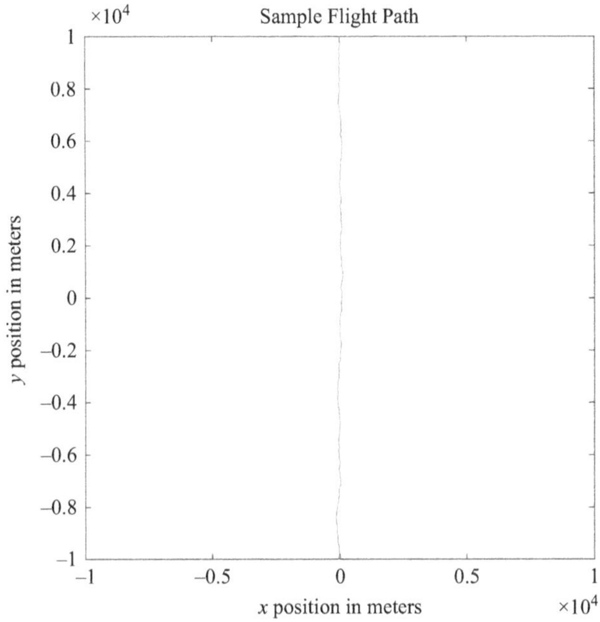

Figure 15.3 Sample flight trajectory consisting of a nominal y-velocity of +100 m/s plus random walk processes on each of x and y driven by discrete-time Gaussian white noise with variance of 400 m² (Note: 1-second step-time utilized)

user and both DMEs are all in a line on the x-axis (see Figure 15.2). As a result, the y geometry (y dilution of precision, YDOP) becomes infinite (conversely, the XDOP is less than 1 since we have redundancy in the x direction). Given the H matrix defined in Equations (15.7) and (15.8), the DOPs can be computed as follows:

$$XDOP = \sqrt{G(1,1)} \qquad\qquad YDOP = \sqrt{G(2,2)} \tag{15.26}$$

where $G = (H^T H)^{-1}$

Figures 15.6 and 15.7 illustrate the DOPs for the nominal flight path that starts at (0,−10,000) and ends at (0,+10,000) meters with the DME transponders located at (−10,000,0) and (+10,000,0).

At both the beginning and the end of the run, the crossing angle between the lines-of-sight from the user to the DME transponders is 90 degrees and thus the XDOP and YDOP are both equal to 1. In the middle, XDOP has decreased to its minimum value of $\frac{1}{\sqrt{2}}$ and YDOP approaches infinity. At the very middle, we note the two DOP plots have a gap. This is due to the fact that when the nominal position and two DME transponders are all in a row, the G matrix is singular and its inverse does not exist. Regarding the minimum XDOP, recall that the standard deviation of the estimation error of the sample mean is inversely proportional to the square-root of the number

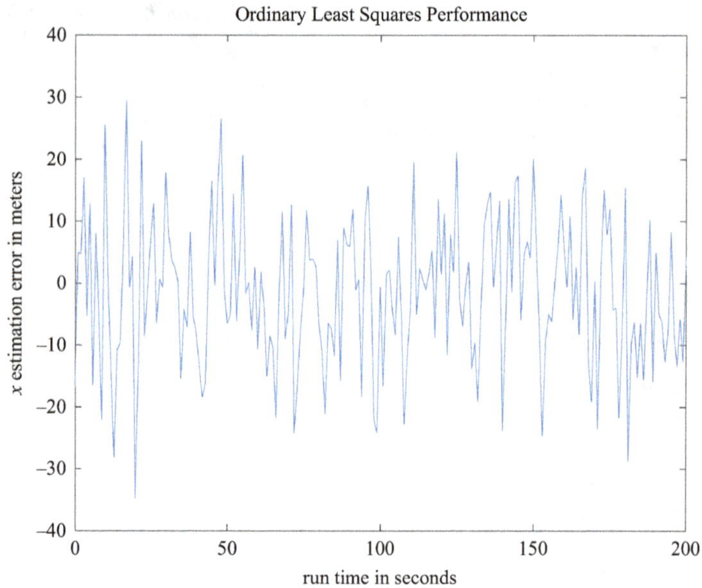

Figure 15.4 X-coordinate estimation errors with ordinary least squares for the trajectory shown in Figure 15.3 with the DME transponder arrangement illustrated in Figure 15.2 and ranging measurement noise with a standard deviation of 15 m

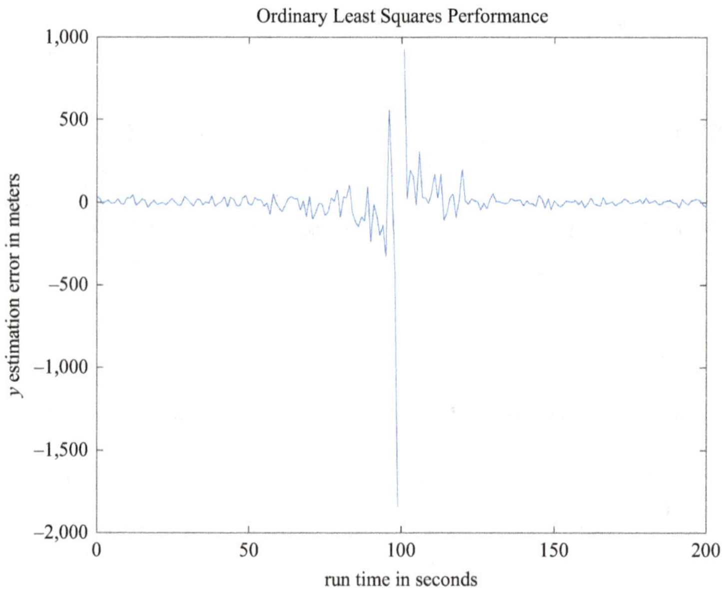

Figure 15.5 Y-coordinate estimation errors with ordinary least squares for the trajectory shown in Figure 15.3 with the DME transponder arrangement illustrated in Figure 15.2 and ranging measurement noise with a standard deviation of 15 m

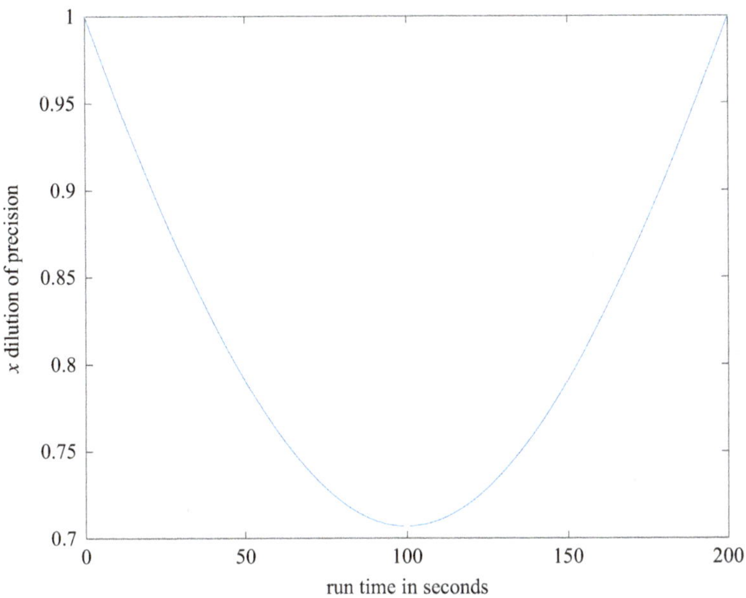

Figure 15.6 X dilution of precision for the flight path illustrated in Figure 15.2

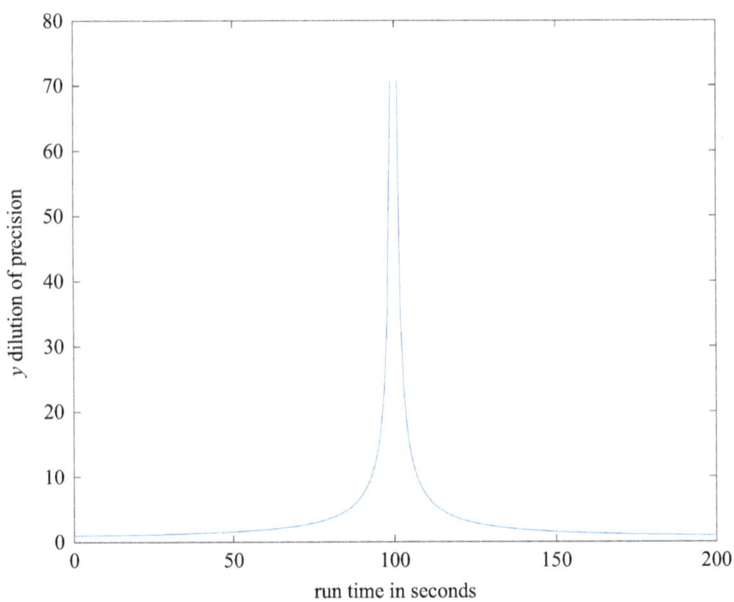

Figure 15.7 Y dilution of precision for the flight path illustrated in Figure 15.2

of measurements. In the middle of the run, the two DME range measurements essentially constitute redundant measurements of the x-coordinate of the user.

With independent white Gaussian measurement errors, the standard deviations of the position errors are related to the standard deviation of the ranging measurement errors as follows:

$$\sigma_{x\ err} = XDOP \cdot \sigma_{rng\ err}, \qquad \sigma_{y\ err} = YDOP \cdot \sigma_{rng\ err} \tag{15.27}$$

Since the measurement error standard deviation was 15 m and the XDOP ranged from about 0.7 to 1, the x position errors shown in Figure 15.4 are consistent. Similarly, given the YDOP shown in Figure 15.7, the y position errors (Figure 15.5) are also consistent.

Given the horrible y geometry, however, one wonders how much better the Kalman filter will do? Remarkably well as a matter of fact. The x and y estimation errors are illustrated in Figures 15.8 and 15.9. The upper dashed line in Figure 15.8 is the square-root of the (1,1) element of the estimation error covariance matrix (P^+) and the lower dashed line is its negative. Similarly, the dashed lines in Figure 15.9

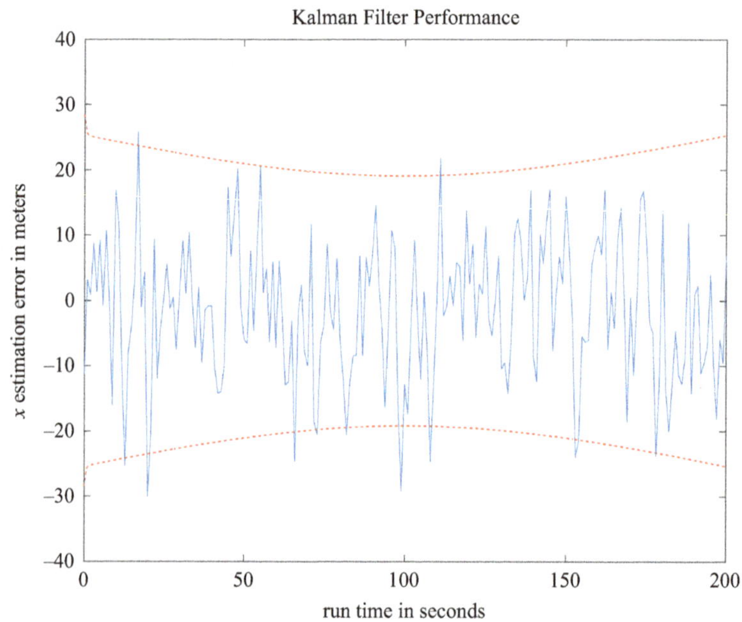

Figure 15.8 X coordinate Kalman filter estimation errors for the trajectory shown in Figure 15.3 with the DME transponder arrangement illustrated in Figure 15.2. The upper dashed line is the square-root of the (1,1) element of the estimation error covariance matrix and the lower dashed line is its negative. The dashed lines thus illustrate the Kalman filter's estimate of the standard deviation of its own estimation error

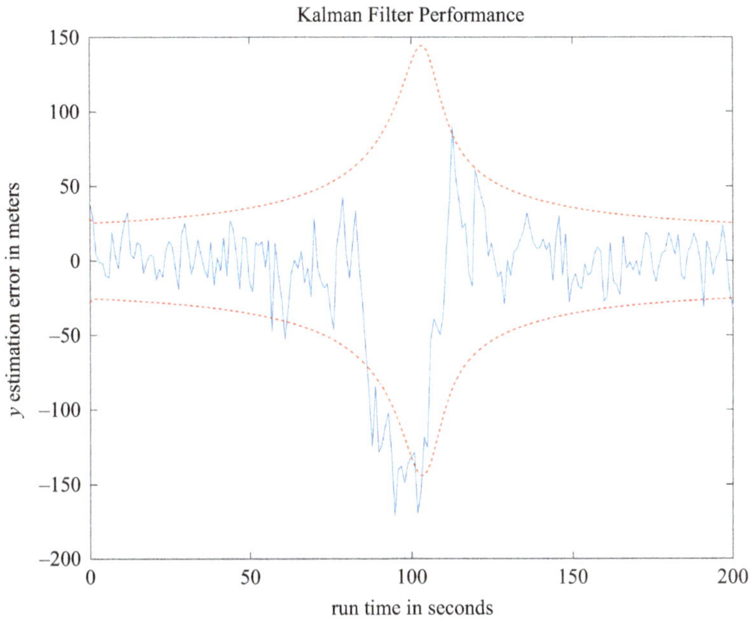

Figure 15.9 Y coordinate Kalman filter estimation errors for the trajectory shown in Figure 15.3 with the DME transponder arrangement illustrated in Figure 15.2. The dashed lines are the filter's estimated error standard deviation

are derived from the (2,2) element of P^+. The dashed lines thus illustrate the Kalman filter's estimate of the standard deviation of its own estimation error.

Comparison of the OLS performance (Figures 15.4 and 15.5) with the Kalman filter (Figures 15.8 and 15.9) reveals very similar results for x but an order of magnitude difference for y. The Kalman filter is affected by the poor geometry but not nearly as much as the OLS estimator. At this point, we must pose the obvious question: "Why?" What is the magic the Kalman filter brings to the situation?

The answer is that the Kalman filter is exploiting its ability to make and utilize predictions. We recall that estimation boils down to weighted sums of predictions and measurements. Least squares only processes the measurements, and if the geometry between the measurements and the solution is bad then least squares is stuck with it. On the other hand, the Kalman is not stuck with just the measurements. The Kalman has its own prediction capability, and because of that it is able to do much better than least squares.

We recall that the filter is weighting the measurements and the predictions according to the statistical error measures on each, and somehow it is able to know that it should de-weight the measurements when the geometry is bad. At the point where the vehicle crosses the x-axis and the YDOP approaches infinity, the Kalman filter somehow is aware and places weight on the prediction and not on the measurement.

How does this happen? Recall the Kalman update equation:

$$\hat{\underline{x}}^{+} = \hat{\underline{x}}^{-} + K(\underline{z} - H\hat{\underline{x}}^{-}) = (I - KH)\hat{\underline{x}}^{-} + K\underline{z} \tag{15.28}$$

where the time indices have been dropped for the purposes of clarity.

The quantity $(K \cdot H)$ is the effective Kalman gain in "state space" (i.e., the solution domain, which in this example is the position domain). Recall also that the Kalman gain itself is a function of the H matrix. Thus the Kalman filter gain and update are inherently accounting for the geometry. We see in (15.28) that $(I - KH)$ weights the prediction. If $(K \cdot H)$ is zero, all weight is on the prediction. If $(K \cdot H)$ is 1, all weight is on the measurement. Thus, $(K \cdot H)$ is effectively the relative weighting placed on the measurement.

For this example, both K and H, and thus their product, are all 2×2 matrices. Designating the product of the two matrices as KH, the (1,1) and (2,2) elements are plotted in Figure 15.10. The plot demonstrates that, indeed, in the middle of the run the filter places zero weight on the measurements (KH(2,2)) for the estimation of y while simultaneously maximizing its weight on the measurements (KH(1,1)) for the

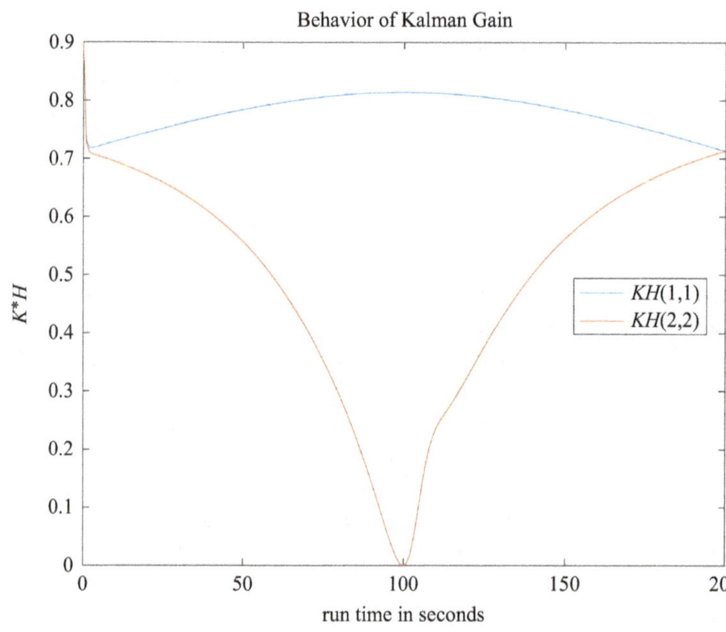

Figure 15.10 Effective Kalman gain for the trajectory shown in Figure 15.3 with the DME transponder arrangement illustrated in Figure 15.2. The (1,1) and (2,2) elements represents the weights being placed on the measurements in the formation of the estimates of the x-coordinate and y-coordinate, respectively

estimation of x. The filter automatically accounts for the geometry in an optimum manner.

A close examination of Figure 15.9 also reveals the estimated error standard deviation peaks *after* the crossing of the x-axis at time $t = 100$ sec. This is the result of non-zero system noise (Q) during the time when the filter is "coasting" on its prediction.

15.3 Case study: DME–DME positioning with non-linear vehicle dynamics via the extended Kalman filter

In the previous case study, an a priori known nominal trajectory, with true position deviations described by a simple random walk process, allowed us to construct a linear system equation. In general, however, a nominal vehicle trajectory is not known in advance. This complicates matters since general motion is a non-linear function of acceleration. To put it simple, turns cause the vehicle dynamics to be non-linear. In later chapters, we will see how integration with an inertial system allows us to form a so-called "complementary filter" and eliminate the non-linear system dynamics. First, however, it is good to explore the trade-offs involved in designing a Kalman filter for a stand-alone radionavigation system.

Since we have no a priori known reference trajectory, we will form one "on the fly" (i.e., in real-time) by extrapolating our current estimate. We will estimate both position and velocity (offsets) in the filter and will model some noise on the velocity states to account for disturbances. This will, of course, not work well during turns but we will explore that shortly. First, let us design the filter. Then we will explore its performance. The linearized system equation is similar to what we had in a previous chapter for the simplified radar problem:

$$\begin{bmatrix} \Delta xpos \\ \Delta xvel \\ \Delta ypos \\ \Delta yvel \end{bmatrix}_{k+1} = \begin{bmatrix} 1 & \Delta t & 0 & 0 \\ 0 & 1 & 0 & 0 \\ 0 & 0 & 1 & \Delta t \\ 0 & 0 & 0 & 1 \end{bmatrix} \begin{bmatrix} \Delta xpos \\ \Delta xvel \\ \Delta ypos \\ \Delta yvel \end{bmatrix}_{k} + \begin{bmatrix} 0 \\ w_{xvel} \\ 0 \\ w_{yvel} \end{bmatrix} \tag{15.29}$$

where the state vector consists of the offsets of x and y position and velocity from the nominal position and velocity and perturbing noise is modeled on the velocity states. Given a current estimate of the total state (position and velocity), we use the state transition matrix from the system equation to derive the algorithm to form the nominal state for the next moment in time:

$$\begin{bmatrix} Xpos_{nom} \\ Xvel_{nom} \\ Ypos_{nom} \\ Yvel_{nom} \end{bmatrix}_{k+1} = \begin{bmatrix} 1 & \Delta t & 0 & 0 \\ 0 & 1 & 0 & 0 \\ 0 & 0 & 1 & \Delta t \\ 0 & 0 & 0 & 1 \end{bmatrix} \begin{bmatrix} Xpos_{est} \\ Xvel_{est} \\ Ypos_{est} \\ Yvel_{est} \end{bmatrix}_{k} \tag{15.30}$$

where "nom" refers to nominal and "est" refers to estimate. Note that the quantities in (15.30) are total quantities and not incremental quantities. Similar to Equation

(15.12), the total estimate is obtained by adding the state vector estimate (which consists of incremental quantities) onto the nominal:

$$\begin{bmatrix} Xpos_{est} \\ Xvel_{est} \\ Ypos_{est} \\ Yvel_{est} \end{bmatrix}_k = \begin{bmatrix} Xpos_{nom} \\ Xvel_{nom} \\ Ypos_{nom} \\ Yvel_{nom} \end{bmatrix}_k + \begin{bmatrix} \Delta xpos_{est} \\ \Delta xvel_{est} \\ \Delta ypos_{est} \\ \Delta yvel_{est} \end{bmatrix}_k \tag{15.31}$$

This use of the filter's estimate to form the nominal point of linearization is why this architecture is known as an "extended Kalman filter" commonly abbreviated "EKF." The Kalman filter's results are being extended to form the point of linearization.

The noise vector in Equation (15.29) can be shown [1] to yield a system noise covariance matrix given by:

$$Q = \begin{bmatrix} S\frac{\Delta t^3}{3} & S\frac{\Delta t^2}{2} & 0 & 0 \\ S\frac{\Delta t^2}{2} & S\Delta t & 0 & 0 \\ 0 & 0 & S\frac{\Delta t^3}{3} & S\frac{\Delta t^2}{2} \\ 0 & 0 & S\frac{\Delta t^2}{2} & S\Delta t \end{bmatrix} \tag{15.32}$$

where Δt is the Kalman cycle/update interval and S, for our purposes, is a scaling parameter. Later we will vary it to see the impact on the filter performance.

The initial prediction of the state vector, $\hat{\underline{x}}_1^-$, is set to all zeros since all a priori information about the state has been incorporated into the nominal state. The initial prediction error covariance matrix, P_1^-, may be specified as a diagonal matrix with values sufficiently large so as to ensure the filter does not place too much reliance on the initial prediction.

Given the state vector shown in Equation (15.29), we can write the measurement equation by expanding the H matrix defined previously in Equation (15.7):

$$\begin{bmatrix} \Delta\rho_1 \\ \Delta\rho_2 \end{bmatrix} = \begin{bmatrix} \left(\frac{x_o - x_1}{\rho_{1o}}\right) & 0 & \left(\frac{y_o - y_1}{\rho_{1o}}\right) & 0 \\ \left(\frac{x_o - x_2}{\rho_{2o}}\right) & 0 & \left(\frac{y_o - y_2}{\rho_{2o}}\right) & 0 \end{bmatrix} \begin{bmatrix} \Delta xpos \\ \Delta xvel \\ \Delta ypos \\ \Delta yvel \end{bmatrix} + \begin{bmatrix} \nu_1 \\ \nu_2 \end{bmatrix} \tag{15.33}$$

where ν_1 and ν_2 are the DME ranging measurement noise/errors. Thus the H matrix has the same non-zero elements as before but the columns of zeros are needed to accommodate the velocity states. For the current case, we will again simulate the DME ranging measurement error as being independent Gaussian noise with a standard deviation of 15 m and thus our R matrix is the same as it was in the previous case (defined in Equation (15.25)).

Finally, since our estimated state vector is used to predict ahead to form the nominal state (Equations (15.31) and (15.30)), the state vector prediction is reset to zeros before going back to the top of the loop:

$$\hat{\underline{x}}_{k+1}^- = \begin{bmatrix} 0 & 0 & 0 & 0 \end{bmatrix}^T \tag{15.34}$$

At first, this seems a bit odd but it is needed since all of the information in the estimated state was already propagated ahead to form the predicted nominal state. [You may also note that forming the Kalman update (e.g., Equation (15.28)) with the state prediction being always a vector of zeroes is numerically inefficient. This is true and alternative formulations have been derived accordingly [1]. Specifically, the algorithm can be reformatted such that the filter estimates total quantities (i.e., position and velocity) directly rather than estimating incremental quantities and having to add them to the nominal in order to form the total estimate.] However, even though the state prediction is zeros, *the prediction error covariance matrix is not*. It continues to be computed as normal.

We will now investigate the impact of non-linear vehicle dynamics on the EKF. A simulated flight path is constructed consisting of a straight path to the east followed by a 180 degree left turn and then a straight path back to the west as shown in Figure 15.11.

In order to focus on the non-linear dynamics and not the radionavigation system geometry, we will simulate the DME transponders at (−50 km, 50 km) and (50 km, 50 km). This makes both XDOP and YDOP approximately equal to 1 throughout the entire flight path.

With the scaling parameter (S) in Equation (15.32) set to 225 (i.e., $(15)^2$), the Kalman filter performance is illustrated in Figures 15.12–15.14. The results are clearly noisy. The velocity is especially noisy but this is not surprising given the noise level on the ranging measurements. The results are not particularly inspiring and we wonder if we could possibly do better. By decreasing the magnitude of the values in the Q matrix, we can achieve more filtering by getting the filter to place more weight on its predictions. The results of setting the S parameter to 1 are shown in Figures 15.15–15.17. Although better performance is achieved in the straight segments, there

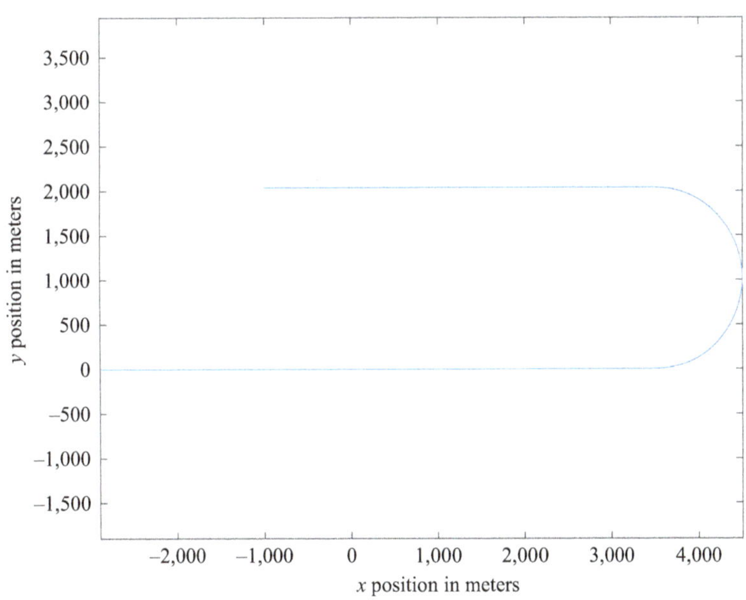

Figure 15.11 Simulated flight path originating at (−2900,0), proceeding in the positive x direction (e.g., East) at 100 m/s, executing an approximately 1 g left turn and then continuing in the negative x direction (e.g., West)

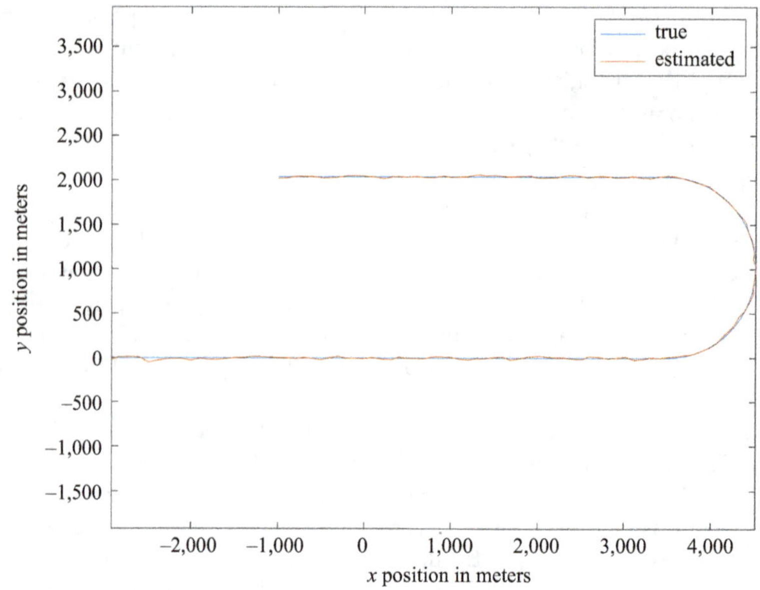

Figure 15.12 True simulated flight path and Kalman filter results with Q set to 'large' values

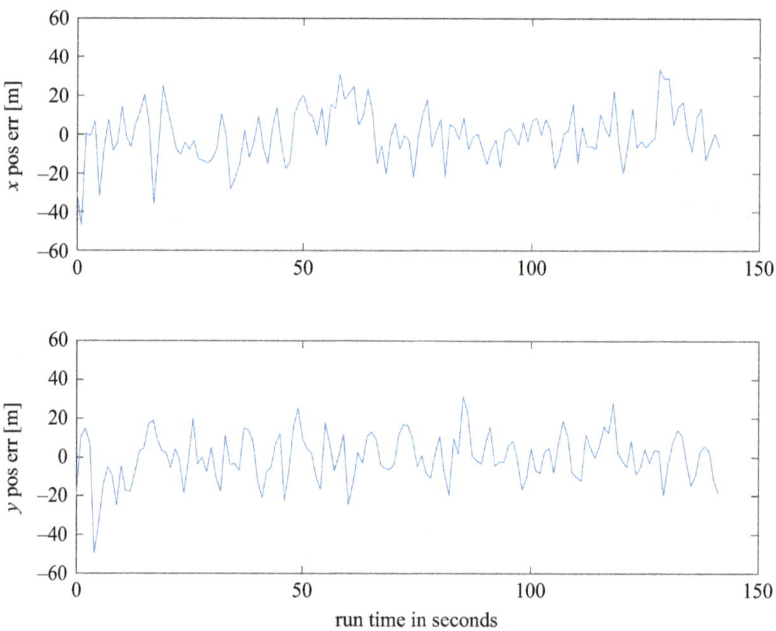

Figure 15.13 Kalman filter errors in estimated position with Q set to 'large' values

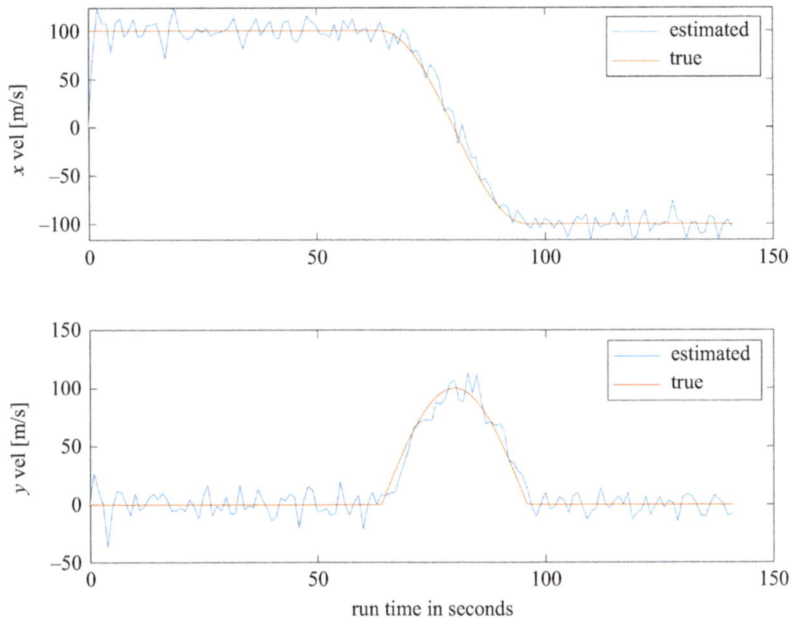

Figure 15.14 True and Kalman filter estimated velocity with Q set to 'large' values

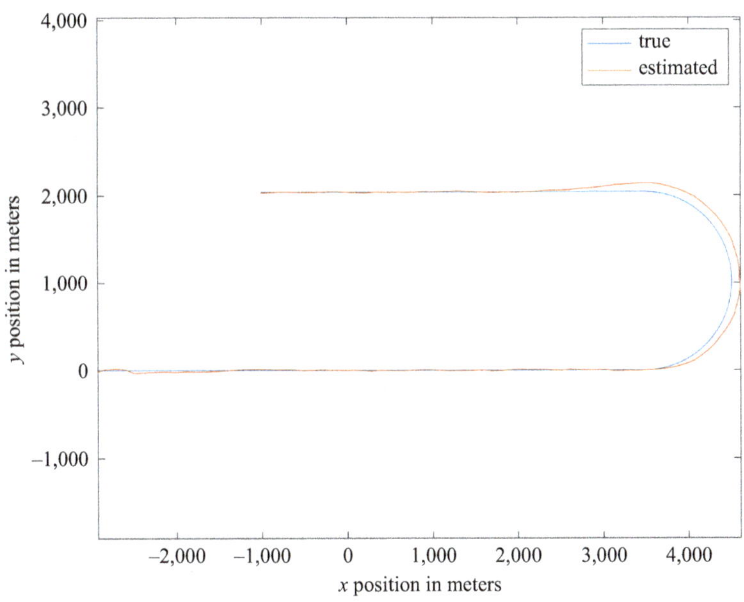

Figure 15.15 True flight path and Kalman filter estimates with Q set to 'small' values

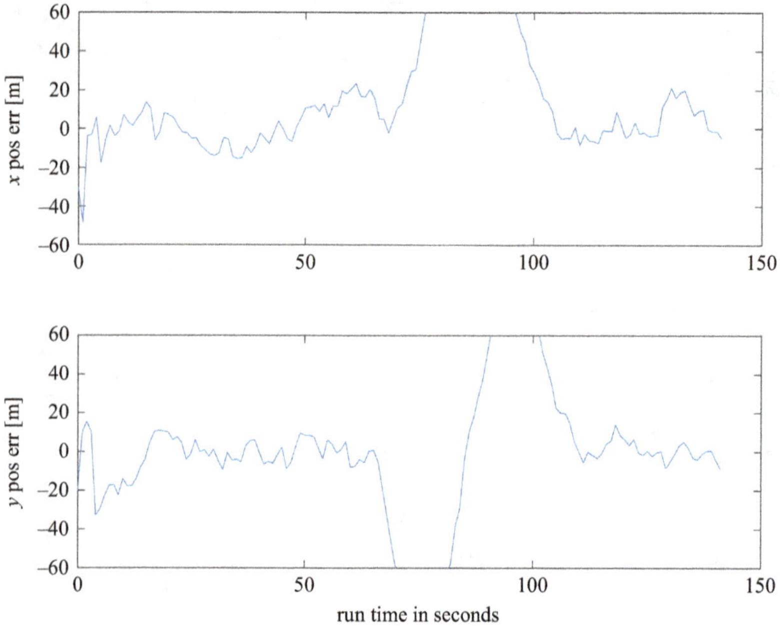

Figure 15.16 Kalman filter errors in estimated position with Q set to 'small' values

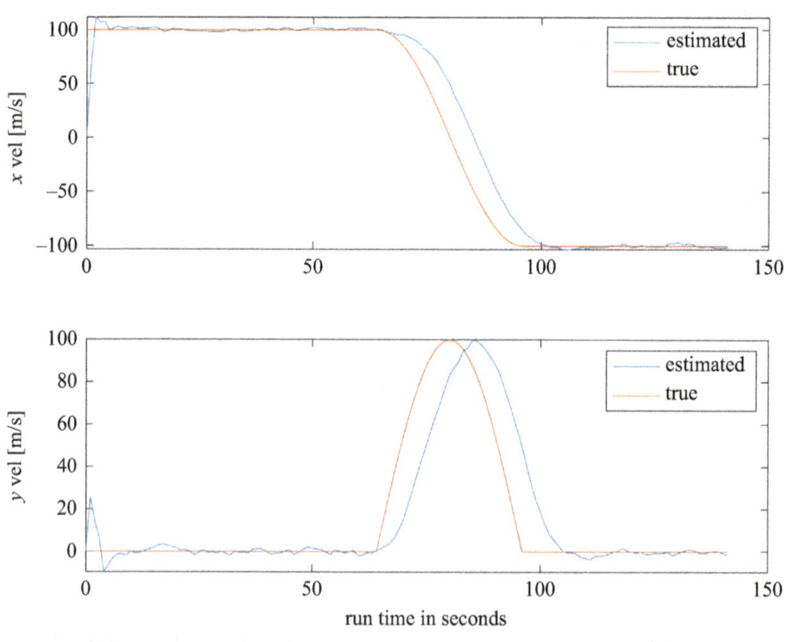

Figure 15.17 True and Kalman filter estimated velocity with Q set to 'small' values

is a large lag in the filter response during the turn that results in large bias errors. By placing too much weight on the (constant velocity) prediction, the filter ignores the measurements when they start to tell the filter that the vehicle is turning. Since we have set Q too small, the filter thinks the "diverging" measurements are just noise. If we optimally tune Q (in between these extremes), we would find that the Kalman filter performs just slightly better than an OLS estimator. As shown earlier, the Kalman filter has the advantage of optimally accounting for the geometry, but in this case, the geometry is good throughout the entire flight path and thus the filter's advantage would not be apparent. It would certainly be desirable to achieve low-noise estimates throughout the entire flight trajectory. As we will see later, the best way to achieve this is to integrate the radionavigation system with an inertial navigation system. We have a couple of major topics that need to be covered before we get there and in the next chapter we will extend the DME EKF to GNSS.

Reference

[1] Brown RG, Hwang PYC. *Introduction to Random Signals and Applied Kalman Filtering: with MATLAB® exercises*. 4th ed. Hoboken, NJ: John Wiley & Sons, Inc.; 2012.

Chapter 16
GPS-only Kalman filtering

In the previous chapter, we introduced the linearized Kalman filter and the extended Kalman filter in the context of two-dimensional radionavigation-based positioning (specifically, multilateration DME). We can now apply these concepts to GNSS. It is useful to study GNSS-only Kalman filtering to understand its performance and limitations. Many GNSS receivers output a position solution produced by a Kalman filter and, within some inertial integration schemes, the system outputs three separate solutions: a GNSS-only solution, an inertial-only solution, and a GNSS-inertial integrated solution. Since our purpose here is not to delve into the details that differentiate the various GNSSs, we will focus our development primarily on GPS.

16.1 GPS versus DME

GPS is distinguished from DME in several ways. First, the GPS measurements are pseudoranges not ranges. In the case of DME, a measurement is formed of the range between the ground transponder and the airborne interrogator. In GPS, on the other hand, the receiver does not make a direct range measurement. Instead, the receiver forms a pseudorange measurement that is a measure of the total range plus whatever the receiver clock offset is. As a result, we must take the receiver clock offset into account when designing the Kalman filter.

In addition, GPS provides simultaneous independent measurements of pseudorange-rate as well as pseudorange. DME, by contrast, does not provide a range-rate measurement. Recall that the pseudorange-rate is derived from the Doppler-shifted carrier-frequency of the received signal and includes the receiver's clock drift rate (i.e., frequency offset). Since GPS provides simultaneous, independent, measurements of pseudorange-rate, the receiver can output a velocity solution that is independent of (i.e., not derived from) the position solution output.

Of course, another distinction between GPS and DME is that GPS provides a three-dimensional solution whereas DME only provides a horizontal solution. Finally, we also note that the GPS transmitters are constantly in motion whereas the DME ground transponders are fixed in concrete. This adds a bit of complexity since the real-time locations of the GPS satellites must be calculated by the receiver and/or integration software (as we will see later).

16.2 GPS receiver clock model

Since the GPS receiver clock offset (sometimes referred to as the clock phase offset or simply as the clock bias) is an unknown that must be solved for along with the three components of position, it must be included as a state in the Kalman filter. Since it is included as a state, we must determine a linear model that characterizes how the clock offset changes over time. In other words, we must determine the receiver clock offset portion of the overall Kalman system equation. One of the most commonly used clock models approximates the clock frequency offset as a random walk process and the clock phase offset (i.e., bias) as the sum of the integrated (frequency offset) random walk process and a separate random walk process. Figure 16.1 illustrates the block diagram of the continuous-time process for this so-called "two-state" model [1].

Recalling that s^{-1} is an integration block, the input to the block is thus the derivative of the output of the block. The continuous-time block diagram thus makes it convenient to write the continuous-time state equation by inspection:

$$\begin{bmatrix} \frac{d}{dt}\left\{clk_{ph}(t)\right\} \\ \frac{d}{dt}\left\{clk_{fr}(t)\right\} \end{bmatrix} = \begin{bmatrix} 0 & 1 \\ 0 & 0 \end{bmatrix} \begin{bmatrix} clk_{ph}(t) \\ clk_{fr}(t) \end{bmatrix} + \begin{bmatrix} w_{cf} \\ w_{ca} \end{bmatrix} \tag{16.1}$$

where clk_{ph} is the clock phase offset, clk_{fr} is the clock frequency offset, w_{ca} is the clock acceleration white noise, and w_{cf} is the clock frequency white noise. From this continuous-time state equation, we need to determine the discrete-time state vector, state transition matrix and system noise covariance matrix. These will subsequently be utilized within the larger Kalman filters.

It is common for the clock states to be expressed (and estimated) in range and range-rate units. Recall the pseudorange bias due to the receiver clock offset is simply the receiver clock phase offset multiplied by a constant (the speed of light: c); similarly the pseudorange-rate bias due to clock frequency offset is simply the clock frequency offset multiplied by another constant (the carrier wavelength: $\lambda = \frac{c}{f_c}$). Thus, the clock bias is typically expressed in units of meters.

$$b_u = c \cdot clk_{ph} \tag{16.2}$$

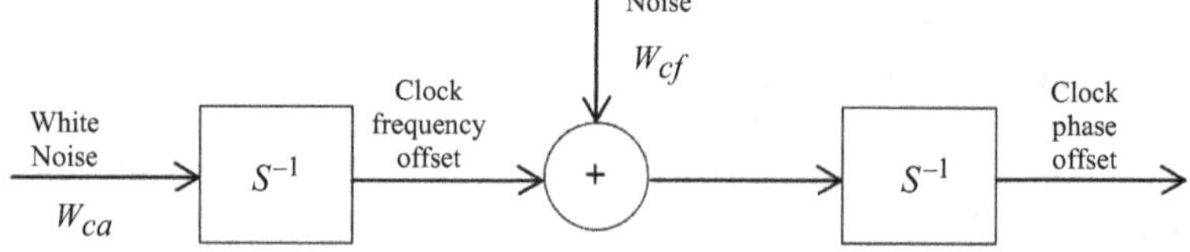

Figure 16.1 Continuous-time model for the GPS receiver clock phase and frequency offset

and the clock frequency offset (which is also the clock bias *rate*) in units of meters-per-second:

$$\dot{b}_u = \frac{c}{f_c} clk_{fr} \tag{16.3}$$

where the "*u*" represents "user" (i.e., the receiver). The discrete-time state vector for the two-state clock model is thus:

$$\underline{x}_k = \begin{bmatrix} b_u & \dot{b}_u \end{bmatrix}^T \tag{16.4}$$

From the system dynamics matrix specified in Equation (16.1), it can be shown that the corresponding discrete-time state transition matrix is given by:

$$\Phi_k = \begin{bmatrix} 1 & \Delta t \\ 0 & 1 \end{bmatrix} \tag{16.5}$$

The system noise covariance matrix is derived in the books by Brown and Hwang [1] and Groves [3] (see, e.g., lower-right 2 × 2 submatrix in Equation (9.152) of Groves). The result is:

$$Q_k = \begin{bmatrix} s_{ph}\Delta t + \frac{(\Delta t)^3}{3} s_{fr} & \frac{(\Delta t)^2}{2} s_{fr} \\ \frac{(\Delta t)^2}{2} s_{fr} & s_{fr}\Delta t \end{bmatrix} \tag{16.6}$$

where s_{ph} is the spectral density of the phase offset and s_{fr} is the spectral density of the frequency offset. Groves ([3, p. 418]) notes that typical values for the temperature-compensated crystal oscillators (TCXOs) commonly used in GNSS receivers are $s_{ph} \approx 0.01 \frac{\text{m}^2}{\text{s}}$ and $s_{fr} \approx 0.04 \frac{\text{m}^2}{\text{s}^3}$.

16.3 State selection for GPS-only Kalman filters

Depending upon the dynamics of the platform on which the GPS receiver is mounted, one of three GPS-only Kalman filters can be selected. The 5-state model estimates three dimensions of position (x, y, z) along with the two clock states (phase and frequency). This model is appropriate only for stationary and extremely low-dynamic (e.g., pedestrian) applications. For low dynamic applications (e.g., ground vehicles), an 8-state model should be used that estimates three dimensions of velocity along with the position and clock states. For medium dynamic applications (e.g., aircraft), an 11-state filter can be used which extends the 8-state filter by estimating three dimensions of acceleration. However, as we will see shortly, the 8-state filter works quite well if pseudorange-rate measurements are used in addition to pseudorange measurements. For high dynamic applications (e.g., rockets and missiles), specialized receivers are required that provide higher than normal data rates and lower than normal data latencies. We will not consider these specialized applications.

It is important not to lose sight of the fact that these are *statistical models* of how the states behave; they are *not* physical models. Remember the filter uses the Kalman system equation (specifically, the state transition matrix) to form a prediction of the states that is then weighted and combined with the weighted measurements to form the estimate. We will discuss this further as we explore some examples.

16.4 5-State GPS-only Kalman filter

The 5-state model estimates three dimensions of position along with the clock phase and frequency offsets. The clock model has already been described above. The model for each component of position is the random walk process illustrated in Figure 16.2. We are *not* assuming the actual position varies according to a random walk. Rather we are statistically modeling the difference between the true position and the nominal (predicted) position as a random walk. By combining the position and clock models, the continuous-time state equation for the 5-state system is given by:

$$\begin{bmatrix} \frac{d}{dt}\{\Delta x(t)\} \\ \frac{d}{dt}\{\Delta y(t)\} \\ \frac{d}{dt}\{\Delta z(t)\} \\ \frac{d}{dt}\{\Delta b(t)\} \\ \frac{d}{dt}\{\Delta \dot{b}(t)\} \end{bmatrix} = \begin{bmatrix} 0 & 0 & 0 & 0 & 0 \\ 0 & 0 & 0 & 0 & 0 \\ 0 & 0 & 0 & 0 & 0 \\ 0 & 0 & 0 & 0 & 1 \\ 0 & 0 & 0 & 0 & 0 \end{bmatrix} \begin{bmatrix} \Delta x(t) \\ \Delta y(t) \\ \Delta z(t) \\ \Delta b(t) \\ \Delta \dot{b}(t) \end{bmatrix} + \begin{bmatrix} w_x(t) \\ w_y(t) \\ w_z(t) \\ w_{cf}(t) \\ w_{ca}(t) \end{bmatrix} \tag{16.7}$$

where w_x, w_y, and w_z are the white noise driving the random walk processes of the delta position states. The states are delta quantities (i.e., incremental quantities) rather than whole-value quantities. The discrete-time system equation is then given by:

$$\begin{bmatrix} \Delta x \\ \Delta y \\ \Delta z \\ \Delta b \\ \Delta \dot{b} \end{bmatrix}_{k+1} = \begin{bmatrix} 1 & 0 & 0 & 0 & 0 \\ 0 & 1 & 0 & 0 & 0 \\ 0 & 0 & 1 & 0 & 0 \\ 0 & 0 & 0 & 1 & \Delta t \\ 0 & 0 & 0 & 0 & 1 \end{bmatrix}_k \begin{bmatrix} \Delta x \\ \Delta y \\ \Delta z \\ \Delta b \\ \Delta \dot{b} \end{bmatrix}_k + \begin{bmatrix} w_{px} \\ w_{py} \\ w_{pz} \\ w_{ph} \\ w_{fr} \end{bmatrix}_k \tag{16.8}$$

where the subscripts indicate the quantities are all in discrete-time. Equation (16.8) has the usual form of the Kalman system equation:

$$\underline{x}_{k+1} = \Phi_k \underline{x}_k + \underline{w}_k \tag{16.9}$$

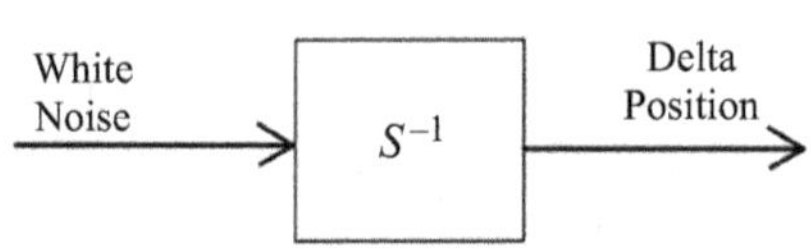

Figure 16.2 Continuous-time model for each dimension of position in the 5-state model. The white noise input is in the velocity domain and the position output is a random walk process. The output is actually "delta position" since at each update the filter is estimating the position offset from the nominal position

The 5-state model is *only* appropriate for stationary or extremely low dynamic platforms since the system equation does not account for any vehicle dynamics.

The purpose of using a random walk model for stationary position components is to create non-zero Q values such that the filter does not completely ignore the measurements. If Q is zero, eventually the filter will place all weight on its own predictions. In the process of doing so, numerical instabilities may also result. Since the delta position states are modeled as being statistically independent, the corresponding portion of the covariance matrix is diagonal only and consists of the integration of the noise spectral density over the update interval. Since the noise spectral density is also modeled as a constant, the integration is simply the spectral density multiplied by the time interval. The total Q matrix is then given by combination of the position state components with (16.6) [1]:

$$Q_k = \begin{bmatrix} s_p\Delta t & 0 & 0 & 0 & 0 \\ 0 & s_p\Delta t & 0 & 0 & 0 \\ 0 & 0 & s_p\Delta t & 0 & 0 \\ 0 & 0 & 0 & s_{ph}\Delta t + \frac{(\Delta t)^3}{3}s_{fr} & \frac{(\Delta t)^2}{2}s_{fr} \\ 0 & 0 & 0 & \frac{(\Delta t)^2}{2}s_{fr} & s_{fr}\Delta t \end{bmatrix} \tag{16.10}$$

where s_p is the spectral density of the position noise. Since the purpose of non-zero Q is to maintain numerical stability and prevent the filter from completely ignoring the measurements, s_p is typically small. Values on the order of $0.1\,\frac{\text{m}^2}{\text{s}}$ are acceptable.

The system equation has thus provided us with the state vector and state transition matrix (Equation (16.8)) and the system noise covariance matrix (Equation (16.10)). To determine the measurement equation, we will proceed in a manner very similar to that which we used for DME-based positioning in the previous chapter. We start with the non-linear equation that relates position to the pseudorange between the user and the ith satellite:

$$PR_i = \sqrt{(x_u - x_i)^2 + (y_u - y_i)^2 + (z_u - z_i)^2} + b_u \tag{16.11}$$

where, again, "u" refers to user (or whatever platform the receiver is on), x_i, y_i, and z_i are the coordinates of the ith satellite and b_u is the receiver clock phase offset (clock "bias") in units of range ($b_u = c \cdot clk_{ph}$). Expanding Equation (16.11) in a Taylor Series expansion about a nominal user state (x_o, y_o, z_o, b_o) and retaining only the zeroth and first-order terms results in the following approximation:

$$\begin{aligned} PR_i \approx PR_{io} &+ \frac{x_o - x_i}{r_{io}}(x_u - x_o) + \frac{y_o - y_i}{r_{io}}(y_u - y_o) \\ &+ \frac{z_o - z_i}{r_{io}}(z_u - z_o) + 1 \cdot (b_u - b_o) \end{aligned} \tag{16.12}$$

where

$$PR_{io} = \sqrt{(x_o - x_i)^2 + (y_o - y_i)^2 + (z_o - z_i)^2} + b_o \tag{16.13}$$

and

$$r_{io} = \sqrt{(x_o - x_i)^2 + (y_o - y_i)^2 + (z_o - z_i)^2} \tag{16.14}$$

Define the difference between the nominal and true user state as:

$$\begin{bmatrix} \Delta x \\ \Delta y \\ \Delta z \\ \Delta b \end{bmatrix} = \begin{bmatrix} x_u - x_o \\ y_u - y_o \\ z_u - z_o \\ b_u - b_o \end{bmatrix} \tag{16.15}$$

Substituting (16.15) into (16.12) yields:

$$PR_i \approx PR_{io} + \frac{x_o - x_i}{r_{io}}\Delta x + \frac{y_o - y_i}{r_{io}}\Delta y + \frac{z_o - z_i}{r_{io}}\Delta z + 1 \cdot \Delta b \tag{16.16}$$

A linear equation in the unknowns (position and clock deltas) is created by moving the zeroth order term (which is a constant) to the left side of the equation. First we define the difference between the measured pseudorange and the pseudorange computed between the satellite and the nominal position:

$$\Delta PR_i = PR_i - PR_{io} \tag{16.17}$$

With (16.17), moving the zeroth order term in (16.16) yields the linear equation:

$$\Delta PR_i = \frac{x_o - x_i}{r_{io}}\Delta x + \frac{y_o - y_i}{r_{io}}\Delta y + \frac{z_o - z_i}{r_{io}}\Delta z + 1 \cdot \Delta b \tag{16.18}$$

Since there are four unknowns in (16.18), we need at least four pseudorange measurements. Gathering N instances of (16.18) yields:

$$\begin{bmatrix} \Delta PR_1 \\ \Delta PR_2 \\ \vdots \\ \Delta PR_N \end{bmatrix} = \begin{bmatrix} \left(\frac{x_o - x_1}{r_{1o}}\right) & \left(\frac{y_o - y_1}{r_{1o}}\right) & \left(\frac{z_o - z_1}{r_{1o}}\right) & 1 \\ \left(\frac{x_o - x_2}{r_{2o}}\right) & \left(\frac{y_o - y_2}{r_{2o}}\right) & \left(\frac{z_o - z_2}{r_{2o}}\right) & 1 \\ \vdots & \vdots & \vdots & \vdots \\ \left(\frac{x_o - x_N}{r_{No}}\right) & \left(\frac{y_o - y_N}{r_{No}}\right) & \left(\frac{z_o - z_N}{r_{No}}\right) & 1 \end{bmatrix} \begin{bmatrix} \Delta x \\ \Delta y \\ \Delta z \\ \Delta b \end{bmatrix} \tag{16.19}$$

which can be written more succinctly as:

$$\underline{\Delta PR} = H_{pos}\underline{\Delta x} \tag{16.20}$$

In reality, there is noise on the measurements and an estimate of the state vector can be obtained using ordinary least squares:

$$\hat{\underline{\Delta x}}_{OLS} = (H_{pos}^T H_{pos})^{-1} H_{pos}^T \underline{\Delta PR} \tag{16.21}$$

However, we are designing a Kalman filter to achieve enhanced performance. Since we are designing a 5-state filter that estimates the receiver clock frequency offset along with the position and clock phase offset states (recall the state vector was defined in Equation (16.8)), Equation (16.19) needs to be expanded to form the Kalman measurement equation:

$$\begin{bmatrix} \Delta PR_1 \\ \Delta PR_2 \\ \vdots \\ \Delta PR_N \end{bmatrix} = \begin{bmatrix} \left(\frac{x_o - x_1}{r_{1o}}\right) & \left(\frac{y_o - y_1}{r_{1o}}\right) & \left(\frac{z_o - z_1}{r_{1o}}\right) & 1 & 0 \\ \left(\frac{x_o - x_2}{r_{2o}}\right) & \left(\frac{y_o - y_2}{r_{2o}}\right) & \left(\frac{z_o - z_2}{r_{2o}}\right) & 1 & 0 \\ \vdots & \vdots & \vdots & \vdots & \vdots \\ \left(\frac{x_o - x_N}{r_{No}}\right) & \left(\frac{y_o - y_N}{r_{No}}\right) & \left(\frac{z_o - z_N}{r_{No}}\right) & 1 & 0 \end{bmatrix} \begin{bmatrix} \Delta x \\ \Delta y \\ \Delta z \\ \Delta b \\ \Delta \dot{b} \end{bmatrix} + \begin{bmatrix} \nu_1 \\ \nu_2 \\ \vdots \\ \nu_N \end{bmatrix} \tag{16.22}$$

where the vector of delta-pseudoranges (defined in Equation (16.17)) constitutes the Kalman observation vector $\underline{z}$ and the Nx5 matrix is the observation matrix (H). Notice the additional column of zeros that distinguishes the 5-state Kalman filter observation matrix from the earlier H_{pos}. The zeros denote the fact that there is no direct relationship between pseudorange and clock-frequency offset.

Since the noise on the pseudorange measurements is independent, the measurement noise covariance matrix is diagonal:

$$R = \begin{bmatrix} \sigma_1^2 & 0 & \cdots & 0 \\ 0 & \sigma_2^2 & \cdots & 0 \\ \vdots & \vdots & \ddots & 0 \\ 0 & 0 & 0 & \sigma_N^2 \end{bmatrix} \tag{16.23}$$

Enhanced fidelity can be achieved by setting the variances of each pseudorange in (16.23) in inverse proportion to the carrier-to-noise ratio (assuming a reliable measurement is provided by the receiver). A very simplified approximation can also be achieved by setting the variances in inverse proportion to the sine of the elevation angle. The presence of correlated errors (e.g., multipath, tropospheric delay, ionospheric delay) are not taken into account by the filter (but will be later when we integrate with the inertial system).

The total estimate is formed by adding the estimated deltas to the nominal states:

$$[\underline{\hat{x}}_{total}]_k = \begin{bmatrix} \hat{x}_u \\ \hat{y}_u \\ \hat{z}_u \\ \hat{b}_u \end{bmatrix} = \begin{bmatrix} x_o \\ y_o \\ z_o \\ b_o \end{bmatrix} + \begin{bmatrix} \hat{\Delta x} \\ \hat{\Delta y} \\ \hat{\Delta z} \\ \hat{\Delta b} \end{bmatrix} \tag{16.24}$$

where the hat symbol indicates an estimate.

Although the 5-state model is only appropriate for stationary or very low-dynamic platforms, we will still implement an extended Kalman filter and thus our current (total) estimate is used to form the next nominal (total) state. The system equation (16.9) is adapted to form the next nominal state from the current total estimate:

$$[\underline{x}_o]_{k+1} = \Phi_k[\underline{\hat{x}}_{total}]_k \tag{16.25}$$

where Φ_k was defined in Equation (16.8). Recall the noise in the system equation is neither known nor predictable and thus is ignored here but it is accounted for in the prediction error covariance matrix (P_{k+1}^-). Since we have used the total estimate to form the next nominal state, there is nothing left to predict and thus the state prediction is reset to zeros for the next moment in time:

$$\underline{\hat{x}}_{k+1}^- = \begin{bmatrix} 0 & 0 & 0 & 0 & 0 \end{bmatrix}^T \tag{16.26}$$

and the associated prediction error covariance matrix is computed in the usual way:

$$P_{k+1}^- = \Phi_k P_k^+ \Phi_k^T + Q_k \tag{16.27}$$

For the very first execution of the Kalman loop, the ordinary least-squares (OLS) solution can be used to determine the initial nominal state (except the clock frequency offset which can be initialized at zero). Accordingly, the initial prediction error covariance matrix can be determined on the basis of the OLS error covariance matrix:

$$P_{OLS} = \sigma_{rng}^2([H_{pos}^T H_{pos}])^{-1} \tag{16.28}$$

where it has been assumed that the variance of all pseudorange measurements (σ_{rng}^2) are equal. [A higher fidelity result can be obtained if a weighted least squares (WLS) solution is formed (instead of an OLS) using the variances of each of the pseudoranges.] The result provided by Equation (16.28) can be used to populate the upper left 4×4 submatrix of the 5-state Kalman filter initial prediction error covariance matrix. The remaining row and column can be set to zeros other than $P_0^-(5,5)$ which should be set according to the expected initial receiver clock frequency error.

One important point to keep in mind when computing quantities such as Equations (16.13) and (16.14) is the satellite motion (which was not an issue for DME since the ground transponders are mounted in concrete). Each satellite position needs to be computed at the signal time-of-transmission but the pseudorange measurements (and the position solution) are valid at the signal time-of-reception. For near-Earth receivers, there is an approximately 70 ms difference between the two. Since the satellites are traveling several kilometers per second, the difference is not trivial. For users on the equator, there will be an east position error of approximately 30 m if the rotation of the Earth during this time period is not taken into account. Thus, the satellite positions used in these computations need to be computed at the time-of-transmission but then expressed in the earth-centered earth-fixed (ECEF) coordinate system at the time-of-reception. This is accomplished with a simple rotation matrix. This earth rotation correction is frequently ignored in computer simulations (obviously both in the generation of the simulated measurements and in the measurement processing).

Figures 16.3 and 16.4 illustrate the results of a simulation of a stationary GPS receiver implementing an OLS position solution and the extended Kalman filter described above. Pseudorange measurements were simulated with noise, multipath, ionospheric delay and tropospheric delay. The pseudorange noise was simulated with a standard deviation of 1 m. With respect to the nominal values, the total simulated tropospheric delay was reduced by 90 per cent and the total simulated ionospheric delay was reduced by 50 per cent to simulate the effects of the application of a standard tropospheric correction model along with the broadcast ionospheric correction model. Multipath error was simulated by a first-order Gauss–Markov process with a standard deviation of 1.6 m and a time-constant of 2 min.

We see that the Kalman filter shows substantial noise reduction over the OLS solution but it does not reduce the "biases" (the multipath and atmospheric delays). This is not surprising since these bias-like errors were not modeled in the filter. Why did we not model them in the filter? In a word, they are not "observable" (a topic we will explore more in a later chapter). Nevertheless the Kalman filter is still providing benefit. In addition to the noise reduction, it is important to remember the filter takes the satellite geometry into account. Before leaving this example, we should also make note of the fact that the atmospheric bias-like errors impact the vertical position and clock solutions significantly more so than the horizontal solution. This is a typical

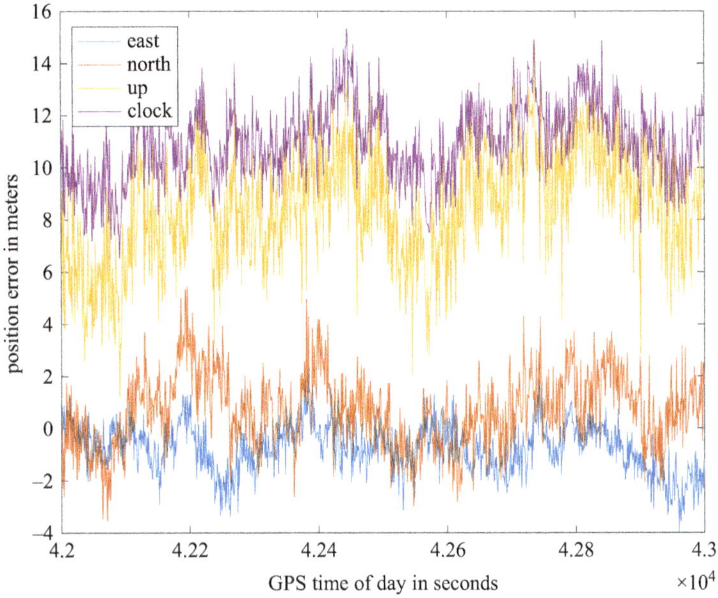

Figure 16.3 Ordinary least-squares position errors for a simulated stationary GPS receiver. The receiver clock offset (also known as a clock phase offset or a clock bias) is expressed in units of distance via implicit multiplication by the speed of light within the position solution

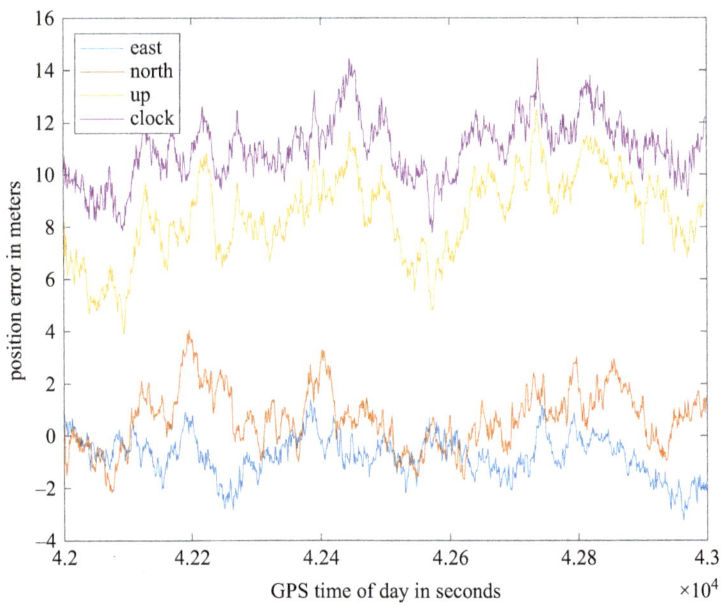

Figure 16.4 Extended Kalman filter position errors for a simulated stationary GPS receiver. The raw measurements are identical to those in Figure 16.3

result when HDOP is small (the satellites are roughly evenly spread in azimuth and thus the bias errors cancel out somewhat).

16.5 8-State GPS-only Kalman filter

The 5-state filter is appropriate for stationary or very low dynamic receivers. The next step up in terms of dynamic responsiveness is to model velocity states within the filter in addition to the position and clock states. In this case, it is the (incremental) velocity states that are modeled as random walk processes and thus the (incremental) position states are modeled as so-called "integrated random walk" processes as illustrated in Figure 16.5.

By inspection of Figure 16.5, along with the previously derived clock model, the continuous-time state equation for the 8-state system is given by:

$$\begin{bmatrix} \frac{d}{dt}\{\Delta x(t)\} \\ \frac{d}{dt}\{\Delta y(t)\} \\ \frac{d}{dt}\{\Delta z(t)\} \\ \frac{d}{dt}\{\Delta \dot{x}(t)\} \\ \frac{d}{dt}\{\Delta \dot{y}(t)\} \\ \frac{d}{dt}\{\Delta \dot{z}(t)\} \\ \frac{d}{dt}\{\Delta b(t)\} \\ \frac{d}{dt}\{\Delta \dot{b}(t)\} \end{bmatrix} = \begin{bmatrix} 0&0&0&1&0&0&0&0 \\ 0&0&0&0&1&0&0&0 \\ 0&0&0&0&0&1&0&0 \\ 0&0&0&0&0&0&0&0 \\ 0&0&0&0&0&0&0&0 \\ 0&0&0&0&0&0&0&0 \\ 0&0&0&0&0&0&0&1 \\ 0&0&0&0&0&0&0&0 \end{bmatrix} \begin{bmatrix} \Delta x(t) \\ \Delta y(t) \\ \Delta z(t) \\ \Delta \dot{x}(t) \\ \Delta \dot{y}(t) \\ \Delta \dot{z}(t) \\ \Delta b(t) \\ \Delta \dot{b}(t) \end{bmatrix} + \begin{bmatrix} 0 \\ 0 \\ 0 \\ w_{ax}(t) \\ w_{ay}(t) \\ w_{az}(t) \\ w_{cf}(t) \\ w_{ca}(t) \end{bmatrix} \tag{16.29}$$

where w_{ax}, w_{ay}, and w_{az} are the acceleration white noise inputs depicted in Figure 16.5 and w_{cf} and w_{ca} are the clock model noise terms described earlier. From the continuous-time state equation, we can derive the Kalman system equation. However, the key elements for the filter design are the state transition matrix and the system noise covariance matrix. Since the system dynamics matrix (F = the 8×8 matrix in

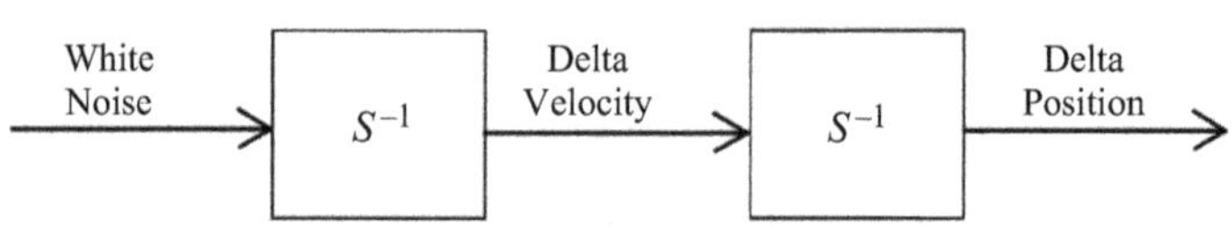

Figure 16.5 Continuous-time model for each dimension of velocity and position in the 8-state model. The white noise input is in the acceleration domain, the velocity is a random walk process and the position is an integrated random walk process

Equation (16.29)) is constant over the update interval (Δt), the state transition matrix can be obtained numerically:

$$\Phi = e^{F\Delta t} \tag{16.30}$$

It can also be shown that in closed form it is given by:

$$\Phi = \begin{bmatrix} 1 & 0 & 0 & \Delta t & 0 & 0 & 0 & 0 \\ 0 & 1 & 0 & 0 & \Delta t & 0 & 0 & 0 \\ 0 & 0 & 1 & 0 & 0 & \Delta t & 0 & 0 \\ 0 & 0 & 0 & 1 & 0 & 0 & 0 & 0 \\ 0 & 0 & 0 & 0 & 1 & 0 & 0 & 0 \\ 0 & 0 & 0 & 0 & 0 & 1 & 0 & 0 \\ 0 & 0 & 0 & 0 & 0 & 0 & 1 & \Delta t \\ 0 & 0 & 0 & 0 & 0 & 0 & 0 & 1 \end{bmatrix} \tag{16.31}$$

As shown in Groves [3, Equation (9.152)], the Q matrix is given by:

$$Q_k = \begin{bmatrix} S_a^e \frac{(\Delta t)^3}{3} & S_a^e \frac{(\Delta t)^2}{2} & 0_{3x1} & 0_{3x1} \\ S_a^e \frac{(\Delta t)^2}{2} & S_a^e \Delta t & 0_{3x1} & 0_{3x1} \\ 0_{1x3} & 0_{1x3} & S_{ph}\Delta t + \frac{(\Delta t)^3}{3} S_{fr} & \frac{(\Delta t)^2}{2} S_{fr} \\ 0_{1x3} & 0_{1x3} & \frac{(\Delta t)^2}{2} S_{fr} & S_{fr}\Delta t \end{bmatrix} \tag{16.32}$$

where 0_{ixj} refers to a matrix of zeros with i rows and j columns. The lower right 2×2 submatrix in (16.32) is the same as was previously discussed. As shown in Groves [3], the acceleration noise spectral density matrix, S_a^e, expressed in the GPS earth (e) frame (i.e., earth-centered earth-fixed coordinates) is given by:

$$S_a^e = C_n^e \begin{bmatrix} S_{aH} & 0 & 0 \\ 0 & S_{aH} & 0 \\ 0 & 0 & S_{aV} \end{bmatrix} C_e^n \tag{16.33}$$

where C_e^n is the earth-to-nav direction cosine matrix (DCM), $C_n^e = (C_e^n)^T$, S_{aH} is the horizontal acceleration spectral density and S_{aV} is the vertical acceleration spectral density. As noted by Groves [3], S_{aH} is on the order of $1\frac{\mathrm{m}^2}{\mathrm{s}^3}$ for marine vessels, $10\frac{\mathrm{m}^2}{\mathrm{s}^3}$ for automobiles and $100\frac{\mathrm{m}^2}{\mathrm{s}^3}$ for fighter aircraft. S_{aV} is very small for marine and land vehicles and is tuned primarily for numerical stability. For low to medium dynamic vehicles such as most civil aircraft, S_{aV} should be tuned according to the expected vertical acceleration profile.

We now go on to the Kalman measurement equation to obtain the observation vector, data matrix and measurement noise covariance matrix. Equation (16.22) can be easily expanded to incorporate the three velocity states:

$$\begin{bmatrix} \Delta PR_1 \\ \Delta PR_2 \\ \vdots \\ \Delta PR_N \end{bmatrix} = \begin{bmatrix} \left(\frac{x_o - x_1}{r_{1o}}\right) & \left(\frac{y_o - y_1}{r_{1o}}\right) & \left(\frac{z_o - z_1}{r_{1o}}\right) & 0 & 0 & 0 & 1 & 0 \\ \left(\frac{x_o - x_2}{r_{2o}}\right) & \left(\frac{y_o - y_2}{r_{2o}}\right) & \left(\frac{z_o - z_2}{r_{2o}}\right) & 0 & 0 & 0 & 1 & 0 \\ \vdots & \vdots & \vdots & \vdots & \vdots & \vdots & \vdots & \vdots \\ \left(\frac{x_o - x_N}{r_{No}}\right) & \left(\frac{y_o - y_N}{r_{No}}\right) & \left(\frac{z_o - z_N}{r_{No}}\right) & 0 & 0 & 0 & 1 & 0 \end{bmatrix} \begin{bmatrix} \Delta x \\ \Delta y \\ \Delta z \\ \Delta \dot{x} \\ \Delta \dot{y} \\ \Delta \dot{z} \\ \Delta b \\ \Delta \dot{b} \end{bmatrix} + \begin{bmatrix} \nu_1 \\ \nu_2 \\ \vdots \\ \nu_N \end{bmatrix} \tag{16.34}$$

The data matrix (H) is thus given by the $N \times 8$ matrix in (16.34) and the observation vector (z) on the left-hand side of equation (16.34) is the same N-element vector derived previously for the 5-state filter from the pseudorange measurements and the nominal state (see Equations (16.17) and (16.19)). Accordingly, the measurement noise covariance matrix is the same as before (Equation (16.23)).

In the same manner as was used with the 5-state filter, the total estimate is formed by adding the estimated deltas to the nominal states. Also, as before, the state transition matrix can be used to extrapolate the current estimate in order to form the next nominal state. Finally, again as before, since the total estimate was used to form the next nominal state, the state prediction is reset to zeros for the next moment in time and the associated prediction error covariance matrix is computed in the usual way.

For the very first execution of the Kalman loop, the ordinary least-squares (OLS) solution can be used to determine the position and clock phase values of the initial nominal state. The clock frequency offset and velocity states can be initialized at zero or a few consecutive OLS position and clock estimates can be used to form crude initial estimates. As before, the initial prediction error covariance matrix can be determined on the basis of the OLS error covariance matrix. The main diagonal elements associated with the velocity states ($P^-(4,4)$ through $P^-(6,6)$) can be conservatively set to the square of the maximum possible velocity of the platform (at the time when the filter is initialized).

Figures 16.6–16.8 show the results of the 8-state filter with a simulated F-16 flight path. The dog-leg flight path simulates the F-16 initially flying east at 200 m/s, executing a 180 degree 1-g left turn and then proceeding back to the west (all at constant velocity). For the purposes of computing the system noise covariance matrix (Equations (16.32) and (16.33)), the horizontal acceleration spectral density (S_{aH}) was set to $50\frac{\text{m}^2}{\text{s}^3}$ and the vertical acceleration spectral density (S_{aV}) was set to $10\frac{\text{m}^2}{\text{s}^3}$. The simulated noise on the pseudoranges was independent white Gaussian with a variance of $1\,\text{m}^2$, and thus R was set as a diagonal matrix with the main diagonal elements all set to 1.

Reasonable performance is obtained given the noisy and biased C/A-code measurements. The horizontal velocity error excursions observed in Figure 16.8 occur during the turn and are the direct result of mismodeling in the filter. Specifically, the model (Figure 16.5) does *not* account for deterministic acceleration in the flight profile. The presence of sustained non-zero-mean acceleration is inappropriately

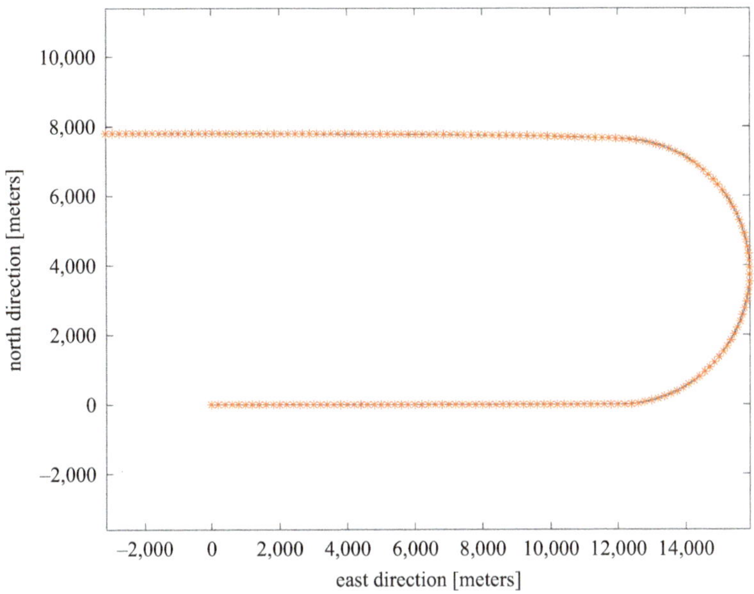

Figure 16.6 True and estimated trajectories with a simulated F-16 flight path. The trajectory starts at the origin. True velocity is 200 m/s throughout and the left turn is executed at 1-g acceleration

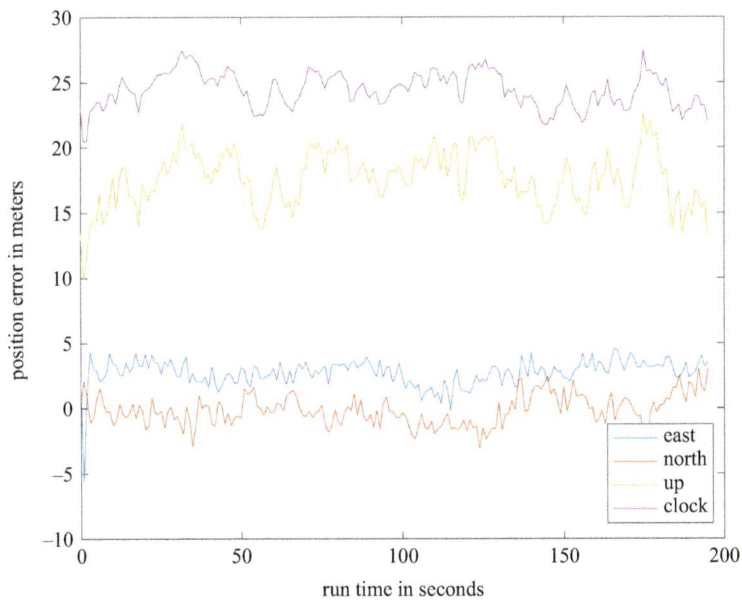

Figure 16.7 8-state Kalman filter position errors for the flight path depicted in Figure 16.6. GPS C/A-code measurements were simulated with thermal noise, multipath, ionosphere, and tropospheric delay (iono reduced by 50 per cent and tropo reduced by 90 per cent to simulate the effects of standard correction models)

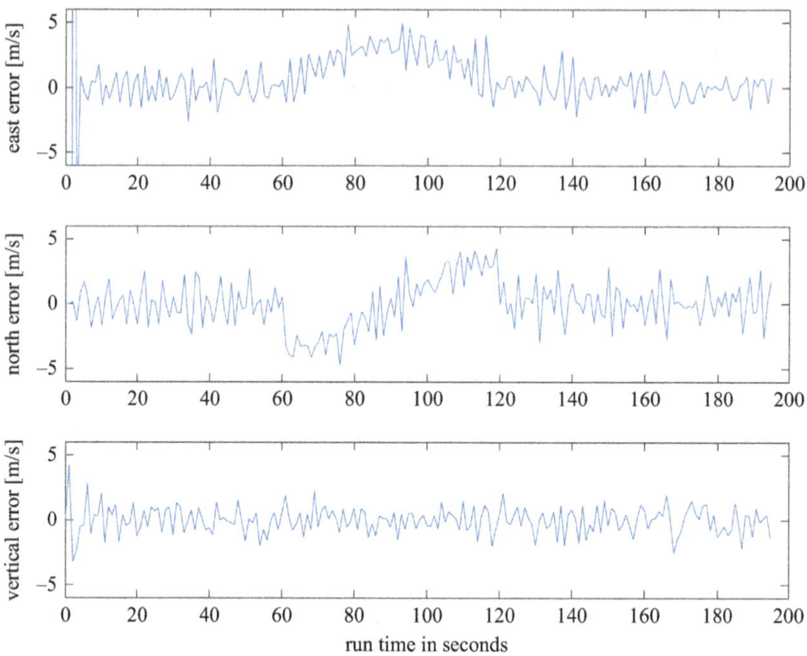

Figure 16.8 8-State Kalman filter velocity errors for the flight path depicted in Figure 16.6

treated as noise by the filter and divergence results for the duration of the turn. Thus the filter is only appropriate for low-dynamic vehicles (e.g., automobiles, but not race cars). There are two potential solutions when considering medium-dynamic trajectories. One is to expand the filter by another three states to model acceleration explicitly. As described in [1], a first-order Gauss–Markov model is needed for the acceleration states since acceleration is not sustained in any practical vehicle (i.e., a random walk process would be inappropriate since it leads to long-term non-zero values). Thus, an 11-state model can be derived that significantly reduces the velocity error excursions exhibited by the 8-state filter. However, for any vehicle capable of medium dynamics, a superior solution can be obtained by utilizing the Doppler (e.g., pseudorange-rate) and/or carrier-phase measurements provided by the receiver. GPS receivers measure the pseudo-Doppler-shifted carrier of the received signal. The term "pseudo" is used since the observable consists of the true Doppler shift plus the receiver's clock frequency error. The pseudo-Doppler can be converted to units of velocity through multiplication by the nominal carrier wavelength. The result is thus a pseudo-range-rate:

$$PRR = -\lambda_c f_{pseudo-Doppler} \tag{16.35}$$

where λ_o is the nominal carrier wavelength, the pseudo-Doppler frequency is expressed in units of Hz and the pseudo-range-rate is expressed in distance-units per second. The negative sign is needed since a positive Doppler shift corresponds to a *decreasing* range between the satellite and the receiver. The true Doppler component of pseudo-range-rate consists of the velocity vectors of the receiver and satellite projected onto the line-of-sight between the receiver and satellite. The pseudo-range-rate observation equation is thus given by:

$$PRR_i = -(\underline{v}_u \cdot \underline{\hat{u}}_i - \underline{v}_i \cdot \underline{\hat{u}}_i) + b^{\cdot}_u \tag{16.36}$$

where $\underline{v}_u$ is the velocity vector of the user, $\underline{v}_i$ is the velocity vector of the ith satellite, $\underline{\hat{u}}_i$ is the unit vector from the receiver to the ith satellite, and $b^{\cdot}_u$ is the receiver clock frequency offset in units of velocity. Although counterintuitive at first, a positive receiver clock frequency offset (i.e., clock frequency error) corresponds to a positive component of pseudo-range-rate by way of Equation (16.35) since it induces a negative component of the pseudo-Doppler measurement.

A linear expression for the unknowns can be formed by moving the known satellite velocity term of (16.36) to the left side of the equation. Let us define:

$$\delta PRR_i = PRR_i - \underline{v}_i \cdot \underline{\hat{u}}_i \tag{16.37}$$

Substituting (16.37) into (16.36) yields:

$$\delta PRR_i = -\underline{v}_u \cdot \underline{\hat{u}}_i + b^{\cdot}_u \tag{16.38}$$

Equation (16.38) can be rewritten in component form as:

$$\delta PRR_i = -v_{ux}\hat{u}_{ix} - v_{uy}\hat{u}_{iy} - v_{uz}\hat{u}_{iz} + b^{\cdot}_u \tag{16.39}$$

$$= -v_{ux}\left(\frac{x_i - x_u}{r_i}\right) - v_{uy}\left(\frac{y_i - y_u}{r_i}\right) - v_{uz}\left(\frac{z_i - z_u}{r_i}\right) + b^{\cdot}_u \tag{16.40}$$

which can be simplified:

$$\delta PRR_i = v_{ux}\left(\frac{x_u - x_i}{r_i}\right) + v_{uy}\left(\frac{y_u - y_i}{r_i}\right) + v_{uz}\left(\frac{z_u - z_i}{r_i}\right) + b^{\cdot}_u \tag{16.41}$$

where (x_u, y_u, z_u) are the receiver position components and r_i is the computed distance from the receiver to satellite i. As long as the nominal position is close to the truth, the same nominal position discussed earlier for the position solution can be utilized. Gathering N instances of (16.41) yields:

$$\begin{bmatrix} \delta PRR_1 \\ \delta PRR_2 \\ \vdots \\ \delta PRR_N \end{bmatrix} = \begin{bmatrix} \left(\frac{x_u - x_1}{r_1}\right) & \left(\frac{y_u - y_1}{r_1}\right) & \left(\frac{z_u - z_1}{r_1}\right) & 1 \\ \left(\frac{x_u - x_2}{r_2}\right) & \left(\frac{y_u - y_2}{r_2}\right) & \left(\frac{z_u - z_2}{r_2}\right) & 1 \\ \vdots & \vdots & \vdots & \vdots \\ \left(\frac{x_u - x_N}{r_N}\right) & \left(\frac{y_u - y_N}{r_N}\right) & \left(\frac{z_u - z_N}{r_N}\right) & 1 \end{bmatrix} \begin{bmatrix} v_{ux} \\ v_{uy} \\ v_{uz} \\ b^{\cdot}_u \end{bmatrix} \tag{16.42}$$

which can be written more succinctly as:

$$\underline{\delta PRR} = H_{vel}\underline{v}_u \tag{16.43}$$

Comparison of (16.19) and (16.20) with (16.42) and (16.43) shows that $H_{vel} = H_{pos}$. If we incorporate the presence of measurement noise, we have:

$$\underline{\delta PRR} = H_{vel}\underline{v}_u + \underline{\varepsilon} \tag{16.44}$$

Obviously (16.44) can be solved using ordinary least-squares as we did earlier with the position solution (see Equation (16.21)). However, our main goal here is to improve the performance of the 8-state filter by adding the pseudorange-rate measurements along with the pseudorange measurements. Equation (16.42) cannot be incorporated directly into the Kalman measurement equation by simply expanding Equation (16.34). This is due to the fact that the state variables in (16.34) are incremental (rather than whole) quantities since the filter is an extended Kalman filter.

Equation (16.43) can be rewritten in terms of nominal and incremental velocity quantities:

$$\underline{\delta PRR} = H_{vel}\underline{v}_u = H_{vel}(\underline{v}_o + \underline{\Delta v}_u) \tag{16.45}$$

Since the nominal quantities are known, they can be moved to the left-hand side to form a new observable:

$$\underline{\Delta PRR} = \underline{\delta PRR} - H_{vel}\underline{v}_o \tag{16.46}$$

With $\underline{\delta PRR}$ computed using (16.37), Equation (16.46) thus provides the observation needed for the Kalman filter (see the bottom half of $\underline{z}$ in Equation (16.48) below). In order to obtain the full Kalman measurement equation, we substitute (16.46) into (16.45) yielding

$$\underline{\Delta PRR} = H_{vel}\underline{\Delta v}_u \tag{16.47}$$

With the unknowns given in terms of incremental quantities, Equation (16.47) can be used to expand (16.34) to determine the Kalman measurement equation as follows:

$$\begin{bmatrix} \Delta PR_1 \\ \Delta PR_2 \\ \vdots \\ \Delta PR_N \\ \Delta PRR_1 \\ \Delta PRR_2 \\ \vdots \\ \Delta PRR_N \end{bmatrix} = \begin{bmatrix} (\frac{x_o-x_1}{r_{1o}}) & (\frac{y_o-y_1}{r_{1o}}) & (\frac{z_o-z_1}{r_{1o}}) & 0 & 0 & 0 & 1 & 0 \\ (\frac{x_o-x_2}{r_{2o}}) & (\frac{y_o-y_2}{r_{2o}}) & (\frac{z_o-z_2}{r_{2o}}) & 0 & 0 & 0 & 1 & 0 \\ \vdots & \vdots & \vdots & \vdots & \vdots & \vdots & \vdots & \vdots \\ (\frac{x_o-x_N}{r_{No}}) & (\frac{y_o-y_N}{r_{No}}) & (\frac{z_o-z_N}{r_{No}}) & 0 & 0 & 0 & 1 & 0 \\ 0 & 0 & 0 & (\frac{x_o-x_1}{r_{1o}}) & (\frac{y_o-y_1}{r_{1o}}) & (\frac{z_o-z_1}{r_{1o}}) & 0 & 1 \\ 0 & 0 & 0 & (\frac{x_o-x_2}{r_{2o}}) & (\frac{y_o-y_2}{r_{2o}}) & (\frac{z_o-z_2}{r_{2o}}) & 0 & 1 \\ \vdots & \vdots & \vdots & \vdots & \vdots & \vdots & \vdots & \vdots \\ 0 & 0 & 0 & (\frac{x_o-x_N}{r_{No}}) & (\frac{y_o-y_N}{r_{No}}) & (\frac{z_o-z_N}{r_{No}}) & 0 & 1 \end{bmatrix} \begin{bmatrix} \Delta x \\ \Delta y \\ \Delta z \\ \Delta \dot{x} \\ \Delta \dot{y} \\ \Delta \dot{z} \\ \Delta b \\ \Delta b^{\cdot} \end{bmatrix} + \begin{bmatrix} \nu_1 \\ \nu_2 \\ \vdots \\ \nu_N \\ \varepsilon_1 \\ \varepsilon_2 \\ \vdots \\ \varepsilon_N \end{bmatrix} \tag{16.48}$$

The $2N \times 8$ matrix in (16.48) is the H matrix for the 8-state filter with pseudorange and pseudorange-rate measurements. The vector of delta pseudorange and delta pseudorange-rate observations is the measurement vector $\underline{z}$. The measurement noise covariance matrix is the covariance of the noise vector on the right side of (16.48). In most practical implementations, the covariances between the observations is ignored thus yielding a diagonal matrix:

$$R = \begin{bmatrix} \sigma_{PR_1}^2 & 0 & \cdots & 0 & 0 & \cdots & 0 \\ 0 & \ddots & \cdots & 0 & 0 & \cdots & 0 \\ 0 & \ddots & \sigma_{PR_N}^2 & 0 & 0 & \cdots & 0 \\ 0 & \ddots & 0 & \sigma_{PRR_1}^2 & 0 & \cdots & 0 \\ 0 & \ddots & 0 & 0 & \ddots & \cdots & 0 \\ 0 & \cdots & 0 & 0 & \cdots & \cdots & \sigma_{PRR_N}^2 \end{bmatrix} \quad (16.49)$$

The noise variances are dependent upon the carrier-to-noise ratio and the receiver architecture (i.e., tracking loop orders and bandwidths). Pseudorange noise standard deviations are on the order of 1 m if the code tracking loop is unaided and on the order of 0.2 m if it is aided by the carrier tracking loop. The pseudorange-rate noise standard deviation is on the order of 0.05–0.1 m/s. For the same flight path shown earlier (Figure 16.6), 8-state filter results utilizing pseudorange-rate measurements along with pseudorange measurements are depicted in Figures 16.9 and 16.10. Comparison of Figures 16.8 and 16.10 shows that inclusion of the pseudorange-rate measurements eliminates the acceleration-induced errors in the velocity estimates. For the sake of completeness, it should be noted that the pseudorange-rate measurements were simulated with a higher level of noise than is typical (standard deviation of approximately 0.28 m/s). Nevertheless, the performance is significantly better than with pseudorange measurements alone.

The simulated pseudorange-rate measurements for this simulation were derived from simulated carrier-phase measurements made at the pseudorange epoch and 50 msec prior to the pseudorange epoch. Pseudorange-rate was computed by differencing the two carrier-phase measurements and dividing by the time interval. A true pseudo-Doppler measurement, however, is an average of measured Doppler over some time interval. Regardless of how you compute pseudorange-rate, however, it is not an instantaneous observation. We must remember that it is impossible to make instantaneous frequency measurements. The easiest way to understand this is to consider a sinusoid of unknown frequency. At least two zero-crossings must be timed in order to estimate a frequency. In other words, a single data point (say, one zero-crossing) is insufficient. Thus, the best we can do is to measure frequency over some finite interval of time. In practice, then, we actually measure average frequency over some time interval.

The pseudorange-rate measurement, then, is an average value rather than an instantaneous one. Also, its time of validity is best defined as the middle of the measurement interval. In the simulation just described, the pseudorange-rate is valid

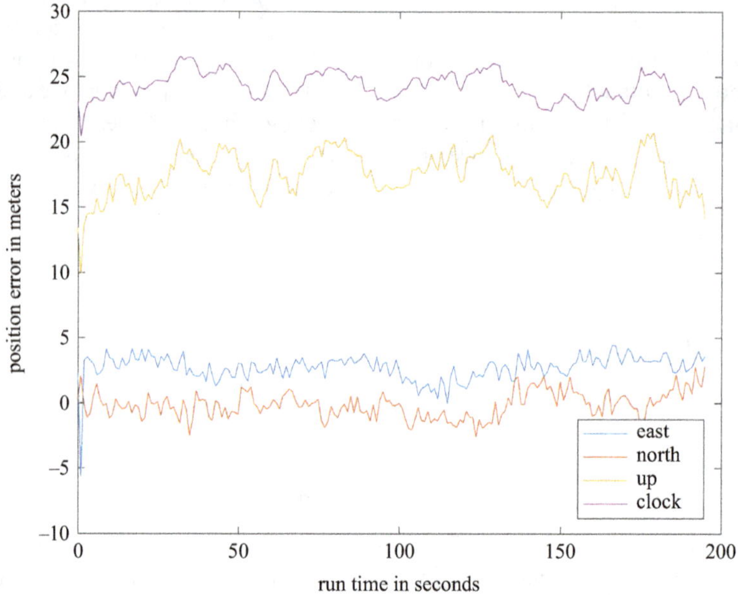

Figure 16.9 8-State Kalman filter results, utilizing both pseudorange and pseudorange-rate measurements, for the flight path depicted in Figure 16.6. The measurements were simulated with the same errors used in the pseudorange-only simulation (Figures 16.7 and 16.8)

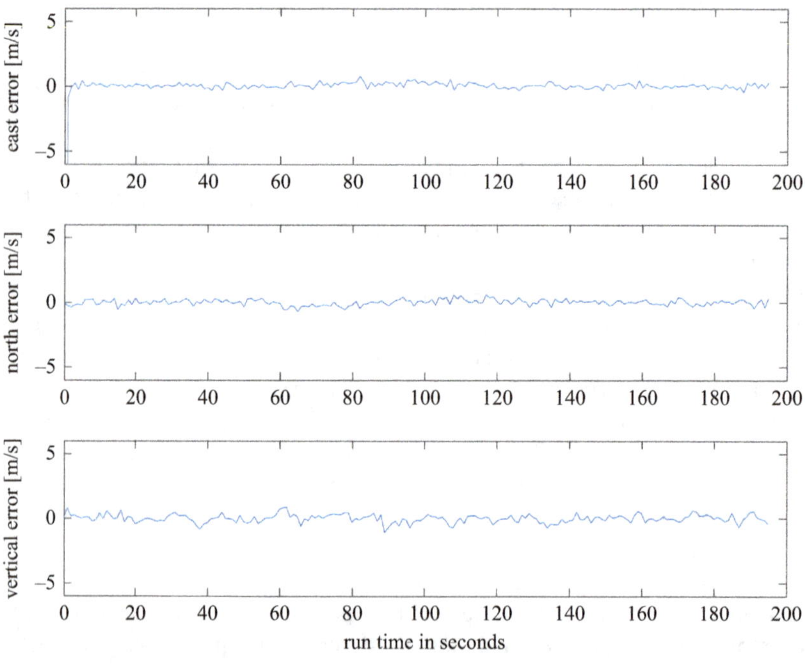

Figure 16.10 Velocity estimation errors for the 8-state filter (Figure 16.9) using both pseudorange and pseudorange-rate measurements

at a time 25 msec prior to the time of validity of the pseudorange measurement. In our filter development, however, we are implicitly assuming that both measurements are valid simultaneously. For the moderate dynamics of the 1-g turn in this scenario, the unmodeled 25 msec lag is negligible. However, we are also missing out on higher quality measurements that could be made if the pseudo-Doppler were measured/averaged over the entire interval between pseudorange measurements (typically 1 sec). These 1-sec interval Doppler measurements are typically known as "delta-range" measurements.

Longer averaging intervals yield lower-noise measurements. However, a half-second lag in the validity of the pseudorange-rate measurements is not negligible for moderate dynamic scenarios. This problem can be approached in at least a couple of different ways. The simplest is to average/interpolate the pseudorange measurements around the pseudorange-rate measurement interval and compute the solution at the pseudorange-rate time of validity. A more complex approach that does not impose a lag in the solution is to incorporate the known lag of the pseudorange-rate measurements into the Kalman measurement model. Finally, it should also be noted that even higher accuracy can be achieved if full carrier-phase measurements are available and used instead of delta-ranges or Doppler.

References

[1] Brown RG, Hwang PYC. *Introduction to Random Signals and Applied Kalman Filtering: with MATLAB® exercises*. 4th ed. Hoboken, NJ: John Wiley & Sons, Inc.; 2012.

[2] Gelb A, editor. *Applied Optimal Estimation*. Cambridge, MA: The M.I.T. Press; 1974.

[3] Groves P. *Principles of GNSS, Inertial, and Multisensor Integrated Navigation Systems*. 2nd ed. Boston, MA: Artech House; 2013.

Chapter 17
Introduction to inertial aiding

17.1 Introduction

We will begin our study of inertial aiding within the context of a one-dimensional problem. Specifically, we are assuming that we are instrumenting a vehicle that is on a perfectly flat, perfectly straight track. Real examples of this are the rocket sleds operated by the United States military in the desert southwest of the country. The sleds are equipped with rockets that enable the sled to accelerate very quickly to very high speeds and travel a few miles before running into water that causes braking action to slow it down. They test various sensors with these sleds. Although real rocket sleds have some lateral motion (sway and vibration), we are going to consider an idealized simplified sled that only moves in one dimension: on a perfectly flat, perfectly straight track (as you will see, we will also take extensive liberties with the length of the track).

Although the real tracks have sensors on the tracks themselves for highly-precise position determination, we are going to assume we have an inertial navigation system on the sled itself. Since it is moving in just one dimension, we only need one sensor: an accelerometer. As long as the accelerometer is mounted parallel to the track, velocity is the result of integrating specific force once and distance along the track is obtained from an additional integration. We will then make the scenario a bit more interesting by simulating errors in the accelerometer. Finally, we will simulate an external aiding source and will input its measurements into a Kalman filter that will estimate the inertial errors so that they can be corrected.

17.2 Simulated Trajectory 1

We will utilize several trajectories in this chapter but will start by simulating a vehicle trajectory with the following parameters:

- $P_0 = 50$ m (initial distance along the track)
- $V_0 = 10\ \frac{\mathrm{m}}{\mathrm{s}}$ (initial velocity of the sled)

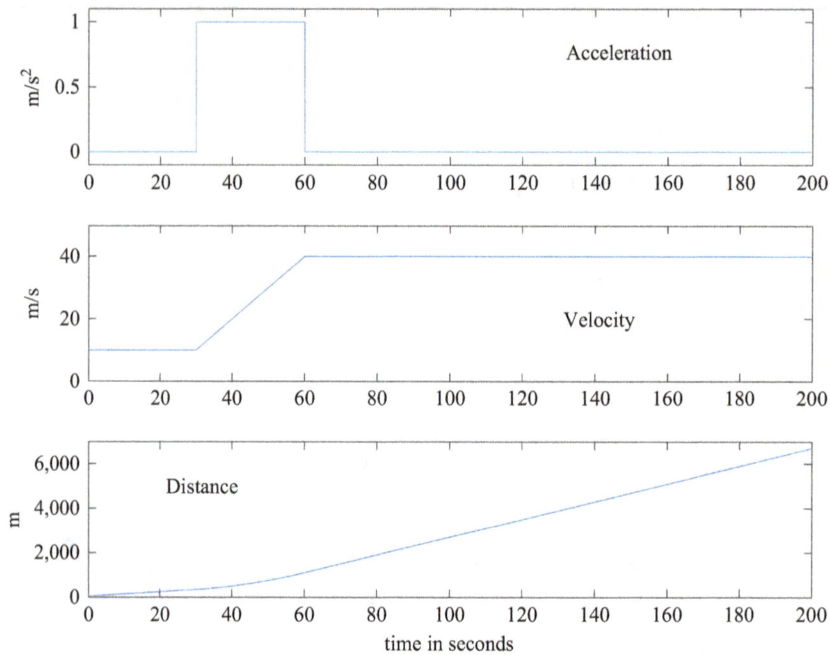

Figure 17.1 Acceleration, velocity and along-track distance profiles of simulated rocket sled Trajectory 1

- $A_0 = 0$; a period of constant acceleration ($1\ \frac{\text{m}}{\text{s}^2}$) occurs from $t = 30$ sec to $t = 60$ sec
- Total simulated trajectory: 200 sec

The acceleration, velocity, and position profiles of Trajectory 1 are depicted in Figure 17.1.

17.3 Inertial navigation simulation for the sled track

The inertial navigation system consists of a single sensor: the accelerometer. One of the dominant errors in an inertial sensor is the uncompensated bias (i.e., the residual bias that exists after all calibrations and compensations have been applied). For the first simulation that we will perform, the accelerometer bias is set approximately to 10 milli-g (specifically, $0.1\ \frac{\text{m}}{\text{s}^2}$). Thus, a "tactical" grade sensor is being simulated. There is no scale factor error, there is no noise and there are no g effects. The total accelerometer error is simply a 10 milli-g bias. The velocity and distance determined by the inertial system is depicted in Figure 17.2 and the errors are shown in Figure 17.3. As expected with a constant acceleration error, the velocity error growth is linear and the distance error growth is quadratic.

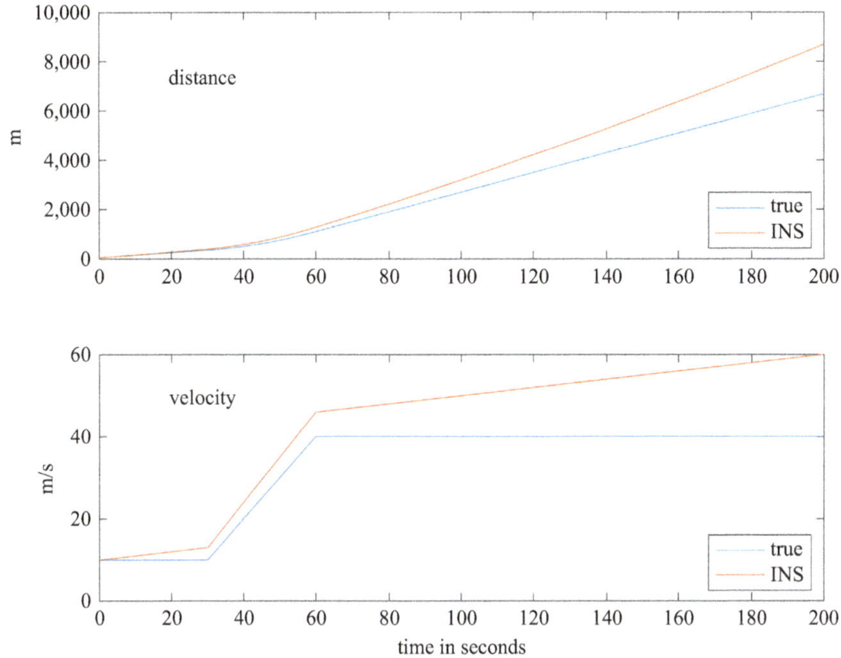

Figure 17.2 Along-track distance and velocity (truth and INS output) for the trajectory depicted in Figure 17.1 for an accelerometer with a 10 milli-g bias

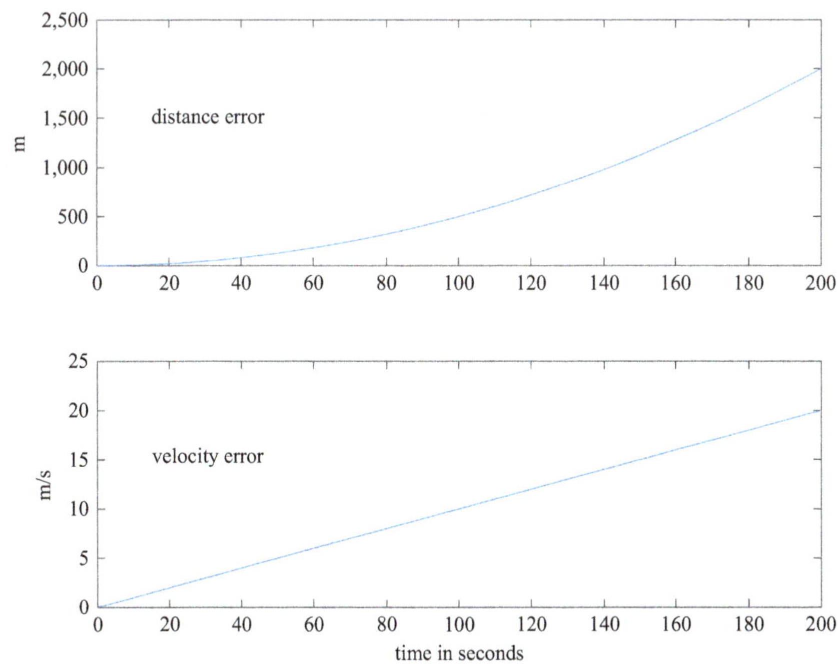

Figure 17.3 Inertial distance and velocity errors with a 10 milli-g accelerometer bias for the simulation depicted in Figure 17.2

17.4 Complementary filtering

As was discussed at length earlier in this book and as the above simulation demonstrates, inertial navigation systems exhibit errors that drift (grow) over time. We seek to incorporate some external measurements in order to constrain the error growth. Since the inertial errors are predominantly low-frequency, ideally the external sensor exhibits predominantly high-frequency errors. Radio-navigation systems approximate this since thermal noise is one of the major components of their errors. We seek to exploit the low noise characteristics of the inertial system with the long-term stability (i.e., no drift) of the radio navigation systems.

It turns out that we can do this and get rid of nonlinear system dynamics *at the same time* through what is called a "complementary filter." In general, the complementary filter assumes that you have two different measurements of the same quantity, and each of the measurements has distinctly different spectral characteristics. One measurement system has predominantly low-frequency errors and the other system has predominantly high-frequency errors:

$$m_1(t) = s(t) - n_1(t) \tag{17.1}$$

$$m_2(t) = s(t) + n_2(t) \tag{17.2}$$

where $s(t)$ is the desired signal (quantity) of interest. The noise on measurement system 1 (n_1) is composed of low-frequency errors and the noise on measurement system 2 is composed of high-frequency errors. The negative sign on n_1 is just a mathematical convenience (as we will see shortly) and does not affect the results. If it bothers you, just remember that the negative of noise is still noise.

We can isolate the noise terms by differencing the measurements:

$$\varepsilon = m_2(t) - m_1(t) = n_1(t) + n_2(t) \tag{17.3}$$

Thus, the observable formed by the difference of the two measurements consists of the *sum* of the errors (this is why n_1 was arbitrarily given a negative sign; although $n_2 - n_1$ is still a "sum" of noise terms, it is easier to think of a sum with a plus sign). Notice that the true signal has canceled out.

Now you may wonder how getting rid of the desired signal can be a good thing. Just hold on, we are not done yet. Since n_1 has a low pass characteristic and n_2 has a high pass characteristic, we can isolate n_1 by running the difference (ε) through a low-pass filter. That allows us to isolate one of the errors. What good is that? Since we have now estimated the error in the measurement of sensor 1, we can use it to correct the measurements from sensor 1 and reduce the overall error considerably.

This is depicted in Figure 17.4. In the upper left is the measurement from sensor 1 (truth minus n_1). In the lower left is the measurement from sensor 2 (truth plus n_2). These two measurements are input to the summing junction that performs the subtraction. The output of the summing junction is the sum of the two measurement errors. The sum is input to a low-pass filter and the output is approximately equal to n_1. Having isolated the sensor 1 error, it can be subtracted off of the raw sensor 1 output in order to correct the output and, ideally, yield the true value.

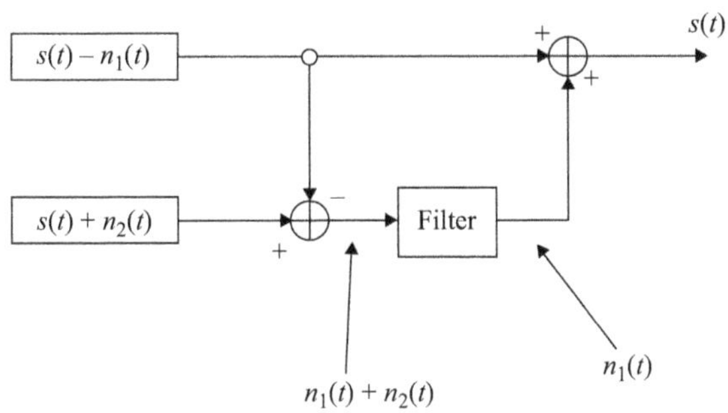

Figure 17.4 Complementary filter architecture

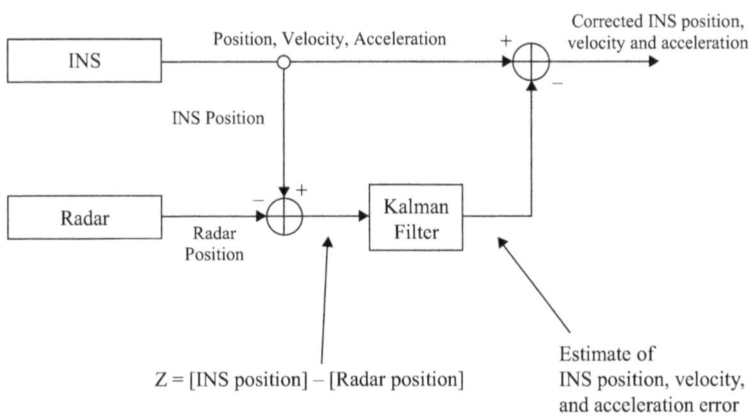

Figure 17.5 The complementary filter architecture as applied to the integration of the inertial system and a radar in the sled track example

17.5 Applying the complementary filter to navigation system integrations sled track case study

We use the complementary filter architecture for navigation system integration but we enhance it significantly by replacing the simple low-pass filter with a multi-state Kalman filter. Instead of just isolating the output errors of one sensor, the Kalman filter is also able to estimate the underlying error sources inside the sensor (or sensors). For our sled track example we will assume that, in addition to the inertial system, there is a radar providing independent measurements of distance along the track. The filter architecture, as applied to our sled track example, is illustrated in Figure 17.5.

The radar will be simulated as having independent white Gaussian measurement error with a standard deviation of 5 m. Additionally, the radar is mounted such that the measurements are perfectly parallel with the track and thus provide a completely independent measurement of the distance (or position or range) of the vehicle along the track.

As shown in Figure 17.5, the Kalman filter is providing estimates of inertial position and velocity error so that the output of the inertial system can be corrected (since this is a one-dimensional example, "position" is the same as distance-along-track). Since the accel bias is the driver of the inertial velocity and position errors, it needs to be estimated as well. The continuous-time state vector is thus:

$$\underline{x} = \begin{bmatrix} p_{err}(t) \\ v_{err}(t) \\ a_{bias}(t) \end{bmatrix} \tag{17.4}$$

The derivatives of the three state variables are given by the following equations (with the assumption that the accel bias is a perfect constant):

$$\frac{d}{dt} a_{bias}(t) = 0 \tag{17.5}$$

$$\frac{d}{dt} v_{err}(t) = a_{bias}(t) \tag{17.6}$$

$$\frac{d}{dt} p_{err}(t) = v_{err}(t) \tag{17.7}$$

The individual derivatives can be combined in the continuous-time state equation:

$$\dot{\underline{x}} = F\underline{x} = \begin{bmatrix} 0 & 1 & 0 \\ 0 & 0 & 1 \\ 0 & 0 & 0 \end{bmatrix} \begin{bmatrix} p_{err}(t) \\ v_{err}(t) \\ a_{bias}(t) \end{bmatrix} \tag{17.8}$$

Since the continuous-time system dynamics matrix (F) is a constant, the discrete-time state transition matrix can be computed via:

$$\Phi = e^{F\Delta t} \tag{17.9}$$

Strictly speaking, F need only be constant over the Kalman filter cycle (loop) interval (Δt) but in this case F is always a constant. In this simple case, there is also a closed form solution and thus the discrete-time Kalman filter system equation for this example is:

$$\begin{bmatrix} p_{err} \\ v_{err} \\ a_{bias} \end{bmatrix}_{k+1} = \begin{bmatrix} 1 & \Delta t & \frac{1}{2}(\Delta t)^2 \\ 0 & 1 & \Delta t \\ 0 & 0 & 1 \end{bmatrix} \begin{bmatrix} p_{err} \\ v_{err} \\ a_{bias} \end{bmatrix}_{k} + \begin{bmatrix} 0 \\ 0 \\ 0 \end{bmatrix} \tag{17.10}$$

where we see that the system noise vector is identically equal to zero. In this simple example where the accel bias is a pure constant and there are no other errors, this noise vector assumption is correct (and implies that Q is all zeros). Later we will explore more realistic scenarios but this simple one is a good place to start.

The system equation gives us half of the terms we need to design the filter. Specifically: the state vector (x), the state transition matrix (Φ) and the system noise covariance matrix (Q). To complete the design, we need to specify the measurement equation. Since this is a complementary filter architecture, the "measurement" or "observable" that is input to the Kalman filter is the difference between the INS position and the "position" (distance) measured by the radar:

$$\underline{z} = pos_{ins} - pos_{radar} \tag{17.11}$$

Equation (17.11) is the observable that will be input to the Kalman filter. For the sake of deriving the measurement equation, however, we *model* the measurements as the sum of the true position and the measurement errors. Thus:

$$\underline{z} = pos_{ins} - pos_{radar} \tag{17.12}$$

$$z = (pos_{true} + ins_{err}) - (pos_{true} + radar_{err}) \tag{17.13}$$

$$z = ins_{err} - radar_{err} \tag{17.14}$$

Given our definition of the state vector in this example, the Kalman filter measurement equation is thus:

$$\underline{z}_k = \begin{bmatrix} 1 & 0 & 0 \end{bmatrix} \begin{bmatrix} p_{err} \\ v_{err} \\ a_{bias} \end{bmatrix}_k + radar_{err} \tag{17.15}$$

where the first matrix in (17.15) is the H matrix (i.e., the data matrix) and the last term in the equation is the measurement noise vector ($\underline{\nu}$). Note that the two terms in (17.15) are scalars in this example. The negative sign on the radar error has been ignored since the sole purpose of the measurement noise vector is to enable us to define the measurement noise covariance matrix R. The sign is irrelevant since it does not have an impact on the covariance (actually, just the variance in this case since the error is a scalar). Since we specified earlier that the radar has Gaussian measurement noise with a standard deviation of 5 m, the measurement noise covariance matrix is simply:

$$R = [(5)^2] = [25] \tag{17.16}$$

In actual practice, the middle expression in Equation (17.16) is a commonly used form since it explicitly shows the standard deviation. Thus, the measurement equation has enabled us to define the other half of the terms we need for the filter. Specifically: the measurement vector ($\underline{z}$), the data matrix (H) and the measurement error covariance matrix (R).

Recalling "Step 0" of the Kalman filter algorithm, we need to determine the initial prediction for the state vector and the initial value of the prediction error covariance matrix. Since the state vector consists of inertial system errors, and since we have no a priori knowledge of these errors, our best initial prediction is simply zeroes:

$$\hat{\underline{x}}_1^- = \begin{bmatrix} 0 \\ 0 \\ 0 \end{bmatrix} \tag{17.17}$$

Obviously the predictions given by (17.17) have uncertainty associated with them. The primary components of the initial prediction error covariance matrix are the variances of the errors of the elements in the initial state prediction. For this simple example, we will assume the radar is used to initialize the position of the inertial and thus the radar error ($\sigma_{radar_{err}} = 5$) dictates the initial inertial position error. The variance of the inertial velocity error can be, very conservatively, set to $(\sqrt{10}\frac{\text{m}}{\text{s}})^2$ and the variance of the accelerometer bias is conservatively set to $(1\frac{\text{m}}{\text{s}^2})^2$. As is conventionally done, the initial prediction error covariances (i.e., the off-diagonal elements) are ignored and thus the matrix is set to:

$$P_1^- = \begin{bmatrix} 25 & 0 & 0 \\ 0 & 10 & 0 \\ 0 & 0 & 1 \end{bmatrix} \tag{17.18}$$

17.6 Coasting (on predictions) between updates

Inertial systems typically perform position/velocity/attitude updating at relatively high rates (e.g., hundreds of times per second). Conversely, external aiding systems typically provide their measurements at much lower rates (e.g., a few times per second). In order to correct the inertial output and maintain the inertial's high data rate and low data latency, the Kalman filter loop (i.e., one pass through steps 1–5) should be executed at the same high rate as the inertial system. This poses a problem since the external aiding data is not available at the high rate.

One solution is to use the Kalman filter's prediction ability to 'coast' through the periods between aiding measurements. The steps within the Kalman recursion are thus modified as follows:

$IF(aiding\ data\ is\ available)$

$$K_k = P_k^- H^T (H_k P_k^- H_k^T + R_k)^{-1}$$

$$\hat{\underline{x}}_k^+ = \hat{\underline{x}}_k^- + K_k(\underline{z}_k - H_k \hat{\underline{x}}_k^-)$$

$$P_k^+ = (I - K_k H_k) P_k^-$$

$ELSE$

$$\hat{\underline{x}}_k^+ = \hat{\underline{x}}_k^-$$

$$P_k^+ = P_k^-$$

END

$$\hat{\underline{x}}_{k+1}^- = \Phi_k \hat{\underline{x}}_k^+$$

$$P_{k+1}^- = \Phi_k P_k^+ \Phi_k^T + Q_k$$

Thus, if no aiding data is available, the estimate is set equal to the prediction and the estimation error covariance is set equal to the prediction error covariance. The prediction steps for the next moment in time (i.e., the steps after "END") are executed at all times. When using Equation (17.9) to compute the state transition matrix, it is important to remember the time interval (Δt) corresponds to the (high) rate at which the Kalman loop is executed (and not the rate of the aiding data). Finally, it should be noted that the above-described coasting procedure in some cases is unduly computationally expensive since the prediction error covariance matrix is only needed at the time of a Kalman update. In such cases alternative formulations can be used which propagate the covariance matrix over the entire update interval.

17.7 Simulation of a constant accel bias in the rocket sled

The inertial navigation simulation depicted in Figure 17.2 processed the inertial sensor data at a 1 kHz rate. For the integrated navigation example, the radar data will be simulated at a 1 Hz rate and thus the Kalman filter will coast for 999 iterations between the radar updates. Figures 17.6–17.8 show the results of the simulation. Figure 17.6 plots the accel bias estimate. There is an initial convergence interval during which the estimate oscillates somewhat. This is typical. For this particular example, the estimate converges on the truth by approximately the 60th Kalman update (i.e., 60 sec). Figure 17.7 plots the error in the Kalman-corrected inertial velocity. Again, there is a convergence period and the error becomes negligible after approximately one minute. Finally, Figure 17.8 plots both the Kalman-corrected inertial position

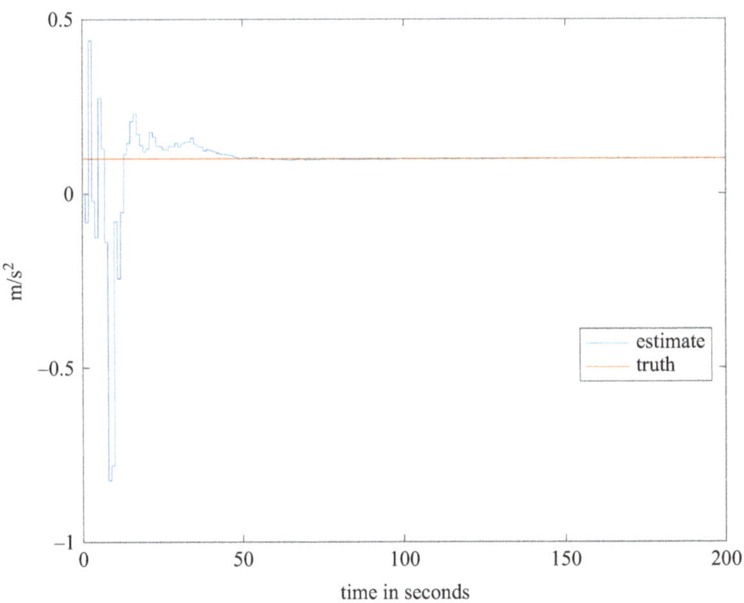

Figure 17.6 Kalman filter accelerometer bias estimate for the rocket sled example. Note that for this simulation, the accelerometer bias is a pure constant and thus Q has been set to all zeros

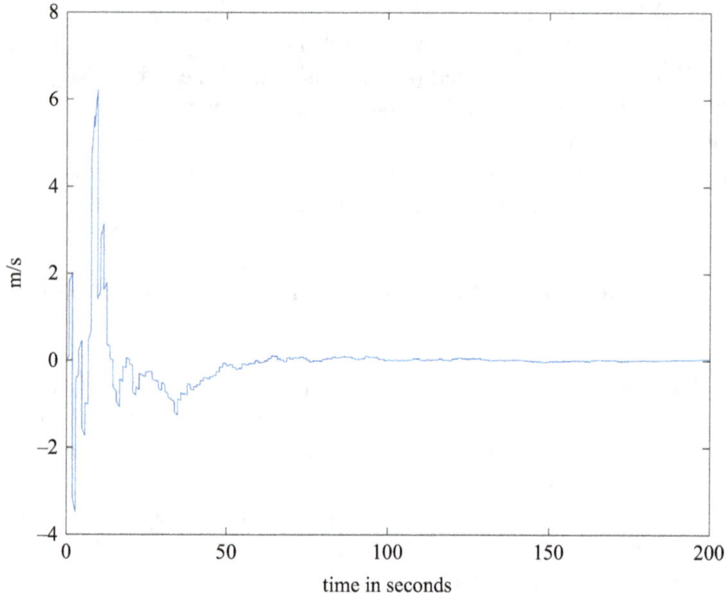

Figure 17.7 Error in the Kalman-corrected inertial velocity for the rocket sled example

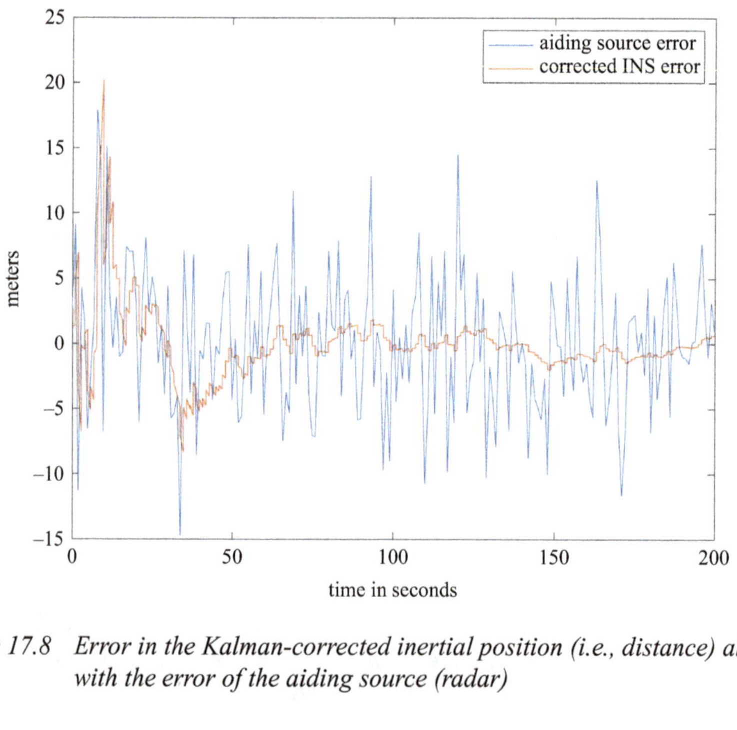

Figure 17.8 Error in the Kalman-corrected inertial position (i.e., distance) along with the error of the aiding source (radar)

(distance) error and the radar measurement error. The filter clearly has eliminated the drift of the inertial-only position error (Figure 17.3) and is significantly less noisy than the radar measurement error. Since our purpose here is to introduce the concept of inertial-aiding, we will not dwell on so-called filter "tuning" issues (i.e., performance optimization).

It is important to remember that the complementary filter structure is allowing the position and velocity output to retain the unbiased characteristics of the aiding source while also retaining the reduced noise characteristics of the inertial system. Remember also that there are nonlinear system dynamics going on at the same time. The vehicle is accelerating in the middle of this profile and yet we are able to run a *linear* Kalman filter and produce good results because of the complementary filter structure. The filter is estimating the errors and *not* the total state (i.e., position and velocity) of the vehicle. To the extent that the errors can be modeled in a linear way, the complementary Kalman filter is very well suited for estimating them.

17.8 Adding more realism to the accelerometer bias simulation and estimation

We set Q to all zeroes in the above example because there was no uncertainty in the system dynamics (recall that in the Kalman filter for this example, the system consists of the state vector of inertial errors). As we discussed before, specifying zero system uncertainty will cause the filter eventually to think that its own predictions are perfect and it would then place zero weight on the measurements. Of course, in reality, an accelerometer bias is not a true bias but rather it is a slowly varying random process. There are other errors as well in an actual accelerometer, but for now we will increase the realism somewhat by showing what happens when the accelerometer bias has some variation over time.

First-order Gauss–Markov processes are commonly used to model inertial sensor errors in the Kalman filter (see the Appendix for a review of this type of random process). For our simple sled-track example, modeling the accel bias as a first-order Gauss–Markov process necessitates that we modify the continuous-time state Equation (17.8) to:

$$\dot{\underline{x}} = F\underline{x} + \underline{\eta} = \begin{bmatrix} 0 & 1 & 0 \\ 0 & 0 & 1 \\ 0 & 0 & \frac{-1}{\tau_{acc}} \end{bmatrix} \begin{bmatrix} p_{err}(t) \\ v_{err}(t) \\ a_{bias}(t) \end{bmatrix} + \begin{bmatrix} 0 \\ 0 \\ \eta_{acc} \end{bmatrix} \tag{17.19}$$

where τ_{acc} is the time-constant of the first-order Gauss–Markov process and η_{acc} is the continuous-time white noise input. There are only two changes that need to be made to the previously described (constant accel bias) Kalman filter to accommodate the first-order Gauss–Markov model of the accel bias: the state transition matrix and the system noise covariance matrix. The state transition matrix can be obtained by evaluating

Equation (17.9) using the system dynamics matrix (F) from Equation (17.19). A first-order approximation can be obtained by using the first two terms of the Taylor Series of the matrix exponential:

$$e^{F\Delta t} \approx I + F\Delta t \tag{17.20}$$

Thus, the first-order approximation of the state transition matrix obtained from (17.19) is:

$$\Phi \approx \begin{bmatrix} 1 & \Delta t & 0 \\ 0 & 1 & \Delta t \\ 0 & 0 & 1 - \frac{\Delta t}{\tau_{acc}} \end{bmatrix} \tag{17.21}$$

The system noise covariance matrix is given by:

$$Q = \begin{bmatrix} 0 & 0 & 0 \\ 0 & 0 & 0 \\ 0 & 0 & q_k \end{bmatrix} \tag{17.22}$$

And the (3,3) element is given by [1]:

$$q_k = \sigma_{GM}^2(1 - e^{-2\beta\Delta t}) \tag{17.23}$$

where σ_{GM}^2 is the variance of the first-order Gauss–Markov process, β is the inverse of the time-constant of the first-order Gauss–Markov process, and Δt is the sampling interval of the first-order Gauss–Markov process. Recall that additional details regarding this process are given in an appendix at the end of this book.

To illustrate the aforementioned concepts, we will modify the previous simulation such that the accelerometer bias is simulated using a first-order Gauss–Markov process with a variance of 0.015 (standard deviation of approximately 0.1225 $\frac{\text{m}}{\text{s}^2}$) and a time-constant of 3,000 sec. To demonstrate optimum performance, the time-constant in the Kalman filter (F(3,3)) is matched to the time constant of the simulated accel bias and the non-zero element of Q is set equal to the value computed by (17.23) with the sampling interval set to 1 milli-sec (note: for this example, q_k is 10^{-8} $\frac{\text{m}}{\text{s}^2}$).

Given the long-time constant of this Gauss–Markov process (3,000 sec), the trajectory we used previously is too short (only 200 sec) to be able to see the long term variations in the accel "bias." Thus, a new simulated vehicle trajectory is needed and we will modify the previous one slightly. The period of 1 $\frac{\text{m}}{\text{s}^2}$ acceleration is started at $t = 130$ sec and ends at $t = 160$ sec and the total length of the profile is extended to 1,500 sec (Note: the length of the run is *not* realistic for an actual sled track).

Figures 17.9–17.11 show the results of the simulation. Although not perfect, the Kalman filter's estimate of the accel bias does quite well in following the true bias. The velocity and position errors both converge but are clearly more noisy than was the case for the previous example.

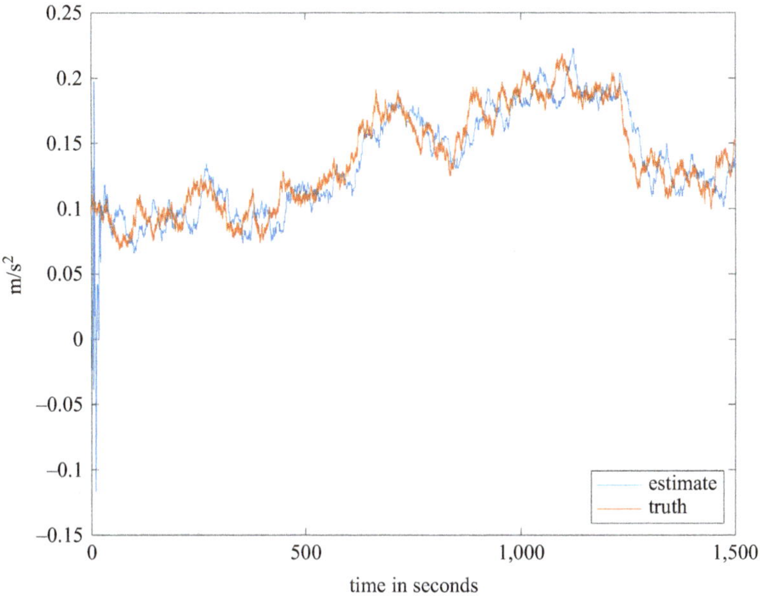

Figure 17.9 Kalman filter accelerometer bias estimate for the rocket sled example with the accelerometer bias simulated and modeled as a first-order Gauss–Markov process with a 3,000 sec time-constant

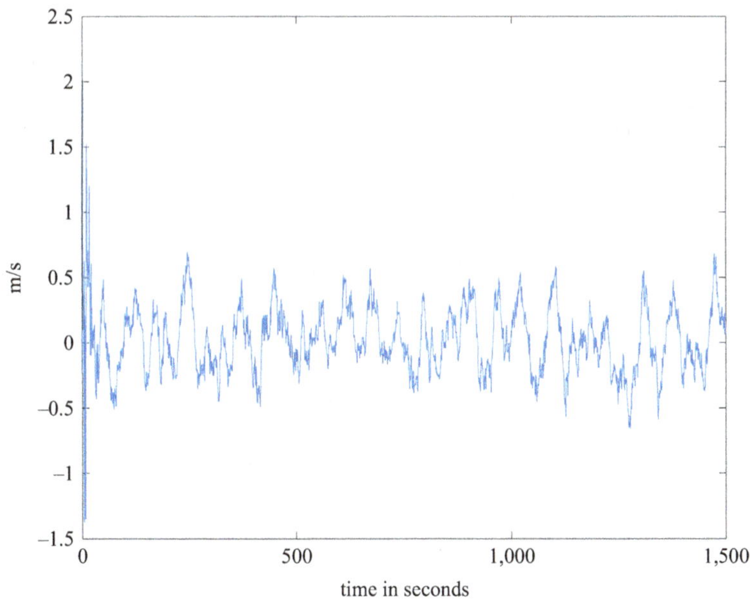

Figure 17.10 Error in the Kalman-corrected inertial velocity for the first-order Gauss–Markov accel bias example

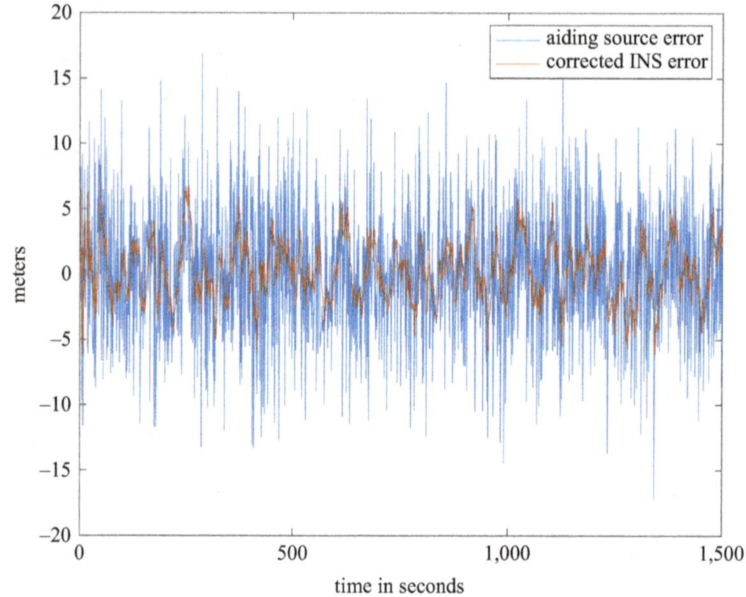

Figure 17.11 Error in the Kalman-corrected inertial position (i.e., distance) along with the error of the aiding source (radar) for the first-order Gauss–Markov accel bias example

The Impact of Non-Zero Q

The increased noise in the velocity and position estimates is a direct result of increasing the value of Q. When we specify a nonzero Q we are telling the filter that there is uncertainty in the way the underlying true system changes over time. As a result, this limits the ability of the filter to reduce noise. If we reran the constant bias case with nonzero Q, the filter would not perform as well. This is the price that we have to pay when we are dealing with uncertainty. We can handle the dynamics of the states we are estimating or we can try to get rid of the noise. We cannot do both perfectly. There is a trade-off.

Scale Factor Errors

Recall that a scale factor provides the conversion from sensor input to sensor output. Low-cost analog accelerometers, for example, will output a voltage that is proportional to the input specific force (e.g., 2 V/g). Although inertial sensors certainly do exhibit some degree of scale factor nonlinearity, the dominant scale factor error component is scale factor stability (frequently referred to simply as scale factor error). Scale factor stability error is a change in the slope of the input–output curve and thus the sensor error is proportional to the true input. For example:

$$\varepsilon_{asf} = s_a f_a \tag{17.24}$$

where ε_{asf} is the accelerometer error, s_a is the scale factor error, and f_a is true specific force.

17.9 Impact of unmodeled scale factor error in the Kalman filter

We will continue to use the vehicle trajectory utilized for the first-order Gauss–Markov accelerometer bias estimation. Recall that profile had a duration of 1,500 sec with a period of constant acceleration (1 $\frac{\text{m}}{\text{s}^2}$) from 130 to 160 sec.

Now before we allow the Kalman filter to deal with this new error, it is good first of all to see what happens if we ignore it. We are going to simulate a scale factor error of 0.1. This would also be referred to as a scale factor error of 100,000 parts-per-million (ppm). Navigation-grade accelerometers typically have scale factor errors on the order of *tens* of parts per million so the simulated error is enormous but it helps to illustrate the impact.

In order to focus on the effect of the scale factor error, we will revert to simulating the accelerometer bias as a pure constant and will go back to the 3-state Kalman filter that models the accel bias as a pure constant (and thus Q is zero). The Kalman filter will thus not do anything to deal with the presence of the scale factor error. Given the presence of unmodeled errors, we would expect the filter to perform poorly.

The results are shown in Figures 17.12–17.14. Somewhat surprisingly, the accel bias estimate (Figure 17.12) performs quite well. The error in the Kalman-corrected velocity (Figure 17.13), however, converges to zero until the true acceleration occurs in the vehicle trajectory at 130 sec. At this point, the error spikes up, then overshoots,

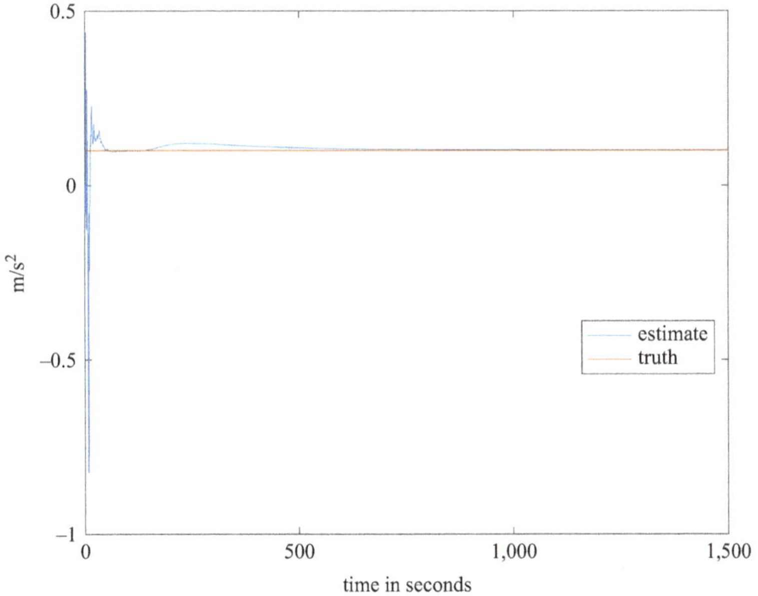

Figure 17.12 Kalman filter accelerometer bias estimate for the rocket sled example with the accelerometer bias simulated as a pure constant but with an unmodeled scale factor error

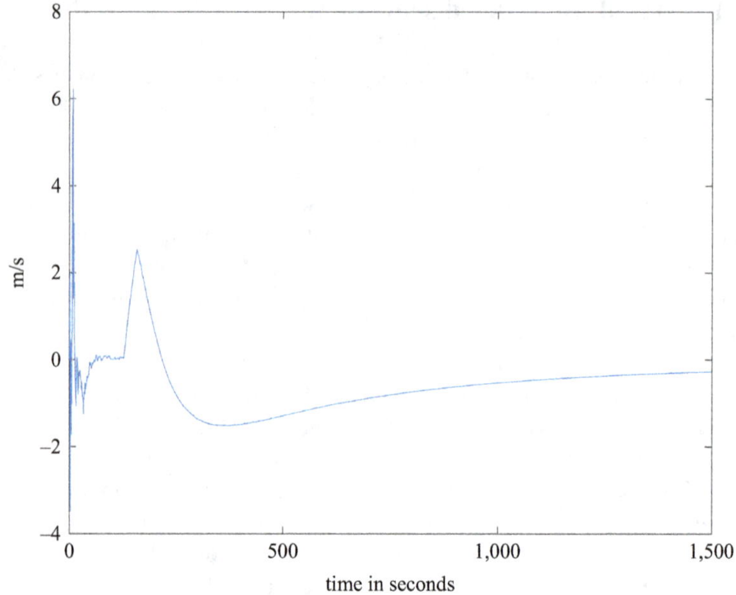

Figure 17.13 *Error in the Kalman-corrected inertial velocity for the unmodeled scale factor error example. The true acceleration in the profile is non-zero from 130 to 160 sec*

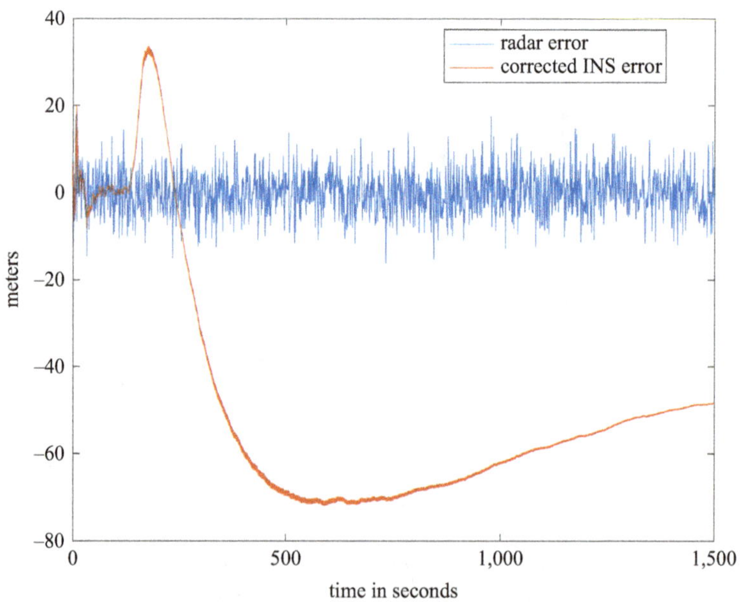

Figure 17.14 *Error in the Kalman-corrected inertial velocity for the unmodeled scale factor error example. The true acceleration in the profile is non-zero from 130 to 160 sec*

then gradually decays. The velocity error has a profound impact on the position error (Figure 17.14). Although the simulated scale factor error was unrealistically large, the simulation clearly indicates the negative impact of unmodeled errors in the Kalman filter.

17.10 Two approaches to deal with unmodeled errors

There are two main approaches to deal with unmodeled errors in the Kalman filter. The first approach is to inflate key elements of the system noise covariance matrix. The second approach is to model explicitly the previously unmodeled errors. We will take a look at each approach in turn.

17.11 Accommodating unmodeled scale factor error by inflating the Q matrix

By inflating the Q matrix, we are telling the filter that there is uncertainty in the system that is not taken into account by the state transition matrix. In the current example, there are three states: position error, velocity error and accelerometer bias. Although the true accelerometer bias is being simulated as a pure constant, we can account for the presence of the scale factor error by adding system noise on the accel bias state. Thus, instead of Q being a matrix of all zeros, we make $Q(3,3)$ a non-zero value (10^{-8} for the simulation results shown here).

The results are shown in Figures 17.15–17.17. We see that the results are noisy. That should not be surprising. When we increase Q, we are telling the filter that there is some inherent system uncertainty and as a result it is not going to be able to filter as much. It is not going to be able to smooth the results as well. The errors exhibit excursions during the time period of the true acceleration (130–160 sec) and thus is the time period when the scale factor error manifests itself (recall Equation (17.24)). However, unlike the previous example where the scale factor error was completely ignored, we see that with the inflated Q the errors converge back to zero instead of diverging even after the true acceleration has stopped.

17.12 Explicit modeling of the scale factor error

An alternative to inflating Q is to include the scale factor error as an *additional* state variable estimated by the filter. We thus expand the system dynamics equation as follows:

$$\dot{\underline{x}} = F\underline{x} + \underline{\eta} = \begin{bmatrix} 0 & 1 & 0 & 0 \\ 0 & 0 & 1 & a(t) \\ 0 & 0 & \frac{-1}{\tau_{ab}} & 0 \\ 0 & 0 & 0 & \frac{-1}{\tau_{asf}} \end{bmatrix} \begin{bmatrix} p_{err}(t) \\ v_{err}(t) \\ a_{bias}(t) \\ SF_{acc}(t) \end{bmatrix} + \begin{bmatrix} 0 \\ 0 \\ \eta_{ab} \\ \eta_{asf} \end{bmatrix} \tag{17.25}$$

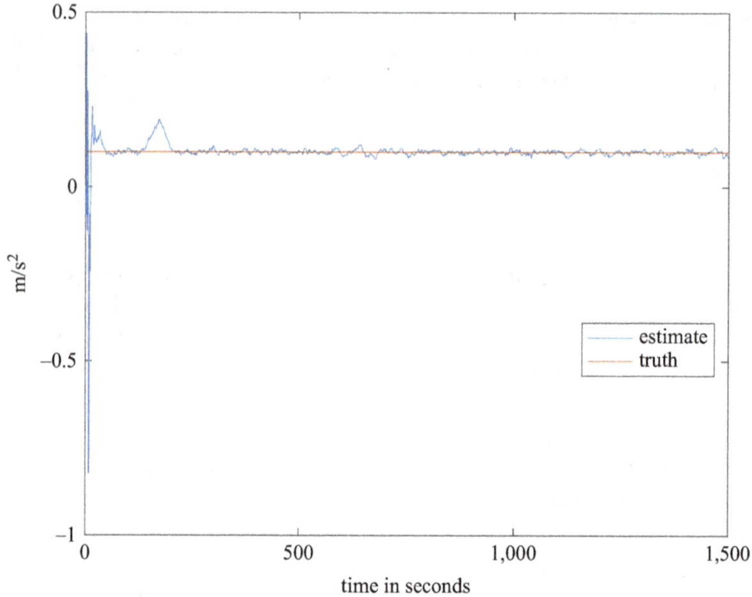

Figure 17.15 Kalman filter accelerometer bias estimate for the rocket sled example with the accelerometer bias simulated as a pure constant but with a scale factor error accommodated through increased system noise covariance

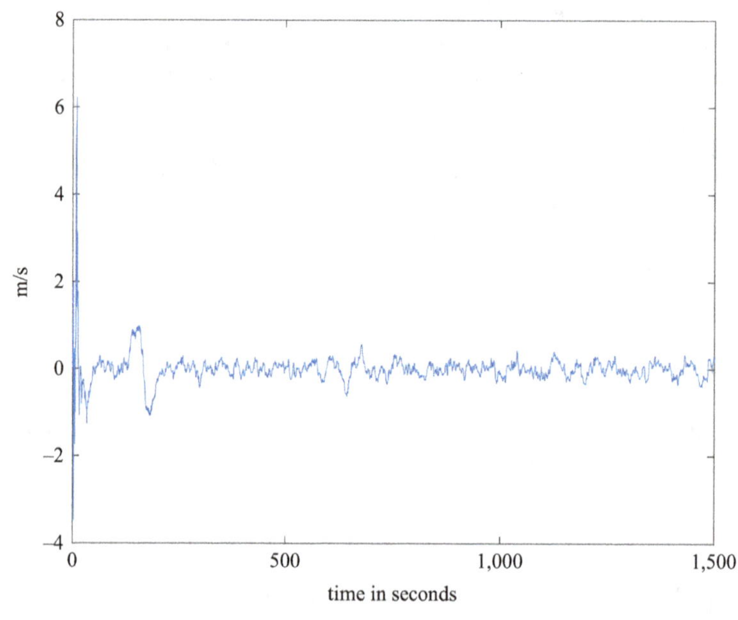

Figure 17.16 Error in the Kalman-corrected inertial velocity for the scale factor error with inflated Q example. The true acceleration in the profile is non-zero from 130 to 160 sec

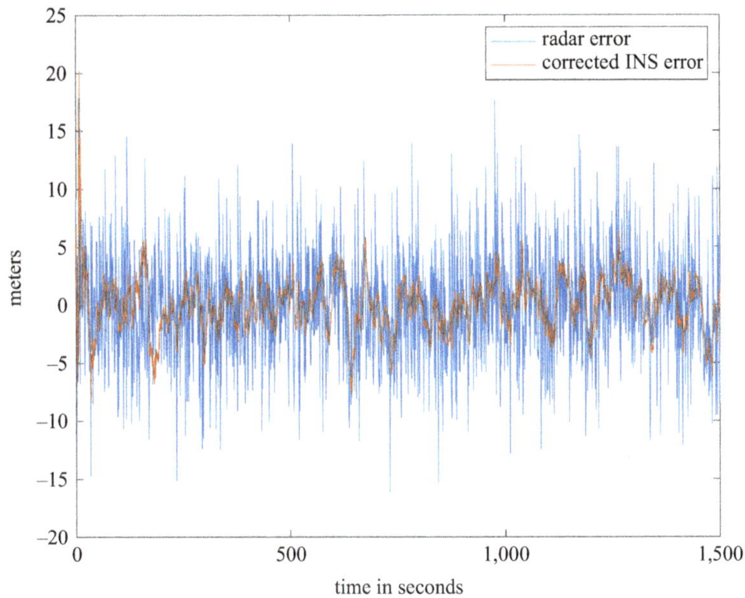

Figure 17.17 Error in the Kalman-corrected inertial position (i.e., distance) along with the error of the aiding source (radar) for the scale factor error with inflated Q example. The true acceleration in the profile is non-zero from 130 to 160 sec

where τ_{ab} and τ_{asf} are the time-constants of the first-order Gauss–Markov process models for the accel bias and accel scale factor error, respectively, and η_{ab} and η_{asf} are the continuous-time white noise inputs for the Gauss–Markov processes.

First consider the system dynamics matrix (the 4×4 matrix in Equation (17.25)). Note that the upper left 3×3 is identical to what we had previously for the case of a time-varying accel bias. What happens if the accel bias is truly a constant? Just make the time constant (τ_{ab}) a very large value. In reality, the accel bias is going to have some variation and it is very commonly modeled as a Gauss–Markov process.

The fourth column and the fourth row of the system dynamics matrix deal with the accel scale factor error. We can see there are only two nonzero elements. As mentioned above, the scale factor error itself is modeled as a first-order Gauss–Markov process with a time constant of τ_{asf}. Note the (2,4) element contains the true value of acceleration. Why? Because the derivative of the velocity error is equal to the accel bias plus the scale factor error multiplied by the true value of acceleration.

Now you are probably thinking, "Wait a minute. If I knew the true value of acceleration then my problem would be solved and I wouldn't need the Kalman filter." That is true. We have to use an estimate of the true acceleration instead. With navigation-grade instruments, it suffices to use the raw accelerometer measurement itself. Alternately, the Kalman-corrected accelerometer values can be used (after the filter has converged, of course).

Since the system dynamics matrix is now time-varying (due to the (2,4) element), the state transition matrix is also time-varying and thus cannot be pre-computed. It has to be computed (using some approximation of Equation (17.9)) each time the Kalman loop is executed. The Q matrix is a 4×4 and the (3,3) and (4,4) elements can be determined from Equation (17.23) based on the estimated variance of the accel bias and scale factor error for the particular sensor being used.

A variation of the previous simulation, created to test the scale factor estimation, utilizes a vehicle trajectory of 1,500 sec duration with a constant 2 $\frac{\text{m}}{\text{s}^2}$ acceleration from 200 to 230 sec and a 1 $\frac{\text{m}}{\text{s}^2}$ acceleration from 1,100 to 1,130 sec. In order to demonstrate the ideal performance of the scale factor error estimation, both the accelerometer bias and scale factor errors are simulated as pure constants (acceleration white noise was added, though, to increase realism). However, the filter models both the accel bias and scale factor errors as first-order Gauss–Markov processes with time-constants of 3,000 sec.

The results are shown in Figures 17.18–17.21. The accelerometer bias estimate converges quickly and remains so for the rest of the run. The accelerometer scale factor error estimate does not converge until the non-zero true acceleration occurs starting at $t = 200$ sec. After convergence, the system noise (Q) driving the Gauss-Markov process of the scale factor error state causes the estimate to gradually drift away from the truth. When the next true acceleration occurs starting at $t = 1{,}100$ sec, the estimate re-converges and then gradually starts to drift again. Figure 17.20 shows

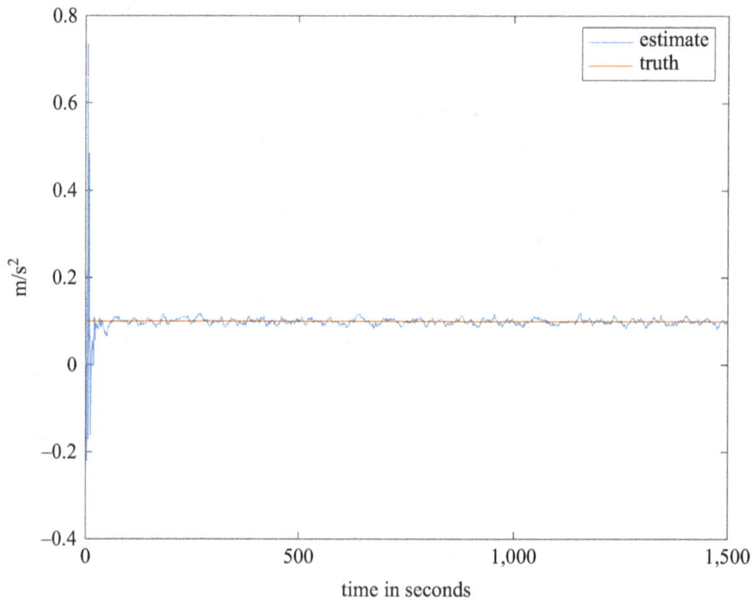

Figure 17.18 Kalman filter accelerometer bias estimate for the rocket sled example with the accelerometer bias and scale factor errors simulated as pure constants. The scale factor error is explicitly estimated by the filter

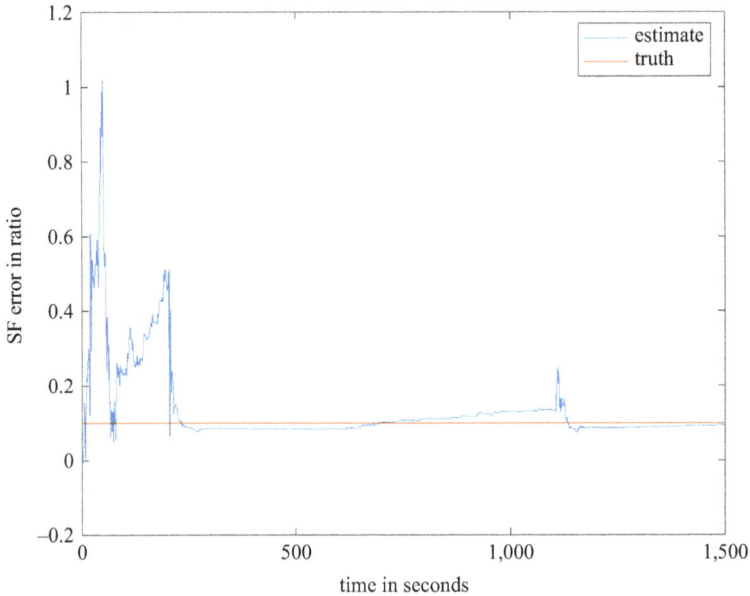

Figure 17.19 Kalman filter accelerometer scale factor error estimate for the rocket sled example with the accelerometer bias and scale factor errors simulated as pure constants. The periods of non-zero true acceleration occur from 200 to 230 sec and from 1,100 to 1,130 sec

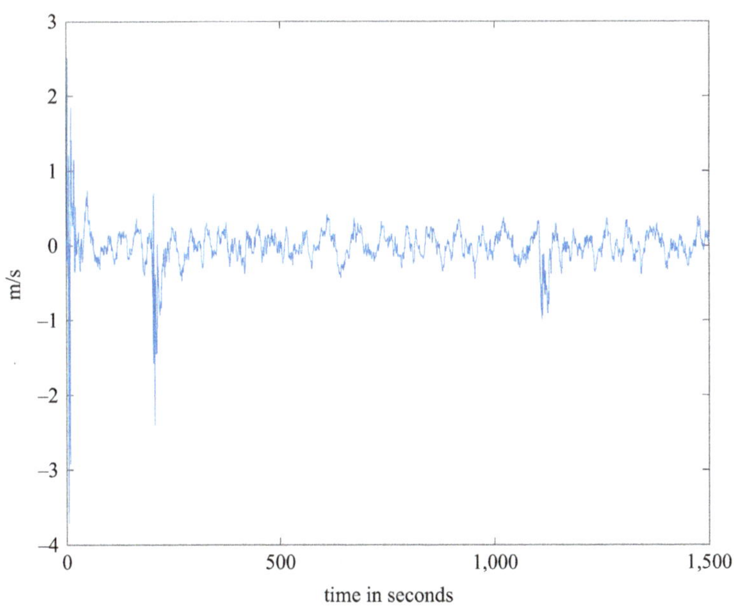

Figure 17.20 Error in the Kalman-corrected inertial velocity. Scale factor error is being estimated by the filter. The periods of non-zero true acceleration occur from 200 to 230 sec and from 1,100 to 1,130 sec

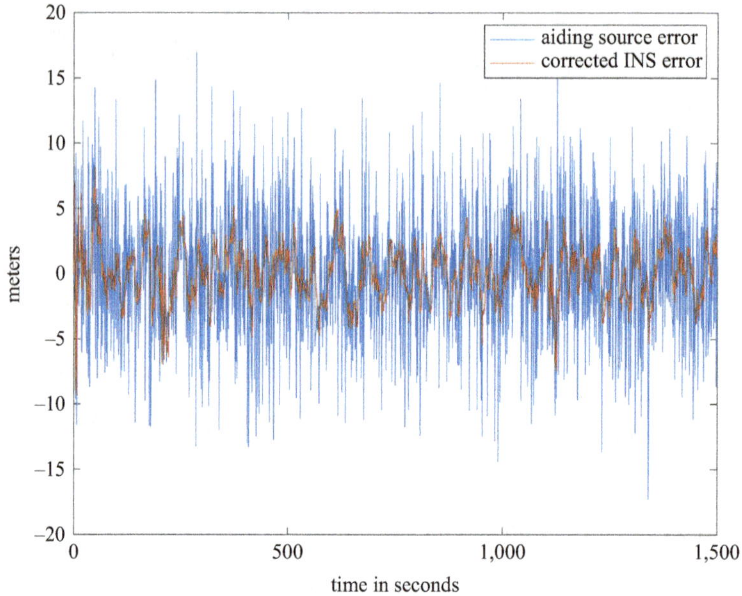

Figure 17.21 Error in the Kalman-corrected inertial position (i.e., distance) along with the error of the aiding source (radar). Accelerometer scale factor error is being estimated by the filter. The periods of non-zero true acceleration occur from 200 to 230 sec and from 1,100 to 1,130 sec

that the velocity converges but does exhibit some increased noise during the periods of acceleration. Finally, Figure 17.21 shows the position error is well behaved.

17.13 Conclusions

We have demonstrated that aiding of the inertial system with external measurements in a complementary Kalman filter architecture works very well if we have good error models. The presence of unmodeled errors, however, can cause filter divergence. We can accommodate unmodeled errors by increasing the system noise covariance matrix or by explicitly modeling and estimating these errors. In practice, however, some unmodeled errors always exist and thus some inflation of the Q matrix is always necessary for stable performance.

References

[1] Gelb A, editor. *Applied Optimal Estimation*. Cambridge, MA: The M.I.T. Press; 1974.

[2] Brown RG, Hwang PYC. *Introduction to Random Signals and Applied Kalman Filtering: with MATLAB® exercises*. 4th ed. Hoboken, NJ: John Wiley & Sons, Inc.; 2012.

Chapter 18
Inertial aiding in two dimensions

18.1 Introduction

In an earlier chapter, we explored how the Kalman filter does an excellent job of taking the radionavigation system geometry into account when forming its estimates but does not do well in handling nonlinear system dynamics. In the previous chapter, we introduced the concept of the complementary filter and applied it to a simple one-dimensional aided-inertial navigation problem. Since the filter in the aided case is estimating inertial errors, instead of the total aircraft state, the non-linearities arising from a dynamic trajectory were not a problem. An aided-inertial architecture is founded on the principle that the inertial navigation system does an excellent job of determining the position, velocity and attitude of the vehicle in the short term and thus the Kalman filter is not needed for that task. The purpose of the aiding is to prevent the inertially determined vehicle state from drifting off over the long term.

In this chapter, we extend the inertial aiding concept to two dimensions. We will use a two-dimensional radionavigation system (DME–DME) to aid a simplified inertial navigation system. We will be estimating inertial errors and one way to view this is that the stand-alone inertial position solution provides the "nominal" trajectory used to linearize the system and measurement equations to make them suitable for the Kalman filter. This implies that the inertial errors may be modeled linearly. In order to introduce some key concepts in this chapter, we will use a very simple inertial error model. In the next chapter, we will develop a significantly more detailed inertial error model but the fact remains that the model must be linear in order to be suitable for the Kalman filter. We will discover in a later chapter that once the inertial error grows sufficiently large the linearity assumption breaks down. We will also discover that the typical solution to this problem is to feedback the estimated errors as corrections to the inertial system thus keeping the inertial errors small.

For the moment, we will keep things simple. Before we can apply the complementary filter architecture to INS and DME, however, we must deal with the fact that the two systems do not provide measurements in the same domain. Recall that in the sled track example both the inertial system and the radar provided measurements of along-track distance. This was important since the Kalman filter observable (i.e., the "z" vector) was formed by subtracting the range measurements from each system. However, we cannot do this directly with INS and DME. The INS provides estimates of position, velocity and attitude. The DME, on the other hand, provides measurements of line-of-sight distance (i.e., range) between the vehicle and each ground transponder. Thus one system provides position-domain information whereas the other system provides range-domain information and of course there is a non-linear relationship between the two. Nevertheless, we must find a way to difference the inertial data with the DME–DME data.

There are two primary ways to do this: (a) convert the INS data into a range-domain equivalent; or (b) first compute a position solution using the DME measurements. The former is referred to as "tight-coupling" and utilizes the aiding system data in its native domain. The later is referred to as "loose-coupling" and utilizes the aiding system data after it has been used to compute a position solution. We will explore each in turn.

18.2 Tightly coupled INS/DME Kalman filter

The tightly coupled INS/DME complementary Kalman filter architecture is depicted in Figure 18.1. The "position-to-range conversion" first involves computing the distance from each DME ground transponder to the *INS-estimated* position:

$$\rho_i^{INS} = \sqrt{(x_{INS} - x_{DME_i})^2 + (y_{INS} - y_{DME_i})^2} \tag{18.1}$$

These INS equivalent-range values are then differenced with the measured DME ranges to form the observation vector that is input to the Kalman filter:

$$\underline{z} = \begin{bmatrix} \Delta\rho_1 \\ \Delta\rho_2 \end{bmatrix} = \begin{bmatrix} \rho_1^{INS} - \rho_1^{DME} \\ \rho_2^{INS} - \rho_2^{DME} \end{bmatrix} \tag{18.2}$$

Of course, measurements from additional DME ground transponders could be incorporated by adding rows to Equation (18.2).

As shown in Figure 18.1, the Kalman filter is estimating the inertial errors. To keep things simple initially, however, we will only estimate horizontal position and velocity errors (i.e., we will ignore vertical position and velocity errors along with

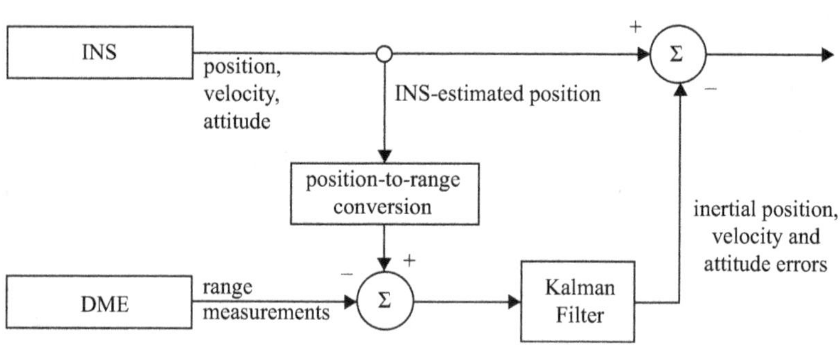

Figure 18.1 Tightly coupled INS/DME complementary Kalman filter architecture

attitude errors and sensor errors). Accordingly, the measurement equation can be adapted from Equation (15.33):

$$\begin{bmatrix} \Delta\rho_1 \\ \Delta\rho_2 \end{bmatrix} = \begin{bmatrix} \left(\frac{x_{INS}-x_{DME_1}}{\rho_1^{INS}}\right) & 0 & \left(\frac{y_{INS}-y_{DME_1}}{\rho_1^{INS}}\right) & 0 \\ \left(\frac{x_{INS}-x_{DME_2}}{\rho_2^{INS}}\right) & 0 & \left(\frac{y_{INS}-y_{DME_2}}{\rho_2^{INS}}\right) & 0 \end{bmatrix} \begin{bmatrix} ins\_x\_pos\_err \\ ins\_x\_vel\_err \\ ins\_y\_pos\_err \\ ins\_y\_vel\_err \end{bmatrix} + \begin{bmatrix} \nu_1 \\ \nu_2 \end{bmatrix} \tag{18.3}$$

where, as before, ν_1 and ν_2 are the DME ranging measurement errors. The measurement error covariance matrix is thus the same as in Chapter 15 (a diagonal matrix with elements of 225 for the case of independent white Gaussian noise with a standard deviation of 15 m).

In order to ease into the topic and keep things initially simple, we will also obtain the system equation by adapting equation (15.29):

$$\begin{bmatrix} ins\_x\_pos\_err \\ ins\_x\_vel\_err \\ ins\_y\_pos\_err \\ ins\_y\_vel\_err \end{bmatrix}_{k+1} = \begin{bmatrix} 1 & \Delta t & 0 & 0 \\ 0 & 1 & 0 & 0 \\ 0 & 0 & 1 & \Delta t \\ 0 & 0 & 0 & 1 \end{bmatrix} \begin{bmatrix} ins\_x\_pos\_err \\ ins\_x\_vel\_err \\ ins\_y\_pos\_err \\ ins\_y\_vel\_err \end{bmatrix}_{k} + \begin{bmatrix} 0 \\ w_{xvel} \\ 0 \\ w_{yvel} \end{bmatrix} \tag{18.4}$$

where w_{xvel} and w_{yvel} account for unmodeled effects on the inertial velocity error.

This linear model of inertial error is a poor approximation. Accelerometer errors, for example, give rise to quadratic position errors. We will live with the approximation for the moment, though, to focus on other aspects of the integration. A more detailed inertial error model will be presented in a subsequent chapter. The system noise covariance matrix (Q) has the same form as before (Equation 15.32) but the scaling term can be greatly reduced since it is only accounting for inertial error growth and not vehicle dynamics.

For the simulation presented herein, we will use the same nonlinear flight path used in the DME–DME EKF (Fig. 15.11). In order to generate non-trivial inertial errors over the very short flight, accelerometer biases of 300–500 micro-gs were simulated along with gyro biases of 0.1–0.3 degree/hour. Gyro white noise of 0.05–0.1 deg/root-hour was also simulated. The resulting inertial east and north position errors are plotted in Figure 18.2. Note that a 200-m bias has been added to both errors in order to simulate the effect of a longer-term inertial drift.

With the scaling parameter (S) in the Q matrix set to 0.01, the INS/DME Kalman filter performance is illustrated in Figures 18.3–18.6. In comparison to the DME-only results presented in Chapter 15, the integration of INS with DME is clearly superior. Low-noise is achieved without vehicle dynamics-induced biases. The only performance issue of concern is the presence of a low-frequency error oscillation that is most noticeable in the x position error (Figure 18.4). This, however, is due to the non-linear inertial position error (Figure 18.2) that is not properly modeled by the Kalman filter (Equation (18.4)). As mentioned earlier, the oversimplified inertial error model represented by Equation (18.4) will be replaced in a subsequent chapter

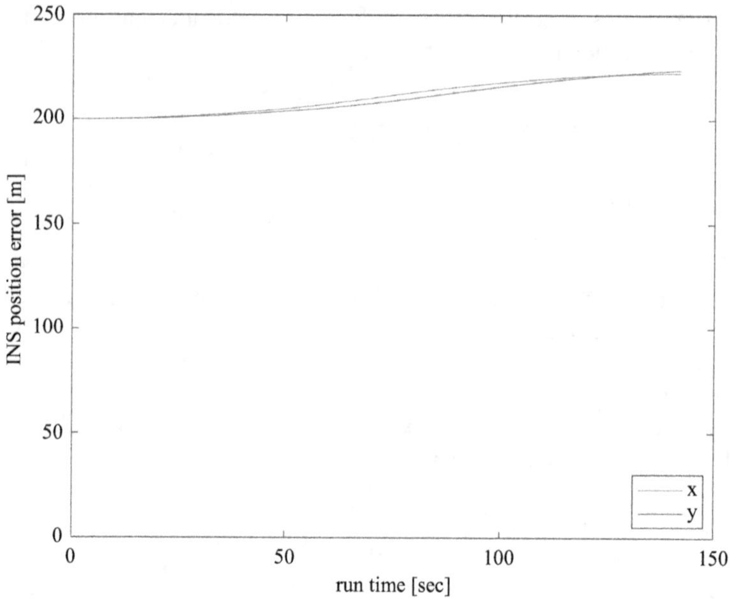

Figure 18.2 Simulated inertial navigation position error components

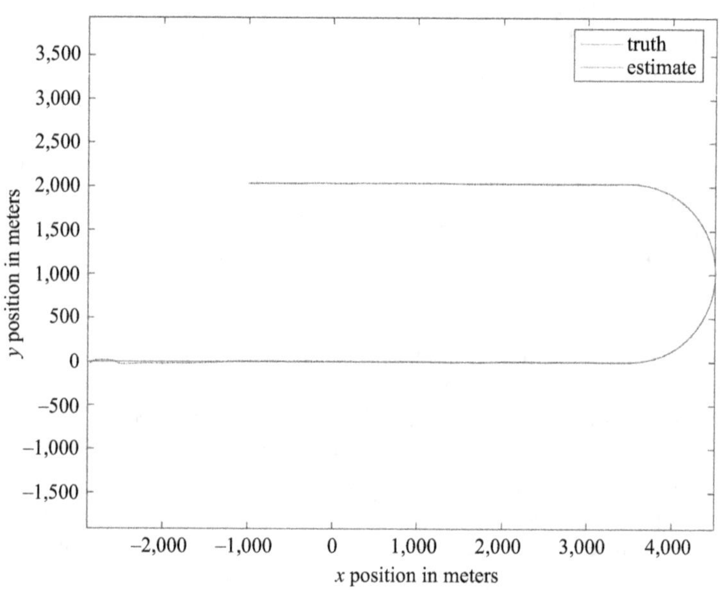

Figure 18.3 True flight path and tightly coupled DME/INS Kalman filter results

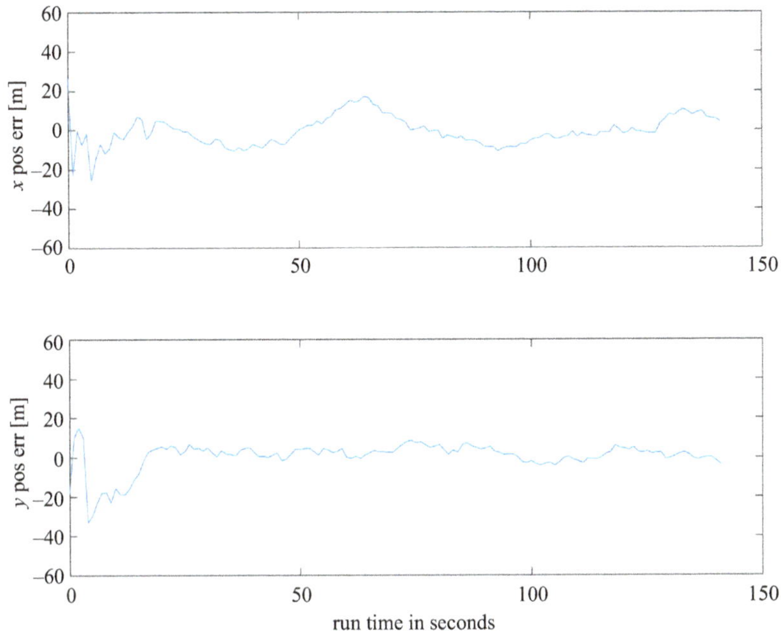

Figure 18.4 Tightly coupled DME/INS Kalman filter position errors

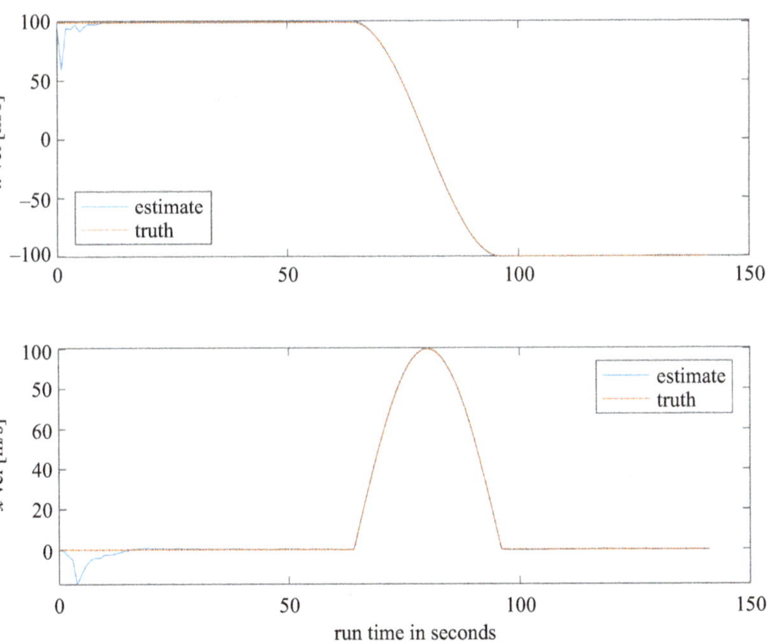

Figure 18.5 True velocity and tightly-coupled DME/INS Kalman filter results

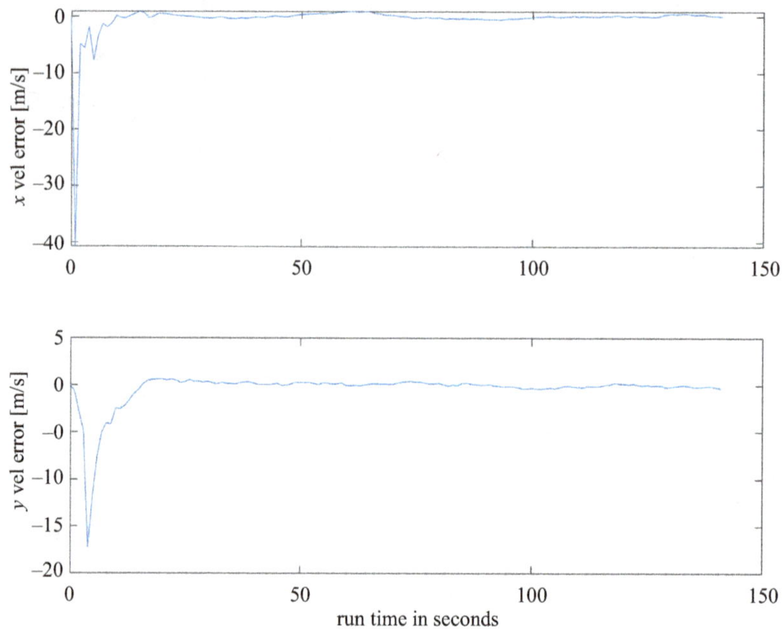

Figure 18.6 Tightly coupled DME/INS Kalman filter velocity errors

with one that is much more detailed. Finally, the transient that occurs in the beginning of the velocity error (Figure 18.6) could be reduced through filter tuning (specifically, optimizing the initial prediction error covariance matrix).

18.3 Loosely coupled INS/DME Kalman filter

In a loosely coupled architecture, the radionavigation system (DME, in our current case) is used to form an independent position (and possibly also velocity) solution that is differenced with the inertial position solution to form the observation vector that is input to the Kalman filter. This forces us to derive a new measurement equation, but note that the system equation remains unchanged since it is not a function of the radionavigation system measurements (this is not true when non-ideal effects are taken into account but we will explore those in later chapters). Since there is no need to convert domains (the INS and DME data are both in the position domain), the measurement equation is straightforward. The observable that we input to the filter is simply the difference between the position components estimated by the INS and the DME–DME solution:

$$\underline{z} = \begin{bmatrix} x_{INS} - x_{DME} \\ y_{INS} - y_{DME} \end{bmatrix} \tag{18.5}$$

The Kalman theoretical measurement equation then follows simply as:

$$\underline{z} = \begin{bmatrix} 1 & 0 & 0 & 0 \\ 0 & 0 & 1 & 0 \end{bmatrix} \begin{bmatrix} ins\_x\_pos\_err \\ ins\_x\_vel\_err \\ ins\_y\_pos\_err \\ ins\_y\_vel\_err \end{bmatrix}_k + \begin{bmatrix} \nu_1 \\ \nu_2 \end{bmatrix} \tag{18.6}$$

where the 2×4 matrix is the data matrix (H) and ν_1 and ν_2 are the multi-DME *position solution* errors.

The R matrix is the covariance of the multi-DME position solution error. For independent white Gaussian noise ranging errors, the position error covariance matrix is given by:

$$R = \sigma_{rng}^2 (H_{DME}^T H_{DME})^{-1} \tag{18.7}$$

where σ_{rng}^2 is the variance of the DME ranging error and H_{DME} was defined in Equation (15.7) for the two-DME case. It is important to note that the data needed to compute (18.7) may not necessarily be available when designing the Kalman filter. For example, the interface on the radionavigation unit (a scanning DME interrogator in our case) may not necessarily provide the data needed to compute the H matrix. In some cases, only a position solution and a figure-of-merit (FOM) are provided. Furthermore, the accuracy of the FOM may be marginal in some cases. As a result, there may be situations in which some crude approximation of R may be all that is realizable.

It is also crucially important that the integrated navigation Kalman filter designer know whether the position solution (out of the radionavigation unit) is computed with an ordinary-least squares estimator or with recursive estimator such as a Kalman filter. If the solution is computed with a least squares estimator, then every data point is statistically independent. This is the ideal situation since the Kalman filter assumes that each data set ($\underline{z}$) input is statistically independent. Conversely, if the radionavigation position solution is formed with a Kalman filter, then the output must be subsampled (i.e., decimated) to a lower data rate. For instance, if the radionavigation filter's time constant is 10 sec, then only one sample can be input to the integrated navigation filter every 20–30 sec (i.e., 2–3 time-constants). This situation of having multiple Kalman filters in sequence is known as the "cascaded filter problem" [1].

Equations (18.5)–(18.7) define the Kalman filter parameters ($\underline{z}$, H and R) that distinguish the loosely coupled architecture from the tightly coupled architecture. Although the radionavigation geometry changes slightly over the flight path, for the simulation results that follow R is approximated with a diagonal matrix with elements of 225 (which assumes a ranging error standard deviation of 15 m, unity DOPs, and ignores the off-diagonal covariance values). With all other aspects of the simulation identical to that described earlier for the tightly coupled architecture, the loosely coupled results are shown in Figures 18.7–18.10. In this example, the loosely coupled architecture provides nearly identical results to those of the tightly coupled architecture.

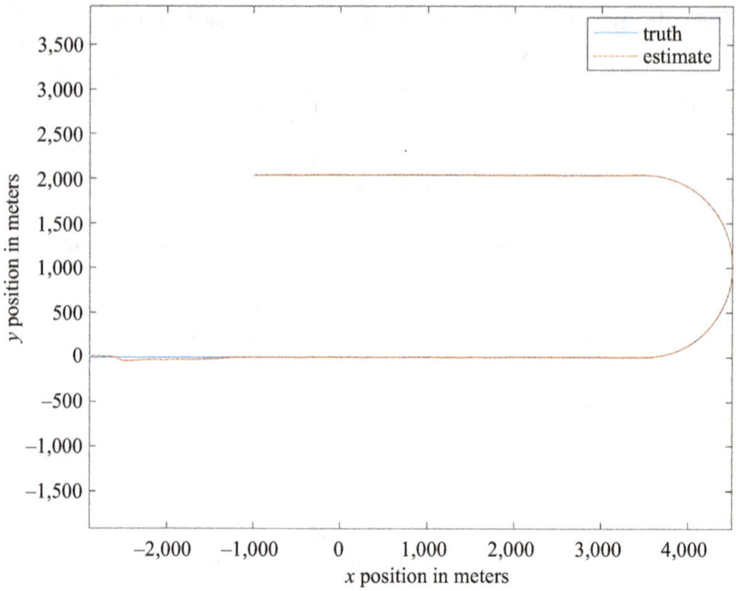

Figure 18.7 True flight path and loosely coupled DME/INS Kalman filter results

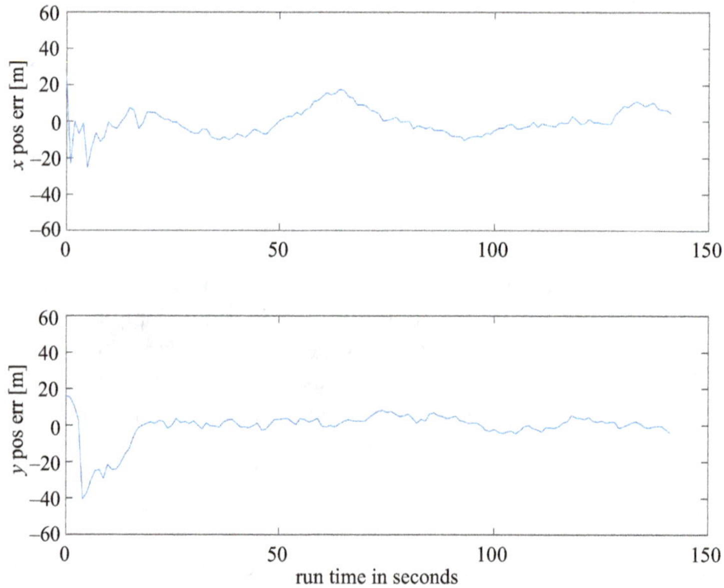

Figure 18.8 Loosely coupled DME/INS Kalman filter position errors

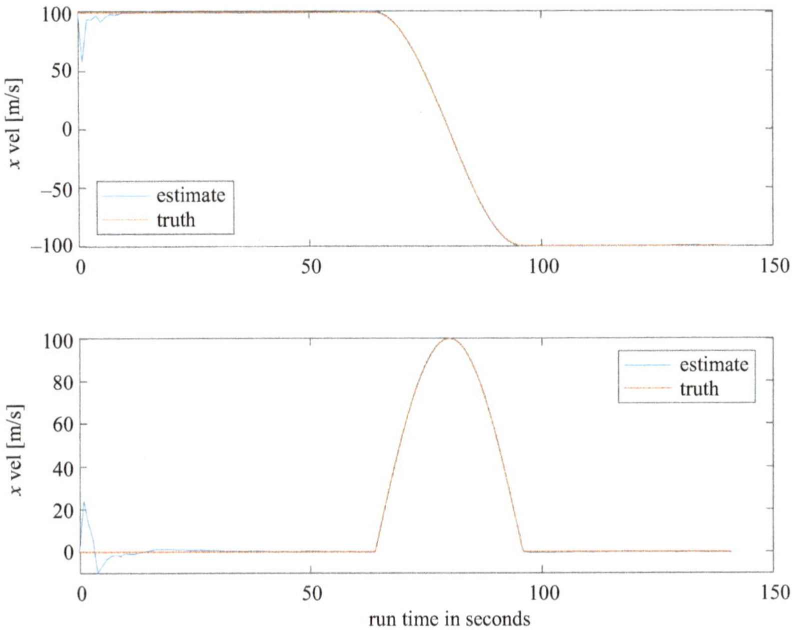

Figure 18.9 True velocity and loosely coupled DME/INS Kalman filter results

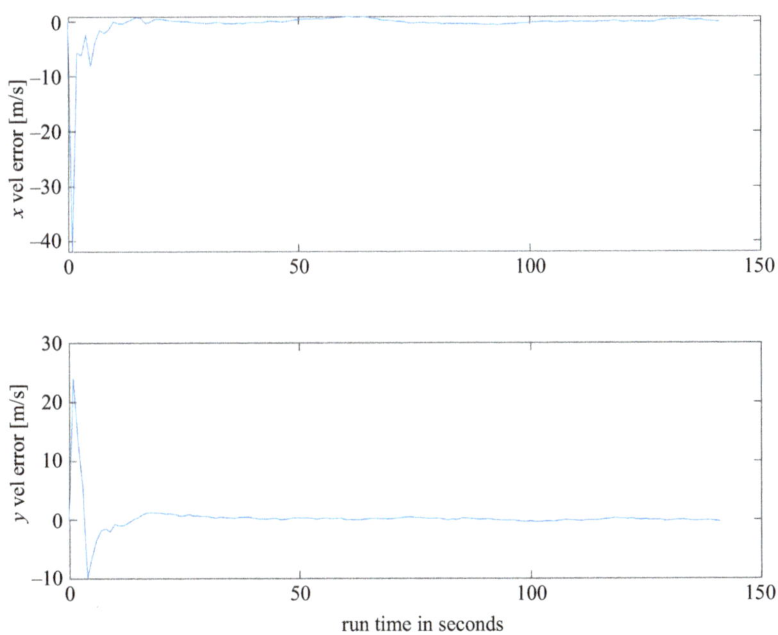

Figure 18.10 Loosely coupled DME/INS Kalman filter velocity errors

18.4 Loose versus tight

The loosely coupled architecture is simpler to implement. The H matrix is constant! However, as mentioned earlier, its performance is dependent upon accurate knowledge of the radionavigation system geometry (for the R matrix) and whether or not the radionavigation system output has been filtered or not. In some cases the designer is forced to integrate an inertial system with an existing radionavigation unit and that unit may only provide 'processed' data (e.g., position, velocity) rather than raw measurements.

The major advantages that tight integration has over loose integration are that: (a) the integration filter takes care of the geometry without the need to construct an elaborate R matrix (with information that may not be available on the interface), and (b) the ability to update the filter at the raw measurement data rate. To illustrate the effects of geometry, an extreme case is simulated. As shown in Figure 18.11, the two DME transponders are located nearly in line with the last straight segment of the flight path. In the previous simulations, they were located far away at angles that provided excellent geometry over the entire flight path. Clearly the YDOP will be horrible although the XDOP will be fine.

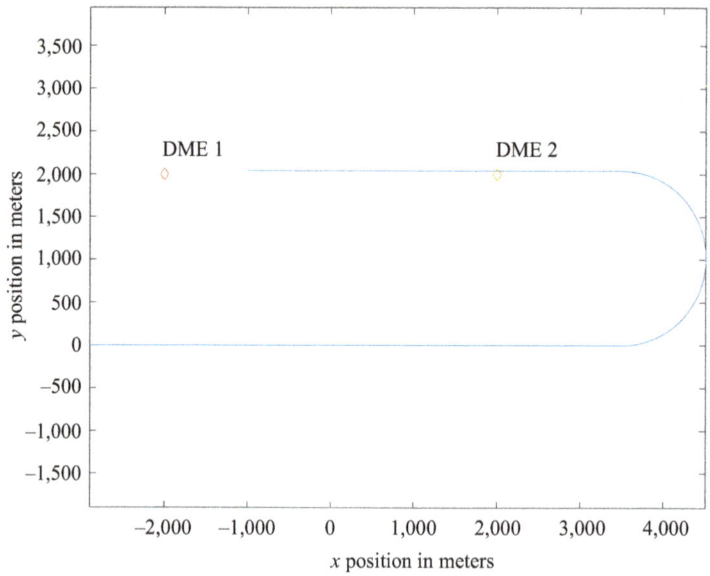

Figure 18.11 Location of the two DME ground transponders to generate extremely high YDOP values over the course of the flight path

The loosely coupled architecture utilizes the position estimate formed by the radionavigation system. Given the horrible geometry, we would expect the OLS position estimator to perform poorly and, as shown in Figure 18.12, this is indeed the case.

Since the accuracy of the OLS solution is so poor, it would be expected that the loosely coupled architecture will not perform well. However the filter utilizes the geometry information provided through the R matrix Equation (18.7) to reduce the weight on the measurements. The results (Figures 18.13–18.16) show that the filter does this rather successfully.

A close analysis of the loosely coupled filter performance would show the results are significantly affected by the use of the poor OLS position results to compute the H matrix in Equation (18.7). The wildly wrong position results (Figure 18.12) cause the computed YDOP to be much lower than the truth. As a result, the filter places too much weight on the measurements. Better performance could have been achieved by using the Kalman-corrected position output to compute H and then R.

Under the same geometry, the tightly coupled filter results are shown in Figures 18.17–18.20. Despite the bad geometry, the filter appropriately weights the measurements and shows good performance.

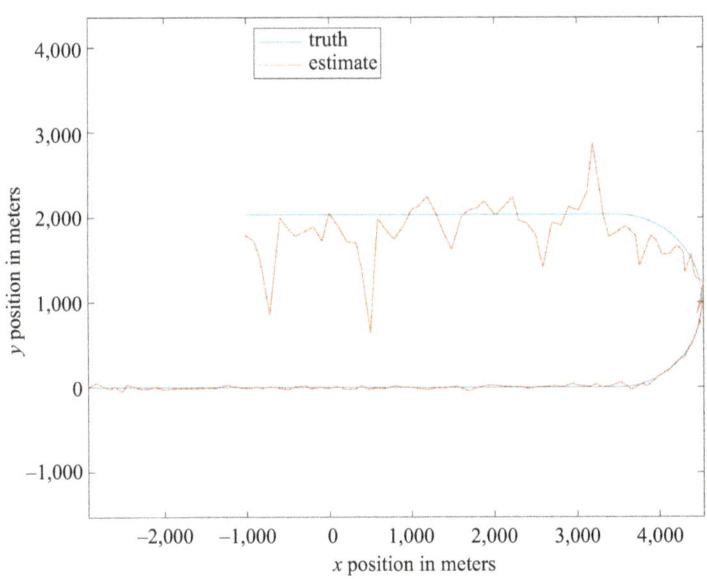

Figure 18.12 True flight path and ordinary least squares position estimation results

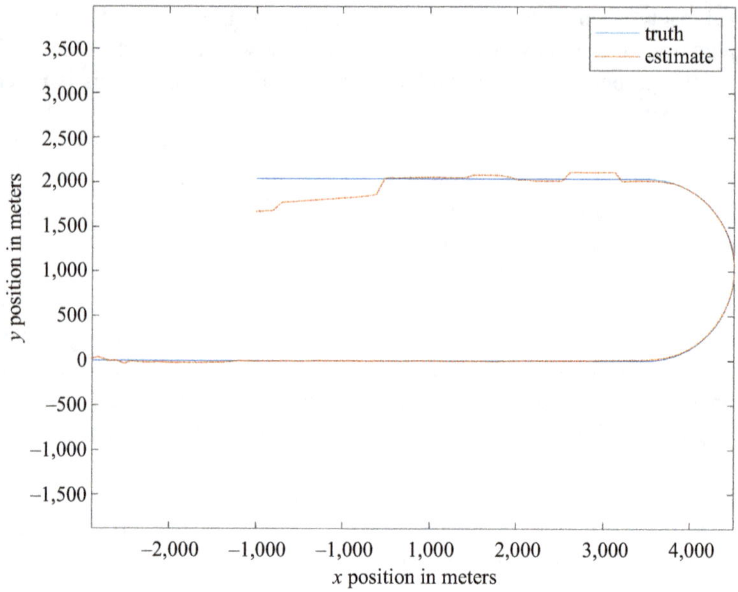

Figure 18.13 True flight path and loosely coupled DME/INS Kalman filter results with bad geometry

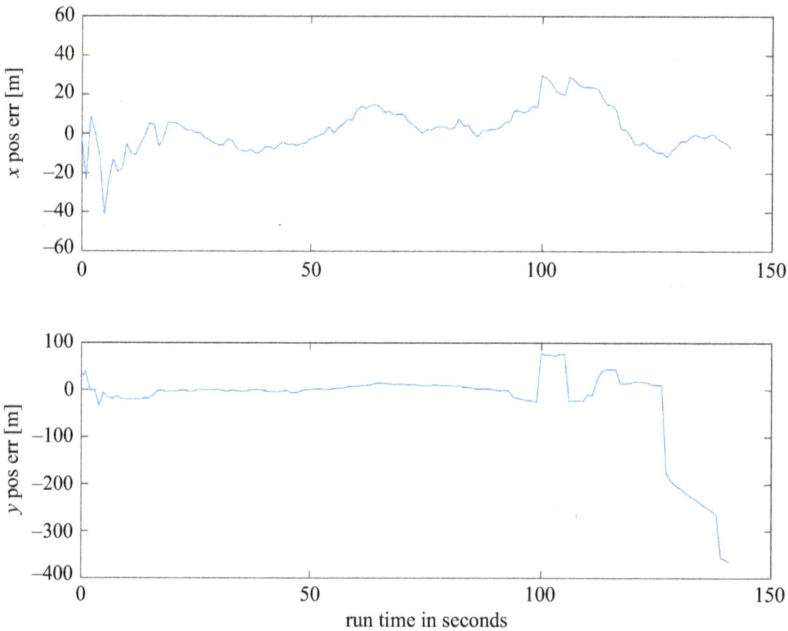

Figure 18.14 Loosely coupled DME/INS Kalman filter position errors with bad geometry

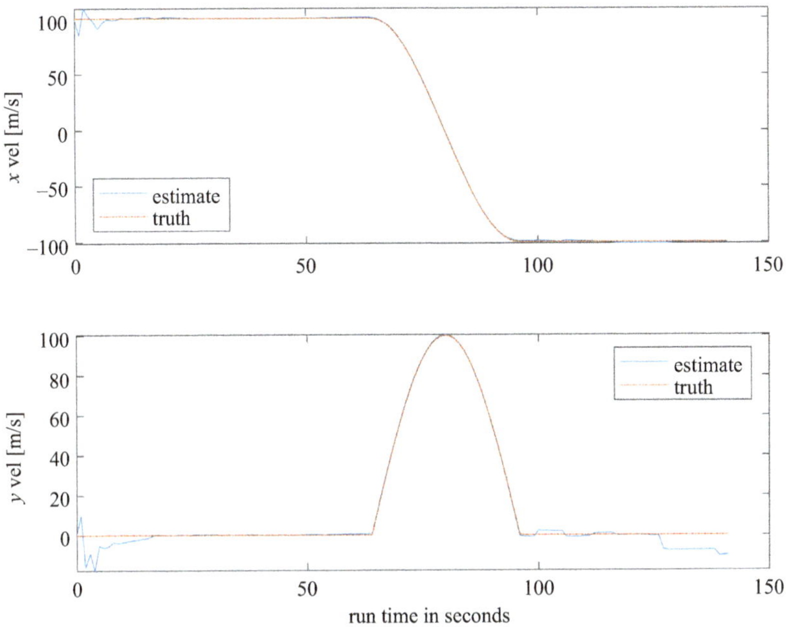

Figure 18.15   True velocity and loosely coupled DME/INS Kalman filter results with bad geometry

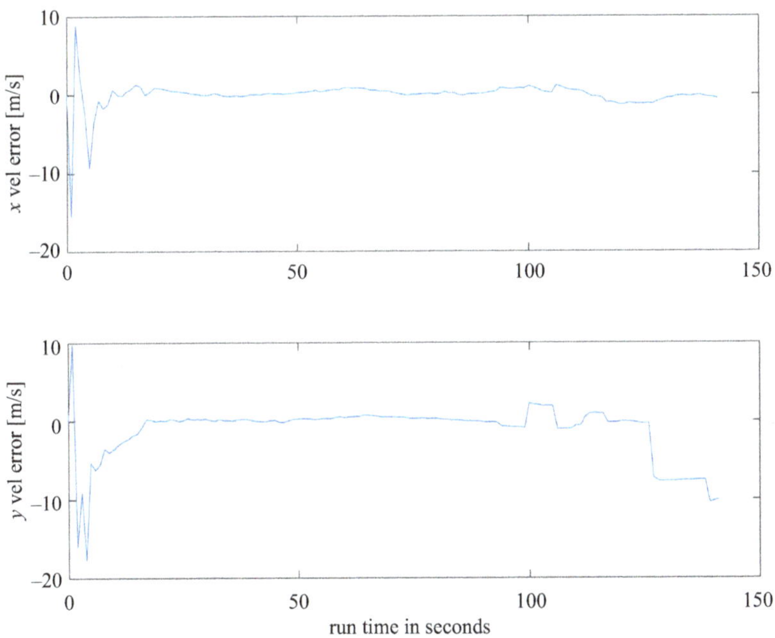

Figure 18.16   Loosely coupled DME/INS Kalman filter velocity errors with bad geometry

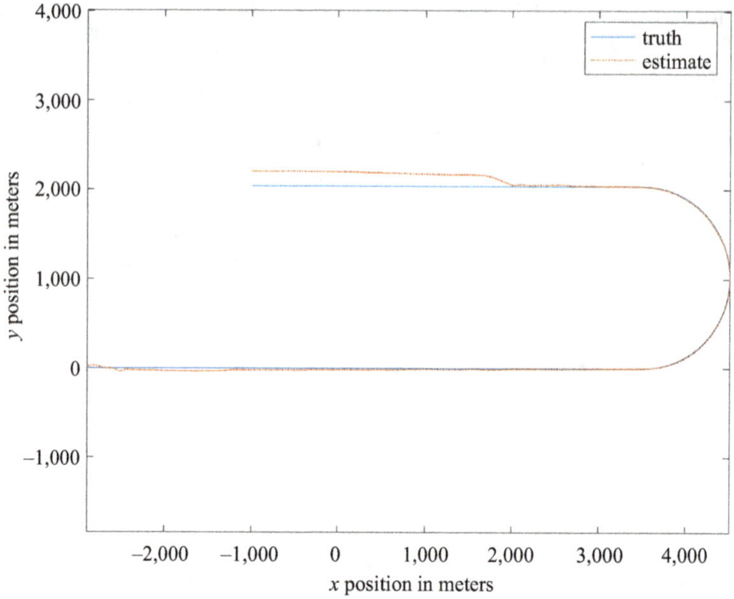

Figure 18.17 True flight path and tightly coupled DME/INS Kalman filter results with bad geometry

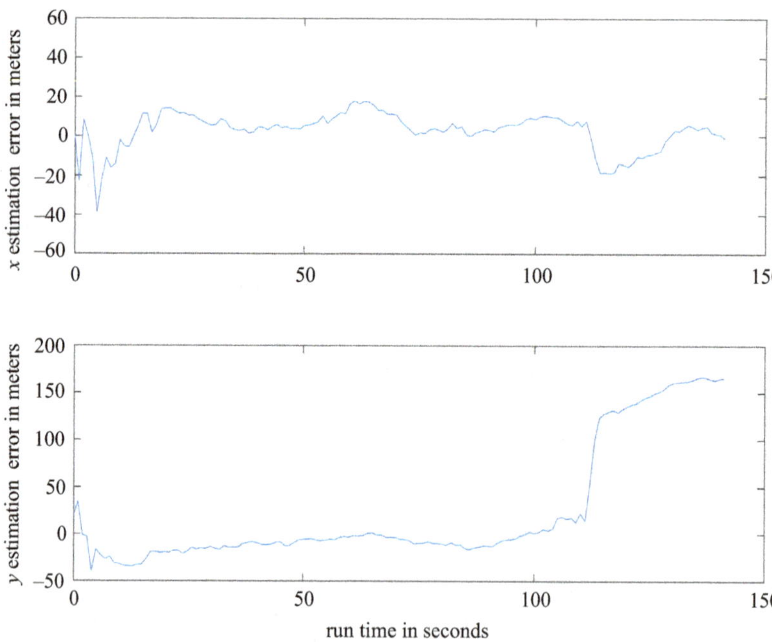

Figure 18.18 Tightly coupled DME/INS Kalman filter position errors with bad geometry

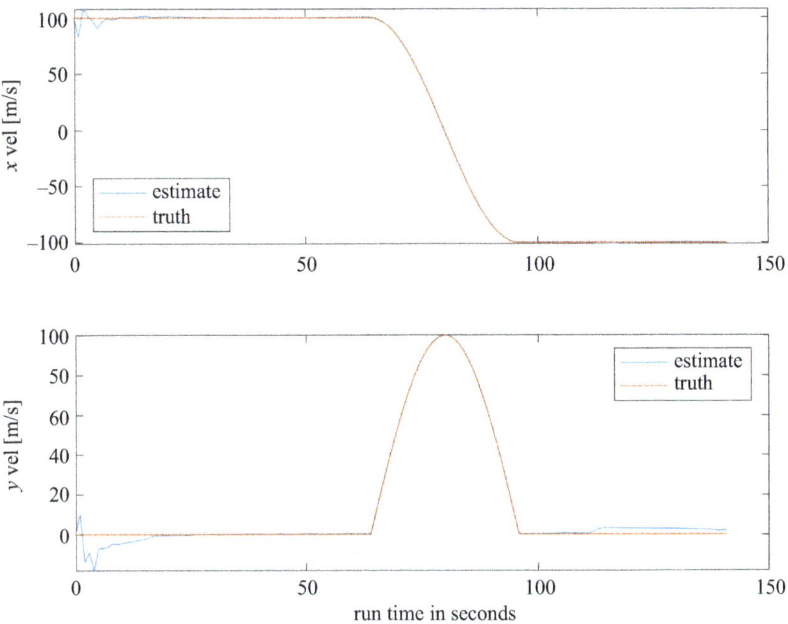

Figure 18.19 True velocity and tightly coupled DME/INS Kalman filter results with bad geometry

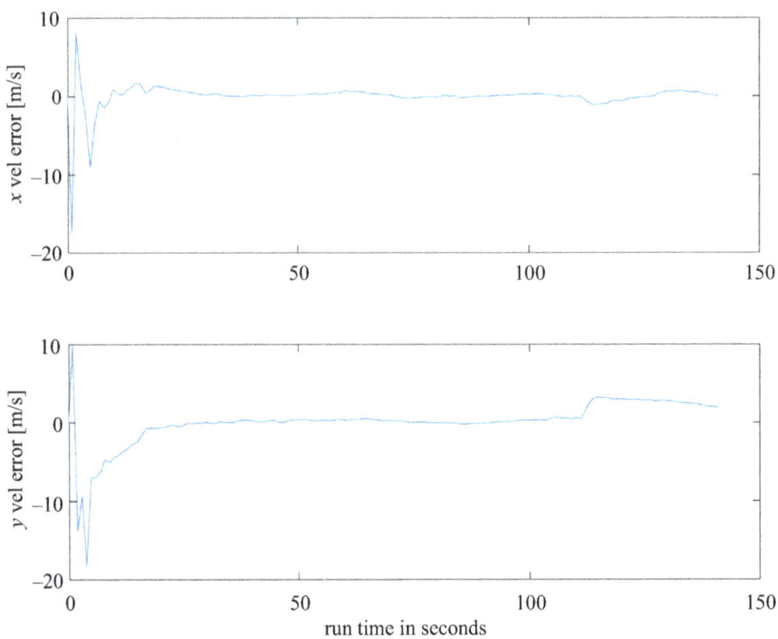

Figure 18.20 Tightly coupled DME/INS Kalman filter velocity errors with bad geometry

18.5 Going forward

In this chapter, we moved from one-dimensional aiding of the INS to two-dimensional aiding. We introduced the concepts of loose-coupling and tight-coupling and along the way discovered three key concepts: (1) the model of the inertial errors (that we are trying to estimate) impacts filter performance and simple models (such as the one used in this chapter) are insufficient for optimum performance; (2) tight-coupling is generally superior to loose-coupling in that the radionavigation system geometry is automatically taken into account whereas it is more difficult to do so in loose-coupling; and (3) since loose-coupling is typically required when the aiding system does not make raw measurement data available on the interface, the cascaded filter problem must be addressed. Specifically, the data from the radionavigation system must be subsampled (decimated) to account for the serial correlation in the output/processed data due to filtering in the radionavigation system itself.

Before we can go any further, we must develop a higher-fidelity inertial error model and that is the topic of the next chapter.

Reference

[1] Brown RG, Hwang PYC. *Introduction to Random Signals and Applied Kalman Filtering: with MATLAB® Exercises*. 4th ed. Hoboken, NJ: John Wiley & Sons, Inc.; 2012.

Chapter 19

Inertial error modeling

19.1 Introduction

At this point, we have established almost all of the necessary background finally to design a Kalman filter to aid the inertial system with GPS. The one remaining item is the determination of the inertial error model. Recall that the GPS/INS complementary Kalman filter estimates the *errors* in the INS. It does not estimate the total position, velocity and attitude of the vehicle. In order to design the Kalman filter, we need to define the system equation that models how the state variables change over time. The states are the inertial position error, inertial velocity error, inertial attitude error, and some related parameters that we will describe later. So before we can start the Kalman filter design, we need a model that describes how the inertial errors behave.

19.2 Reference frames for inertial error analysis

The three commonly used reference frames in inertial navigation error analysis were described as early as 1963 by Pinson [1] and more recently (i.e., in the era of strapdown systems) in 1993 by Siouris [2] and in 2007 by Pue [3]. Before we define them, let us first consider the notional true and computed positions depicted on a portion of the Earth as illustrated in Figure 19.1. The "computed position" is the position computed by the inertial navigation system. The true and computed positions are obviously separated by the inertial system position error.

Figure 19.1 also illustrates two of the reference frames needed for the error analysis. For the sake of illustration convenience, east-north-up (ENU) frames are shown (as opposed to NED frames, just for the sake of visual clarity). Regardless of the convention chosen, the "True" frame is the frame located at the true position and is aligned with the reference directions. The "Computer" frame is the frame located at the computed position and is aligned with the reference directions (*at the computed position*).

A two-dimensional depiction of the three reference frames is shown in Figure 19.2. The third frame is the so-called "Platform" frame or p-frame. In a gimbaled system, the platform frame is the frame defined by the accelerometers that are mounted on the physical gimbaled platform (ignoring non-orthogonality of the accel axes). In a strapdown system, the platform frame is the reference frame *computed*

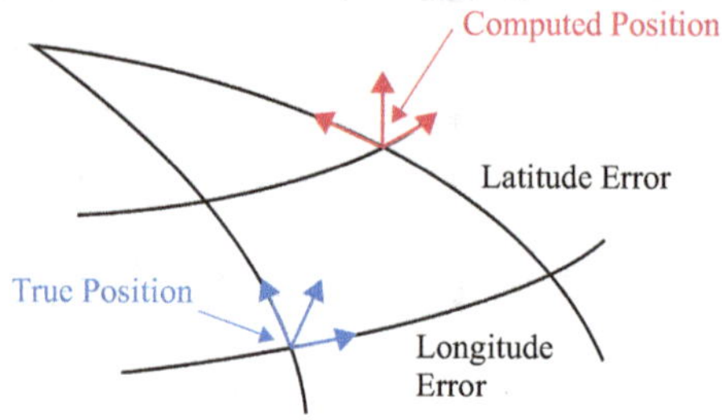

Figure 19.1 Depiction of the true position and inertial system-computed position along with an east-north-up coordinate frame at each [3]

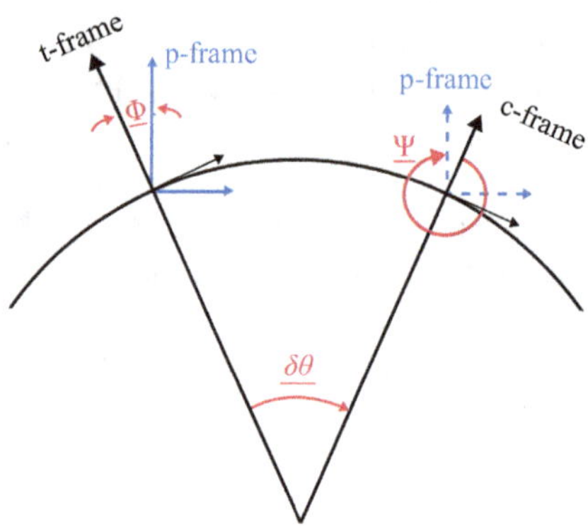

Figure 19.2 Depiction of the angular relationships between the True (t), Platform (p) and Computer (c) frames. The t-frame is located at the true position and the c-frame is located at the computed position. $\underline{\delta\Theta}$ is the vector angular position error. $\underline{\phi}$ is the vector angular difference between the t-frame and the p-frame and thus is the total computed attitude error. $\underline{\psi}$ is the vector angular difference between the c-frame and the p-frame. The total attitude error, $\underline{\phi}$, is the sum of $\underline{\delta\Theta}$ and $\underline{\psi}$ [3]

by the inertial system (e.g., the computed navigation frame or computed n-frame). This is *not* to be confused with the aforementioned c-frame. The p-frame is the *actual* orientation of the *actual* accelerometer triad (in a gimbaled system) and is obviously located at the true position of the vehicle. The c-frame is a frame which is locally-level at the (erroneously) computed position. Figure 19.2 depicts the three angular vectors that relate the three frames:

$$\underline{\delta\Theta} = \begin{bmatrix} \delta\Theta_x & \delta\Theta_y & \delta\Theta_z \end{bmatrix}^T \tag{19.1}$$

$$\underline{\phi} = \begin{bmatrix} \phi_x & \phi_y & \phi_z \end{bmatrix}^T \tag{19.2}$$

$$\underline{\psi} = \begin{bmatrix} \psi_x & \psi_y & \psi_z \end{bmatrix}^T \tag{19.3}$$

where the "T" denotes vector transpose (*not* the True frame). $\underline{\delta\Theta}$ is the vector angular position error and is the angular difference between the True and Computer frames. $\underline{\phi}$ is the vector attitude error and is the angular difference between the True and Platform frames. $\underline{\psi}$ is the angular difference between the Platform and Computer frames and has a physical significance that we will explore shortly. Although all three angular vectors are errors, the "δ" is only used in the case of the angular position error since the true position angles are sometimes denoted by:

$$\underline{\Theta} = \begin{bmatrix} \Theta_x & \Theta_y & \Theta_z \end{bmatrix}^T \tag{19.4}$$

This is the convention we will use but it is not universal, unfortunately, and some authors will denote the position error with $\underline{\Theta}$. Furthermore, Equation (19.1) provides some potential for confusion since a similar symbol is used to denote the incremental angular measurements provided by the gyros (e.g., the "delta thetas"). In this book, the convention is that incremental angular measurements are denoted $\Delta\theta$ (with an upper case "delta" and a lower case "theta"). Generally this potential ambiguity may be resolved by the context in which the symbols are being used.

As mentioned above, the error in the inertially determined attitude ($\underline{\phi}$) is the difference between the true frame and the platform frame. Similarly, the angular position error ($\underline{\delta\Theta}$) is the difference between the true frame and the computer frame. But what is the purpose of the so-called "psi-angle" ($\underline{\psi}$)? Since the psi-angle is the difference between the platform frame and the computer frame, it follows that:

$$\underline{\phi} = \underline{\delta\Theta} + \underline{\psi} \tag{19.5}$$

In other words, the total attitude error can be viewed as consisting of two components. $\underline{\delta\Theta}$ is the component of attitude error that is strictly due to position error. Note that the computer frame is locally level and correctly points in the reference directions (e.g., ENU or NED) but is nevertheless rotated with respect to the true frame. Thus, $\underline{\psi}$ is the remainder of the attitude error. As we will see shortly, $\underline{\psi}$ is driven by the errors in the gyros (whereas $\underline{\delta\Theta}$ is not a direct function of the gyro errors). Thus, it can be said that $\underline{\delta\Theta}$ is the component of attitude error due to position error and $\underline{\psi}$ is the component of attitude error that is due to gyro error. This approach decouples some of the error equations and makes the analysis a little bit simpler. It is a commonly used technique.

If you are still struggling with the notion of "attitude error due to position error," consider the following: assume an inertial navigation system is mounted in a stationary vehicle and is perfect in every way except for one level accelerometer. The gyros are perfect, the initialization is perfect, the knowledge of gravity is perfect, et cetera (for the sake of this example, ignore Earth rotation). But one of the level accelerometers has a bias. So what happens over time? The INS erroneously computes that it is moving in the direction of the accelerometer bias. Due to Earth curvature, transport rate gets applied to the computed navigation frame (i.e., the INS adjusts what it thinks is local level). In a gimbaled system, torquing commands are computed that physically torque the platform (in this case, erroneously away from the true local level). In a strapdown case, the INS adjusts the body-to-nav direction cosine matrix (or quaternion) to account for this perceived nav-frame rotation.

Now look back at Figure 19.2. Although the INS is located at the true position, the accel bias, at some point, causes the INS to think that it is at the (erroneously) computed position. Since the only error in the whole system is the accel bias, the computed navigation frame is the c-frame shown in the figure. In this example, $\underline{\psi} = 0$ since there are no gyro errors. Thus in this example $\underline{\phi} = \underline{\delta\Theta}$ and the c-frame and p-frame are coincident. There is attitude error in this case, but it is due only to position error.

In general, of course, there will be both position errors and gyro errors. As a result, in general the three reference frames are not coincident and all three angle error vectors in (19.5) are non-zero.

19.3 Two primary INS error models

There are two primary approaches to the development of the inertial error model. One, sometimes referred to as the "perturbation" or "phi-angle" approach, derives the equations for $\underline{\phi}$ directly. This obviously has the advantage of providing the equations that describe the total attitude error (it also provides a direct description of velocity error among other advantages). An alternative, known as the "psi-angle" approach, derives separate equations for $\underline{\psi}$ and for $\underline{\delta\Theta}$. The two can then be combined via (19.5) to determine the total attitude error (as we will see, velocity error can then be determined as well).

At first the psi-angle approach appears to be a longer way to achieve the same result as the phi-angle approach. However, the psi-angle approach effectively decouples the error equations in such a way as to make them easier to code and debug. This makes the psi-angle approach a good way to ease into the topic of inertial error modeling. However, it must be emphasized that both approaches are widely used throughout the industry. They are derived and compared in the seminal article by Benson [4]. At this point in our study, the reader should simply be aware that the phi-angle approach consists of a separate set of error equations that can be utilized in the Kalman filter just as well as the psi-angle approach. For now, we will focus on the psi-angle approach.

19.4 Psi-angle INS error model

Recall why we are bothering with all this in the first place: we need a model of how the inertial errors change over time so that the Kalman filter can do a better job of estimating the errors. A differential equation does that since, by definition, it describes the rate-of-change of the given quantity. As derived by Benson [4] and described further by Siouris [2] and Pue [3], the psi-angle error model is given by the following three vector differential equations:

$$\underline{\delta\dot{R}}^t = -\underline{\omega}^t_{et} \times \underline{\delta R}^t + \underline{\delta V}^t \tag{19.6}$$

$$\underline{\delta\dot{V}}^c = C^c_b \underline{\delta f}^b + \underline{\delta g}^c - \underline{\psi}^p \times \underline{f}^c - (2\underline{\omega}^c_{ie} + \underline{\omega}^c_{ec}) \times \underline{\delta V}^c \tag{19.7}$$

$$\underline{\dot{\psi}}^p = -\underline{\omega}^p_{ip} \times \underline{\psi}^p - C^p_b \underline{\delta\omega}^b_{ib} \tag{19.8}$$

where $\underline{\delta R}$ is the vector of linear (not angular) components of position error and $\underline{\delta V}$ is the vector of velocity error components. $\underline{\delta f}$ is the error in measured specific force. $\underline{\delta g}$ is the error in computed gravity. $\underline{\delta\omega}_{ib}$ is the error in measured angular rate. The angular rate term in (19.6) is transport rate and the angular rate term in the cross-product of (19.8) is spatial rate.

The reader is referred to the references for the details of the derivations. Still, we need to achieve some level of comfort with the results. As Benson describes [4], to a first-order approximation, any given error quantity (e.g., $\underline{\psi}$, $\underline{\delta V}$, $\underline{\delta R}$, $\underline{\delta f}$, $\underline{\delta\omega}$) may be interchangeably expressed in any of the three frames (t, p, c). Thus, for example, to first order: $\underline{\delta V}^t = \underline{\delta V}^c$. The first-order approximation holds very well so long as the position angle errors and psi angle components are small (a good assumption for nav-grade inertial navigation systems). Note also that the use of the true frame in equation (19.6) is meant to indicate that the results are valid in any "true" frame (e.g., n-frame, w-frame, et cetera).

The use of the computer frame in Equation (19.7) is the direct result of the use of the psi-angle error model. As shown by Benson [4], the perturbation error model yields an expression directly for the velocity error expressed in the true frame. Strictly speaking, the velocity error determined from the use of the psi-angle model must be converted from the computer frame to the true frame for it to be useful. However, if the position angle errors are small (i.e., the first-order approximation mentioned earlier), there is little difference between the velocity error expressed in the computer frame or the true frame. If the INS is operated over long periods of time without external aiding, however, the position angle errors may no longer be small and the differences between frames may no longer be negligible. Benson [4] provides simulation examples for a high speed aircraft (592 knots) traveling either due east or due west at high latitude (75 deg N) with either an east accelerometer bias of 100 micro-g or a north gyro bias of 0.01 deg/hour. In each case, the difference in the north velocity errors computed with the perturbation approach and the psi-angle approach becomes non-trivial after approximately one- to two-tenths of an hour (the east velocity errors are identical for both approaches for these scenarios). This is somewhat of an extreme case for aircraft applications. Significant differences would take longer to manifest at lower latitudes and/or lower true speed.

Four changes are needed in Equations (19.6)–(19.8) to make them more useful for real-time error estimation in the Kalman filter of an aided-inertial system. The first is the spatial rate term in Equation (19.8). In real-time, the system thinks the INS is located at the computed position and thus computes spatial rate (the sum of earth rate and transport rate) in the computer frame. It is, therefore, helpful to replace (19.8) with an equivalent expression in which the platform-frame terms are replaced with their equivalent computer-frame terms.

The second change is the replacement of $\underline{\psi}^p$ with $\underline{\psi}^c$ in Equation (19.7). As mentioned earlier, to first order the two are the same. Furthermore, the change maintains the explicit coupling between (19.8) and (19.7).

The third change is in Equation (19.6) and involves expressing the equation in the computer frame instead of the true frame. The relation holds as long as the position angle errors are small. Further, it clarifies the direct coupling between (19.6) and (19.7). More specifically, as will be developed later, the Kalman filter exploits (19.7) in its estimation of the velocity error. The velocity error, in turn, is coupled to (19.6) that is exploited in the estimation of position error. Again, the use of the error equations in the Kalman filter will be developed later. With the aforementioned three (of four) changes, the error equations take a form similar to that derived by Pue [3]:

$$\underline{\delta \dot{R}}^c = -\underline{\omega}^c_{ec} \times \underline{\delta R}^c + \underline{\delta V}^c \tag{19.9}$$

$$\underline{\delta \dot{V}}^c = C^c_b \underline{\delta f}^b + \underline{\delta g}^c - \underline{\psi}^c \times \underline{f}^c - (2\underline{\omega}^c_{ie} + \underline{\omega}^c_{ec}) \times \underline{\delta V}^c \tag{19.10}$$

$$\underline{\dot{\psi}}^c = -\underline{\omega}^c_{ic} \times \underline{\psi}^c - C^c_b \underline{\delta\omega}^b_{ib} \tag{19.11}$$

The fourth and final change deals with the fact that, in real time, the system does not have access to the body-to-computer DCM. The presence of gyro errors causes the system only to have access to the body-to-[estimated local level frame] DCM. Frequently the chosen local level reference frame is the navigation (or wander frame) and thus the system only has access to the body-to-estimated-nav DCM. As long as $\underline{\psi}$ is small (a good assumption for nav-grade IMUs), then:

$$C^{\hat{n}}_b \approx C^c_b \tag{19.12}$$

Equation (19.12) can be substituted into Equations (19.10) and (19.11) to yield a set of equations that can be used in real time estimation (as we will learn later).

Having addressed the coordinate frame issues in this error model, we can move on to make some general observations. First, notice the psi-angle (19.11) is driven by the gyro errors ($\underline{\delta\omega}^b_{ib}$). Once a non-zero psi-angle is established, the components propagate and cross-couple through the cross-product of the psi-angle with spatial-rate (the first term in (19.11)). The cross-product accounts for the Coriolis effect due to the rotation of the computer frame relative to the inertial frame.

Velocity error (19.10) is driven by multiple effects: accelerometer errors, gravity computation errors, and the cross-product of psi-angle with true specific force. The error also propagates through the cross-product of velocity error with earth rate and transport rate (note that this term is very similar to a term in the inertial velocity update expression).

Finally, position error (expressed in linear, not angular, units in Equation (19.9)) is driven by velocity error but also propagates with the cross-product of transport rate and position error. Again, the cross-product accounts for the Coriolis effect due to the rotation of the computer frame relative to the Earth frame.

The error in computed gravity (in Equation (19.10)) is, to first order, given by:

$$\underline{\delta g}^c = \begin{bmatrix} -\dfrac{g_0}{R_0} & 0 & 0 \\ 0 & -\dfrac{g_0}{R_0} & 0 \\ 0 & 0 & \dfrac{2g_0}{R_0 + h} \end{bmatrix} \underline{\delta R}^c \tag{19.13}$$

where g_0 is the nominal magnitude of gravity, R_0 is the nominal radius of the Earth, and h is the altitude above the reference ellipsoid. The value of gravity at 45 degrees latitude and zero altitude can be used, for example. Since the Earth is best described by an ellipsoid with two orthogonal radii of curvature, the geometric mean radius of the Earth at 45 degrees latitude can also be used. For more precision, the local values for gravity and earth radii may be used. Strictly speaking, altitude can be included in the denominator of the horizontal terms (in addition to the vertical). However, to first order this is not necessary.

Equation (19.13) can be derived from the horizontal and vertical single-channel error models for the INS described in an earlier chapter. The horizontal model (which yielded Schuler oscillations) shows the feedback path consists of the attitude error multiplied by the nominal value of gravity. Since (19.13) (and 19.10) are being evaluated in the computer frame, however, the attitude error is equal to the position angle error ($\underline{\Delta\Theta}$). To first order, the horizontal components of position angle error are given by $\frac{\delta R}{R_0}$. Recall also that the negative sign is needed to account for the negative feedback of gravity in the horizontal channel. The vertical component of (19.13) may be obtained directly from the INS vertical channel error analysis.

19.5 INS error state-space model

In subsequent chapters, we will utilize the inertial error model (Equations (19.9)–(19.13)) in aided-inertial Kalman filters. Before doing so, however, we will first investigate their behavior in a "stand-alone" fashion. Specifically, we will show how they can be used to simulate the performance of an INS in a much more computationally-efficient manner than the "whole value" simulations utilized earlier in this book. Along the way, we will introduce some additional equations that are needed to take the psi-angle results and convert them into forms that are more familiar. To do this, we first need to rewrite Equations (19.9)–(19.11) in state-space form as follows:

$$\begin{bmatrix} \delta\dot{\underline{R}} \\ \delta\dot{\underline{V}} \\ \dot{\underline{\psi}} \end{bmatrix} = \begin{bmatrix} [-\underline{\omega}^c_{ec}\times] & I & 0 \\ W & [-(2\underline{\omega}^c_{ie} + \underline{\omega}^c_{ec})\times] & [\underline{f}^c\times] \\ 0 & 0 & [-(\underline{\omega}^c_{ie} + \underline{\omega}^c_{ec})\times] \end{bmatrix} \begin{bmatrix} \delta\underline{R} \\ \delta\underline{V} \\ \underline{\psi} \end{bmatrix}$$

$$+ \begin{bmatrix} 0 & 0 & 0 \\ 0 & C^c_b & 0 \\ 0 & 0 & -C^c_b \end{bmatrix} \begin{bmatrix} 0 \\ \delta\underline{f}^b \\ \delta\underline{\omega}^b_{ib} \end{bmatrix} \tag{19.14}$$

where "0" is a 3×3 matrix of zeros, "I" is a 3×3 identity matrix, W is the 3×3 matrix from (19.13) and the terms in square brackets with the × at the end are the skew-symmetric matrix forms (also known as cross-product form) of the respective vectors. The first term after the equal sign is thus a 9×9 matrix. It is the core of the inertial error model. A more compact form of (19.14) is:

$$\begin{bmatrix} \delta\dot{\underline{R}} \\ \delta\dot{\underline{V}} \\ \dot{\underline{\psi}} \end{bmatrix} = \begin{bmatrix} [-\underline{\omega}_{ec}^c \times] & I & 0 \\ W & [-(2\underline{\omega}_{ie}^c + \underline{\omega}_{ec}^c)\times] & [\underline{f}^c \times] \\ 0 & 0 & [-(\underline{\omega}_{ie}^c + \underline{\omega}_{ec}^c)\times] \end{bmatrix} \begin{bmatrix} \delta\underline{R} \\ \delta\underline{V} \\ \underline{\psi} \end{bmatrix}$$

$$+ \begin{bmatrix} 0 & 0 \\ C_b^c & 0 \\ 0 & -C_b^c \end{bmatrix} \begin{bmatrix} \delta\underline{f}^b \\ \delta\underline{\omega}_{ib}^b \end{bmatrix} \tag{19.15}$$

Equation (19.14) or (19.15) can be abbreviated in conventional state-space form as:

$$\dot{\underline{x}} = F_{INS}\,\underline{x} + G\,\underline{u} \tag{19.16}$$

where $\underline{x}$ is the state vector consisting of the position error vector, velocity error vector and psi-angle vector; F_{INS} is the aforementioned 9×9 matrix; G is the input connection matrix consisting of two instances of the body-to-computer DCM; and $\underline{u}$ is the input vector consisting of the accelerometer error vector and gyro error vector.

Under the assumption that F_{INS} is time-invariant, the discrete-time solution of (19.16) is given by [5]:

$$\underline{x}_d[k+1] = \Phi\,\underline{x}_d[k] + G_d\,\underline{u}_d[k] \tag{19.17}$$

where

$$\underline{x}_d[k] = \underline{x}[kT] \tag{19.18}$$

$$\underline{u}_d[k] = \underline{u}[kT] \tag{19.19}$$

$$\Phi = e^{F_{INS}T} \tag{19.20}$$

$$G_d = \int_0^T e^{F_{INS}\lambda} G\,d\lambda \tag{19.21}$$

the "d" refers to the discrete-time and $T =$ the discrete-time interval (i.e., sampling period).

Strictly speaking, the assumption of time-invariance is false. The components of both the earth-rate and transport-rate vectors will change over time. More importantly, the specific-force vector components will change appreciably during any kind of acceleration (e.g., turns). We can, however, approximate F_{INS} as being constant over a sufficiently short time interval. Care must be taken, then, in choosing an appropriate "T" for the given vehicle dynamics. For static cases in which only sensor errors are being modeled, time intervals on the order of 1–10 sec are acceptable.

The two discrete-time matrices in (19.17) can be approximated by:

$$\Phi \approx I + F_{INS}T + \frac{F_{INS}^2 T^2}{2!} + \frac{F_{INS}^3 T^3}{3!} + \cdots \tag{19.22}$$

$$G_d \approx \sum_{j=0}^{N} \frac{F_{INS}^j T^{j+1}}{(j+1)!} G \tag{19.23}$$

where the total number of terms used is a design choice dependent upon the required level of accuracy. Typically the total is less than 10. If T is sufficiently small, the two matrices can be approximated with as few as two terms.

From Equation (19.14), we see that we can specify/simulate certain levels of accelerometer and gyro errors (via $\underline{u}_d$) and we can iteratively compute (19.17) in a simulation to see how these errors impact the inertially derived position, velocity and attitude. However, the results provided by the discrete-time state vector need some modifications to put them into a more usable form.

The total attitude error is given by (19.5) but the first three elements of the state vector are $\underline{\delta R}$ and not $\underline{\delta \Theta}$. For example, to do the conversion assuming $\underline{\delta R}$ has been specified in the ENU frame (and thus no wander angle):

$$\begin{aligned} \delta\Theta(1) &= -\frac{\delta R(2)}{R} \\ \delta\Theta(2) &= \frac{\delta R(1)}{R} \\ \delta\Theta(3) &= \tan(Lat)\delta\Theta(2) \end{aligned} \tag{19.24}$$

Having specified the position angle error vector, the earth-to-nav DCM estimated by the INS is given by:

$$\hat{C}_e^n = C_e^c = C_n^c C_e^n = (I - [\underline{\delta\Theta}\times])C_e^n \tag{19.25}$$

where the "hat" indicates an estimated quantity. From this DCM, the estimated latitude and longitude values can be extracted and then compared to the truth to compute the error.

Equation (19.5) can be used to compute the total attitude angle error vector ($\underline{\phi}$) given $\underline{\delta\Theta}$ (computed from $\underline{\delta R}$) and $\underline{\psi}$ (obtained from the state vector). The total attitude error can then be determined by first computing the estimated body-to-nav DCM:

$$\hat{C}_b^n = C_n^p C_b^n = (I - [\underline{\phi}\times])C_b^n \tag{19.26}$$

Estimated Euler angles can be extracted from this DCM and then compared to the truth to compute the error.

Finally, note also that since the velocity error components are given in local-level coordinates, they can be used directly (with the aforementioned assumption that the position errors are sufficiently small such that the velocity error components are the same in both the computer-frame and the true-frame).

19.6 Example: east gyro bias

The following plots illustrate the results of using the state-space inertial error model to simulate a 0.01 deg/hour east gyro bias. Figure 19.3 shows the position and velocity errors over a 4-hour period. As we would expect with an east gyro bias, the dominant impact is on the latitude. We see a Schuler oscillation that rides on top of a longer term variation. In addition, there is cross coupling due to earth rate. The presence of earth rate causes a cross coupling between the east and north errors and so, although the error starts off dominant in the north-south direction, eventually it cross couples into east-west and we start to pick up longitude error over time as well. This is the identical result we saw with the so-called "whole value" simulation discussed in the earlier Inertial Error Simulation chapter. We also see identical behavior in the attitude error (Figure 19.4). Figure 19.5 depicts the level components of $\underline{\delta\Theta}$. Recall that in addition to its role as the angular position error vector, it is also the component of attitude error due to angular position error.

Figure 19.6 shows the components of $\underline{\psi}$. You can see they slowly vary over the course of this simulation. Recall they are being driven by the gyro error. A gyro bias is being simulated but the components of $\underline{\psi}$ are not increasing at a constant rate because of the Coriolis term in the psi angle Equation (19.11). As mentioned above, a side-by-side comparison of the results produced with the state-space error model and a full inertial system simulation (under various input conditions) would show good agreement. However, the first-order assumptions built into the state-space model will gradually lose validity as the state variables grow over time.

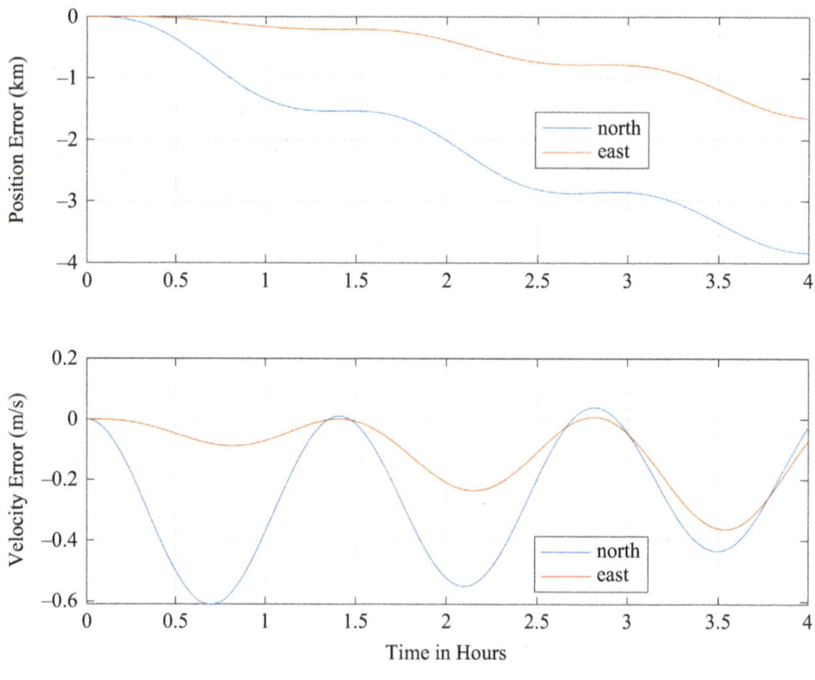

Figure 19.3 Inertial error state-space model position and velocity error results for a stationary platform at 45 degrees North latitude and zero altitude with an east gyro bias of 0.01 deg/hour

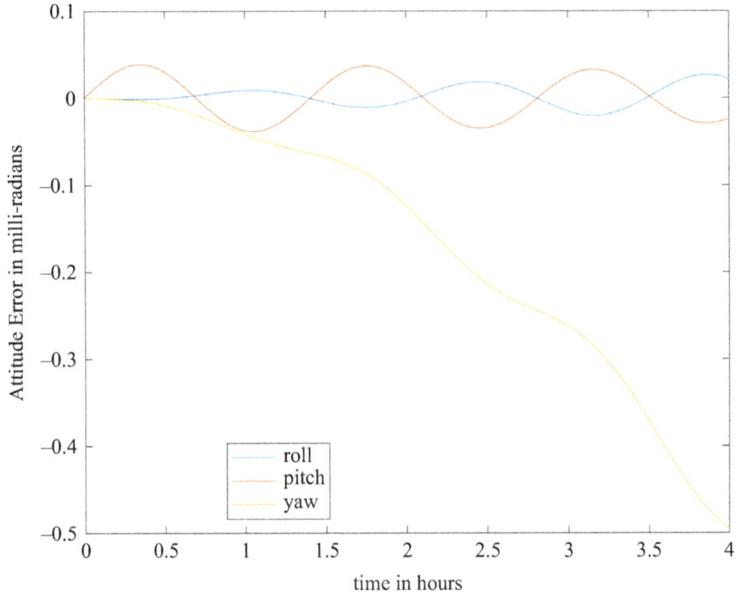

Figure 19.4 *Inertial error state-space model Euler angle components for a stationary platform at 45 degrees North latitude and zero altitude with an east gyro bias of 0.01 deg/hour*

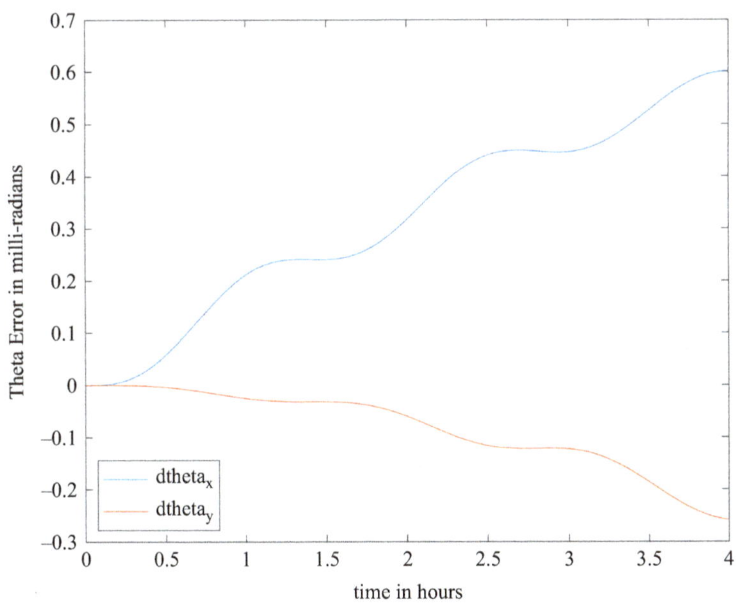

Figure 19.5 *Inertial error state-space model position angle errors for a stationary platform at 45 degrees North latitude and zero altitude with an east gyro bias of 0.01 deg/hour*

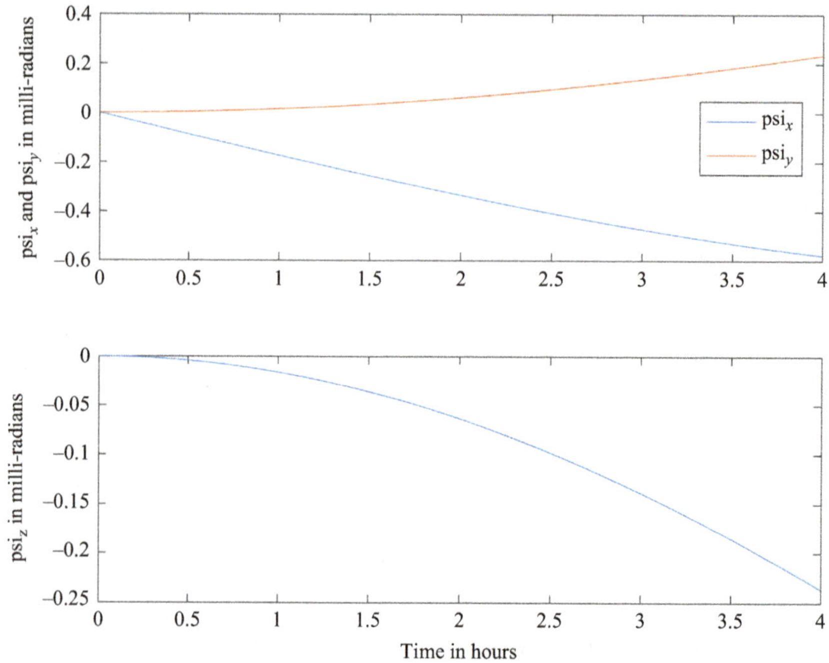

Figure 19.6 Inertial error state-space model psi-angle components for a stationary platform at 45 degrees North latitude and zero altitude with an east gyro bias of 0.01 deg/hour

References

[1] Pinson J. Inertial guidance for cruise vehicles. In: Leondes C, editor. *Guidance and Control of Aerospace Vehicles*. New York, NY: McGraw-Hill; 1963.

[2] Siouris G. *Aerospace Avionics Systems—A Modern Synthesis*. San Diego, CA: Academic Press; 1993.

[3] Pue A. *Integration of GPS with Inertial Navigation Systems [Short Course Notes]*. NavtechGPS; 2007.

[4] Benson D. A comparison of two approaches to pure-inertial and Doppler-inertial error analysis. *IEEE Transactions on Aerospace and Electronic Systems*. 1975;AES-11(4):288–290.

[5] Kamen E, Heck B. *Fundamentals of Signals and Systems using MATLAB®*. Upper Saddle River, NJ: Prentice-Hall; 1996.

Chapter 20
GNSS-aided INS: loose coupling

20.1 Introduction

Having established a state-space model that describes inertial error propagation, we have at last arrived at the point where a complete aided-inertial Kalman filter can be described. Although we showed in a previous chapter that loose coupling is inferior to tight coupling (i.e., loosely coupled filters cannot de-weight aiding measurements under conditions of bad geometry), we will study it nevertheless since the measurement equation has a simpler form than it does in tight coupling and thus it allows us to ease into the complexity of the full aided-inertial filter. In addition, it must be remembered that in some instances, the designer is forced to use loose coupling. This occurs, for example, when the aiding source (e.g., GNSS receiver) that is to be utilized has already been installed on a given platform (e.g., aircraft or land vehicle) but provides only processed data (e.g., position, velocity) on the external interface.

20.2 Loosely coupled Kalman filter with generic position aiding source

We start off considering a simplified case in which the aiding source only provides a position solution and, furthermore, the errors of the aiding source consist of independent white Gaussian noise. As a result, we can derive the continuous-time state equation from the inertial error state-space model presented in the previous chapter (repeated here for convenience):

$$\begin{bmatrix} \delta\dot{\underline{R}} \\ \delta\dot{\underline{V}} \\ \dot{\underline{\psi}} \end{bmatrix} = \begin{bmatrix} [-\underline{\omega}_{ec}^{c}\times] & I & 0 \\ W & [-(2\underline{\omega}_{ie}^{c}+\underline{\omega}_{ec}^{c})\times] & [\underline{f}^{c}\times] \\ 0 & 0 & [-(\underline{\omega}_{ie}^{c}+\underline{\omega}_{ec}^{c})\times] \end{bmatrix} \begin{bmatrix} \delta\underline{R} \\ \delta\underline{V} \\ \underline{\psi} \end{bmatrix} + \begin{bmatrix} 0 & 0 & 0 \\ 0 & C_b^c & 0 \\ 0 & 0 & -C_b^c \end{bmatrix} \begin{bmatrix} 0 \\ \delta\underline{f}^{b} \\ \delta\underline{\omega}_{ib}^{b} \end{bmatrix} \tag{20.1}$$

Since the gyro and accel errors are unknown in a real system, they must be incorporated into the state vector as additional variables that must be estimated. Although complete gyro and accel error models include scale factor errors, temperature effects, acceleration, and jerk-dependencies as well as noise, first-order approximations frequently ignore these and focus only on the so-called "bias" errors. At this point, we should remind ourselves that gyro and accel "bias" errors are more correctly "bias-like" errors (although nobody uses this term). They are bias-like insofar as they vary slowly with time. The slower the variation, the more bias-like they are. All of this is justification for the fact that these "bias" errors are typically modeled with a time-varying stochastic process rather than a pure constant.

As was described in an earlier chapter (see also the appendix on random processes), the gyro and accel bias errors are typically modeled with first-order Gauss–Markov processes. By incorporating the gyro and accel errors into the state vector, and modeling them with first-order Gauss–Markov processes, we can modify Equation (20.1) to yield the following continuous-time system equation:

$$\begin{bmatrix} \delta\dot{\underline{R}} \\ \delta\dot{\underline{V}} \\ \dot{\underline{\psi}} \\ \delta\dot{\underline{f}}^b \\ \delta\dot{\underline{\omega}}_{ib}^b \end{bmatrix} = \begin{bmatrix} [-\underline{\omega}_{ec}^c\times] & I & 0_{3x3} & 0_{3x3} & 0_{3x3} \\ W & [-(2\underline{\omega}_{ie}^c + \underline{\omega}_{ec}^c)\times] & [\underline{f}^c\times] & C_b^c & 0_{3x3} \\ 0_{3x3} & 0_{3x3} & [-(\underline{\omega}_{ie}^c + \underline{\omega}_{ec}^c)\times] & 0_{3x3} & -C_b^c \\ 0_{3x3} & 0_{3x3} & 0_{3x3} & \frac{-1}{\tau_{acc}}I & 0_{3x3} \\ 0_{3x3} & 0_{3x3} & 0_{3x3} & 0_{3x3} & \frac{-1}{\tau_{gyr}}I \end{bmatrix} \begin{bmatrix} \delta\underline{R} \\ \delta\underline{V} \\ \underline{\psi} \\ \delta\underline{f}^b \\ \delta\underline{\omega}_{ib}^b \end{bmatrix} + \begin{bmatrix} 0_{3x1} \\ 0_{3x1} \\ 0_{3x1} \\ \underline{\eta}_{acc} \\ \underline{\eta}_{gyr} \end{bmatrix} \tag{20.2}$$

where $\underline{\eta}_{acc}$ and $\underline{\eta}_{gyr}$ are the Gaussian white noise inputs for the first-order Gauss–Markov models and all Is are 3×3 identity matrices.

The upper left 9×9 submatrix in (20.2) is the core inertial error model and it will be convenient, later, to give it a name. Thus, we define:

$$F_{INS} = \begin{bmatrix} [-\underline{\omega}_{ec}^c\times] & I & 0_{3x3} \\ W & [-(2\underline{\omega}_{ie}^c + \underline{\omega}_{ec}^c)\times] & [\underline{f}^c\times] \\ 0_{3x3} & 0_{3x3} & [-(\underline{\omega}_{ie}^c + \underline{\omega}_{ec}^c)\times] \end{bmatrix} \tag{20.3}$$

Equation (20.2) can thus be rewritten as:

$$\begin{bmatrix} \delta\dot{\underline{R}} \\ \delta\dot{\underline{V}} \\ \dot{\underline{\psi}} \\ \delta\dot{\underline{f}}^b \\ \delta\dot{\underline{\omega}}_{ib}^b \end{bmatrix} = \begin{bmatrix} \multicolumn{3}{c}{} & 0_{3x3} & 0_{3x3} \\ \multicolumn{3}{c}{F_{INS}} & C_b^c & 0_{3x3} \\ \multicolumn{3}{c}{} & 0_{3x3} & -C_b^c \\ 0_{3x3} & 0_{3x3} & 0_{3x3} & \frac{-1}{\tau_{acc}}I & 0_{3x3} \\ 0_{3x3} & 0_{3x3} & 0_{3x3} & 0_{3x3} & \frac{-1}{\tau_{gyr}}I \end{bmatrix} \begin{bmatrix} \delta\underline{R} \\ \delta\underline{V} \\ \underline{\psi} \\ \delta\underline{f}^b \\ \delta\underline{\omega}_{ib}^b \end{bmatrix} + \begin{bmatrix} 0_{3x1} \\ 0_{3x1} \\ 0_{3x1} \\ \underline{\eta}_{acc} \\ \underline{\eta}_{gyr} \end{bmatrix} \tag{20.4}$$

Over sufficiently short intervals of time (i.e., such that the specific-force vector and angular rates can be considered constant), the discrete-time solution of Equation (20.2) yields the usual discrete-time system equation:

$$\underline{x}_{k+1} = \Phi_k \underline{x}_k + \underline{w}_k \tag{20.5}$$

where the state transition matrix is given by

$$\Phi_k = e^{F_k \Delta t} \tag{20.6}$$

and where F_k is the 15×15 system dynamics matrix given in Equation (20.2). Recall that the matrix exponential is frequently approximated by the first few terms of the Taylor series expansion. The system noise vector in Equation (20.2) is unknown but it is assumed that its covariance, Q_k, is known or, at least, can be well approximated.

Equations (20.2)–(20.6) completely specify the system equation and thus specify the state vector, the state transition matrix and the system noise covariance matrix. The measurement equation enables us to specify another three terms. Recall the generic form of the measurement equation:

$$\underline{z}_k = H_k \underline{x}_k + \underline{\nu}_k$$

Similar to the DME/INS example described in an earlier chapter, with a generic position aiding source the measurement vector is given by the difference of the INS-determined position and the position determined by the aiding source:

$$\underline{z} = \begin{bmatrix} x_{INS} - x_{AIDING.SOURCE} \\ y_{INS} - y_{AIDING.SOURCE} \\ z_{INS} - z_{AIDING.SOURCE} \end{bmatrix} \tag{20.7}$$

As we have discussed in earlier chapters, this type of observable consists of the sum of the errors of the two systems. The measurement equation is thus:

$$\underline{z} = \begin{bmatrix} I_{3\times3} & 0_{3\times12} \end{bmatrix} \begin{bmatrix} \underline{\delta R} \\ \underline{\delta V} \\ \underline{\psi} \\ \underline{\delta f^b} \\ \underline{\delta \omega_{ib}^b} \end{bmatrix} + \begin{bmatrix} \nu_x \\ \nu_y \\ \nu_z \end{bmatrix} \tag{20.8}$$

where the data matrix, H, is the first matrix after the equal sign and is a 3×15 matrix consisting mostly of zeros except the identity matrix occupying the first three columns. The measurement noise vector consists of the position error components of the aiding source. Its specific values at any given moment are unknown but its covariance matrix, R, is assumed to be known or well approximated (recall that for the case of DME we estimated R by multiplying the ranging error standard deviation by HDOP). The measurement equation has thus provided us with $\underline{z}$ (formed with Equation (20.7)), H and R.

The filter design is complete once the initial prediction of the state vector and the associated initial prediction error covariance matrix have been specified. Since it is assumed that the initial state vector is completely unknown, but is assumed to be zero mean, the initial prediction is given simply by:

$$\hat{\underline{x}}_1^- = \begin{bmatrix} 0 & 0 & \cdots & 0 \end{bmatrix}^T \tag{20.9}$$

The initial prediction error covariance matrix, P_1^-, can be approximated simply as a diagonal matrix in which the non-zero elements are given by the estimated variance of the state variables. Assuming the GNSS receiver is used to initialize position, the initial position error variances can be roughly estimated based on DOPs and something similar can be done for velocity as well. If the platform is approximately stationary (i.e., ignoring wind buffeting, etc.) the velocity error variances will be quite small. Psi-angle variances can be approximated based on the known leveling/gyrocompassing performance of the INS and, finally, the gyro and accel bias variances can be initialized based either on manufacturer specifications or lab test data. Recall that we are concerned here with the residual bias errors that are left after all calibration/compensations have been applied.

20.3 Loosely coupled GNSS/INS Kalman filter

The generic position-aiding filter described above must be modified to accommodate GNSS aiding. The GNSS position errors are not well approximated by independent white Gaussian noise for two reasons. First, in addition to thermal noise, GNSS errors include the effects of ionospheric delay, tropospheric delay and multipath, none of which are noise-like. The atmospheric delays, in particular, are quite bias-like in behavior. Multipath error tends to span a mid-range of frequencies in between the white noise-like behavior of thermal noise and the bias-like behavior of atmospheric delays. As a result, GNSS position error is not well approximated solely by an independent, white Gaussian measurement noise vector ($\underline{\nu}$).

Second, the computed GNSS position will 'jump' when there is a constellation change. That is, assuming the GNSS-computed position is unfiltered (recall that is what we prefer!), a step error will occur when a satellite measurement is dropped from, or added to, the position solution. Again, such behavior is not well approximated by the measurement noise vector. As a result, additional state variables are needed to model the non-noise components of the GNSS position error. This is typically handled by adding so-called "position-bias" states. We will see shortly how they are to be used and we will see what happens if they are not used. The continuous-time system equation is thus expanded from (20.4) to become [1]:

$$\begin{bmatrix} \delta\dot{\underline{R}} \\ \delta\dot{\underline{V}} \\ \dot{\underline{\psi}} \\ \delta\dot{\underline{f}}^b \\ \delta\dot{\underline{\omega}}_{ib}^b \\ \delta\dot{\underline{R}}_{GNSS} \end{bmatrix} = \begin{bmatrix} & & & 0_{3x3} & 0_{3x3} & 0_{3x3} \\ & F_{INS} & & C_b^c & 0_{3x3} & 0_{3x3} \\ & & & 0_{3x3} & -C_b^c & 0_{3x3} \\ 0_{3x3} & 0_{3x3} & 0_{3x3} & \frac{-1}{\tau_{acc}}I & 0_{3x3} & 0_{3x3} \\ 0_{3x3} & 0_{3x3} & 0_{3x3} & 0_{3x3} & \frac{-1}{\tau_{gyr}}I & 0_{3x3} \\ 0_{3x3} & 0_{3x3} & 0_{3x3} & 0_{3x3} & 0_{3x3} & \frac{-1}{\tau_{pos}}I \end{bmatrix} \begin{bmatrix} \delta\underline{R} \\ \delta\underline{V} \\ \underline{\psi} \\ \delta\underline{f}^b \\ \delta\underline{\omega}_{ib}^b \\ \delta\underline{R}_{GNSS} \end{bmatrix} + \begin{bmatrix} 0_{3x1} \\ 0_{3x1} \\ 0_{3x1} \\ \underline{\eta}_{acc} \\ \underline{\eta}_{gyr} \\ \underline{\eta}_{gnss} \end{bmatrix} \tag{20.10}$$

where F_{INS} was defined in (20.3) and τ_{pos} is the time-constant of the first-order Gauss-Markov process modeling the GNSS position-bias errors ($\underline{\delta R}_{GNSS}$) and $\underline{\eta}_{gnss}$ is the Gaussian white noise driving the GNSS position-bias Gauss–Markov process (the other variables are the same as in Equation (20.4)). The time-constants can be set in accordance with the expected variations in atmospheric (and possibly multipath) errors (e.g., $\sim$ 1,000–3,000 sec). In the case of a receiver utilizing dual-frequency ionospheric corrections, and assuming tropospheric delay is well modeled and corrected, the residual GNSS position errors may be dominated by relatively higher frequency multipath error and the time constant may be reduced accordingly. The values of the diagonal elements of the Q matrix are set in accordance with the modeled Gauss–Markov processes.

Since we have expanded the state vector, the measurement Equation (20.8) will need to be modified accordingly. The measurement error now consists of the sum of the inertial position error, the GNSS position-bias, and the GNSS position noise:

$$\underline{z} = \begin{bmatrix} I_{3\times 3} & 0_{3\times 12} & I_{3\times 3} \end{bmatrix} \begin{bmatrix} \underline{\delta R} \\ \underline{\delta V} \\ \underline{\psi} \\ \underline{\delta f}^b \\ \underline{\delta \omega}_{ib}^b \\ \underline{\delta R}_{gnss} \end{bmatrix} + \begin{bmatrix} \nu_x \\ \nu_y \\ \nu_z \end{bmatrix} \tag{20.11}$$

Thus the H matrix is now 3×18 with identity matrices occupying the first three and last three columns (all other columns are all zeros).

This expansion of the filter to 18 states accommodates the non-noise components of the GNSS position error. However, in order to deal with constellation changes, the software must reset the last three rows and columns of the prediction error covariance matrix at the time of the constellation change. These rows and columns correspond to the GNSS position bias states and resetting them amounts to zeroing them out and resetting the main diagonal components to their original values. This has the effect of causing the filter to apply the "jumps" in the measurement vector to the position bias states. As we will see shortly, if the position bias states are not included in the filter, these measurement vector jumps will be erroneously applied to all of the other state variables.

20.4 Case study: loosely coupled GPS/INS

In order to get a sense for how the filter operates, we will simulate three scenarios. First we will operate the filter with a fixed set of visible satellites. This will provide us with insight into the steady-state operation of the filter. Second, we will simulate a change in the number of visible satellites (in the middle of the run) but we will not reset the last three rows and columns of the prediction error covariance matrix. This will provide us with a depiction of the undesired performance that results. Finally, we will re-run the second scenario but with the prediction error covariance reset. This will allow us to see the dramatic performance improvement that is achieved.

The simulated flight trajectory for this case study involves an F-16 performing a series of s-turns as it makes five step climbs* up from the surface to a maximum altitude of 3,000 m and then makes four step descents down to 1,000 m (Figure 20.1). The climbs, descents, and turns provide "observability" into the states the Kalman filter is trying to estimate. We will discuss the very important topic of observability in a subsequent chapter.

In order to generate some meaningful inertial errors during the relatively short run, a tactical-grade IMU was simulated. Accel biases ranging from 300 to 500 micro-g were simulated on the three axes along with gyro biases of 0.08–0.15 deg/hour. Although the sensor biases were simulated as perfect constants, the time constants of the bias states in the filter were set to 1,000 sec.

A nominal 24-satellite GPS constellation was simulated and there were six satellites visible throughout the run. Ionospheric delay, tropospheric delay, and noise

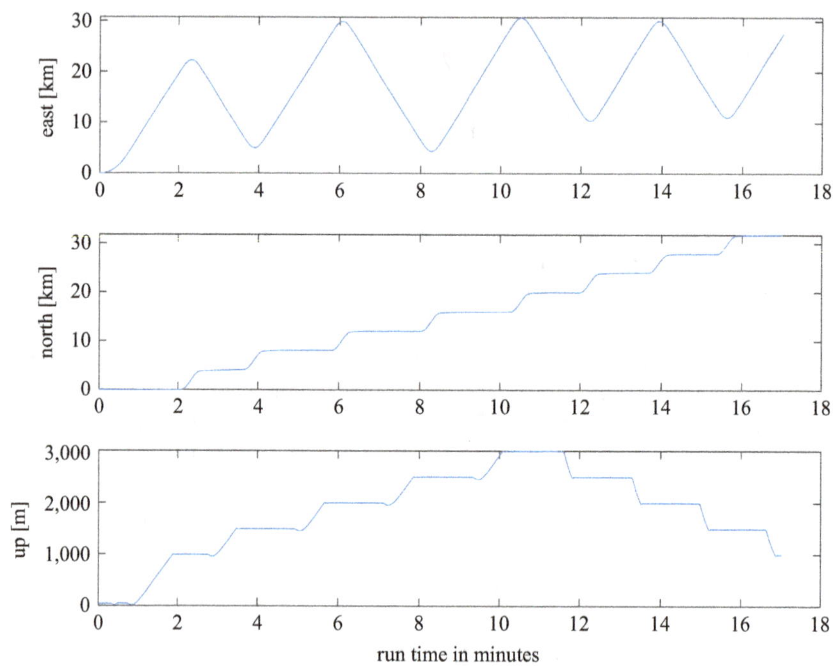

Figure 20.1 Trajectory for the loosely coupled filter simulations

*Step climbs/descents involve a series of climbs/descents to intermediate altitudes with a period of level flight in between each step.

were simulated on the GPS pseudorange measurements. The pseudorange noise was simulated with a standard deviation of one meter.

20.4.1 Loosely coupled GPS/INS: fixed number of visible satellites

The simulation results are shown in Figures 20.2–20.6. The dashed lines are the +/− one standard deviation of the given state computed by taking the square root of the relevant diagonal element of the estimation error covariance matrix. In Figure 20.2, we see the estimated position errors are quite reasonable given the simulated GPS measurement noise level and the fact that the filter does little, if anything, to deal with the GPS multipath and atmospheric delays. Figure 20.3 depicts excellent velocity performance after the filter has converged to steady-state. Figure 20.4 shows the Euler angle estimation error. Particularly for pitch, the error exhibits something like step changes. This occurs at particular points in the turns and climbs and is reflected in the estimated error standard deviations. Zoomed views of the accel and gyro bias estimation errors are depicted in Figures 20.5 and 20.6. The accel bias estimates converge quite quickly whereas the gyro bias estimates take significantly longer.

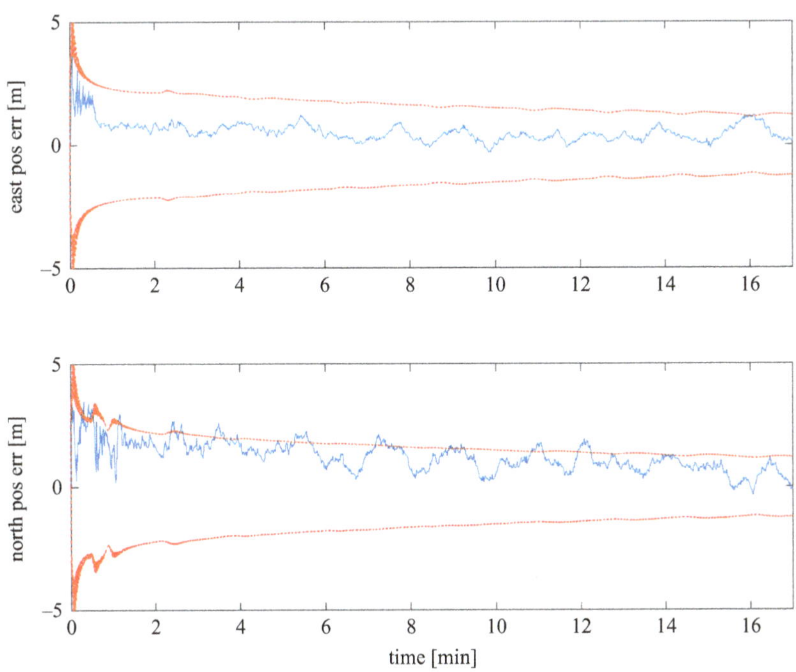

Figure 20.2 Error in Kalman-corrected inertial position. 18-State loosely coupled GPS/INS filter. The dashed lines depict the filter's estimated error covariance (specifically, the standard deviation of the given state)

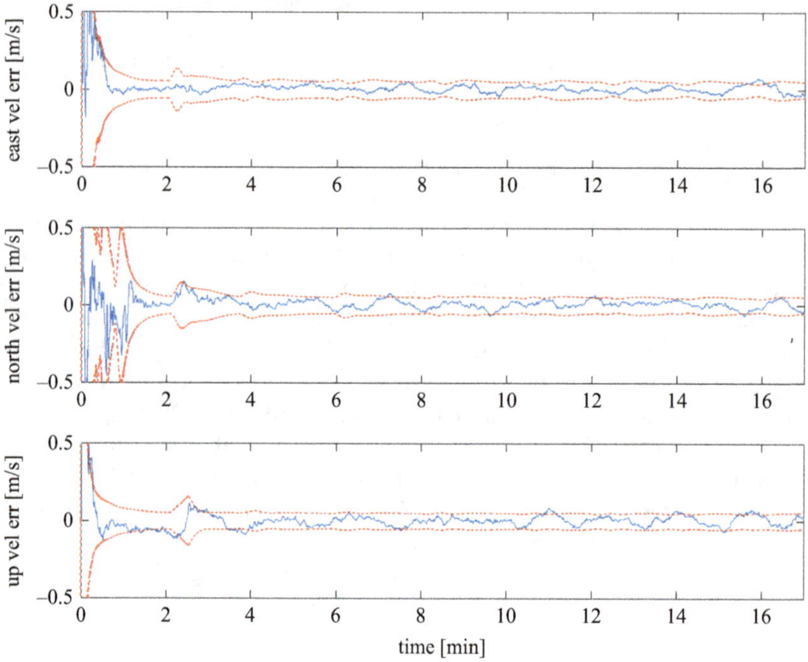

Figure 20.3   Error in Kalman-corrected inertial velocity. 18-State loosely coupled GPS/INS filter

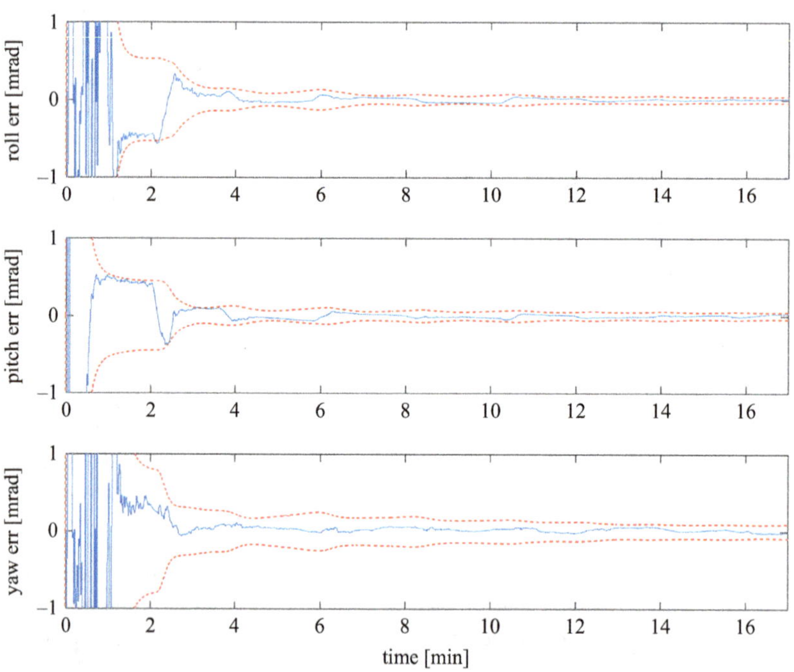

Figure 20.4   Error in Kalman-corrected inertial attitude. 18-State loosely coupled GPS/INS filter

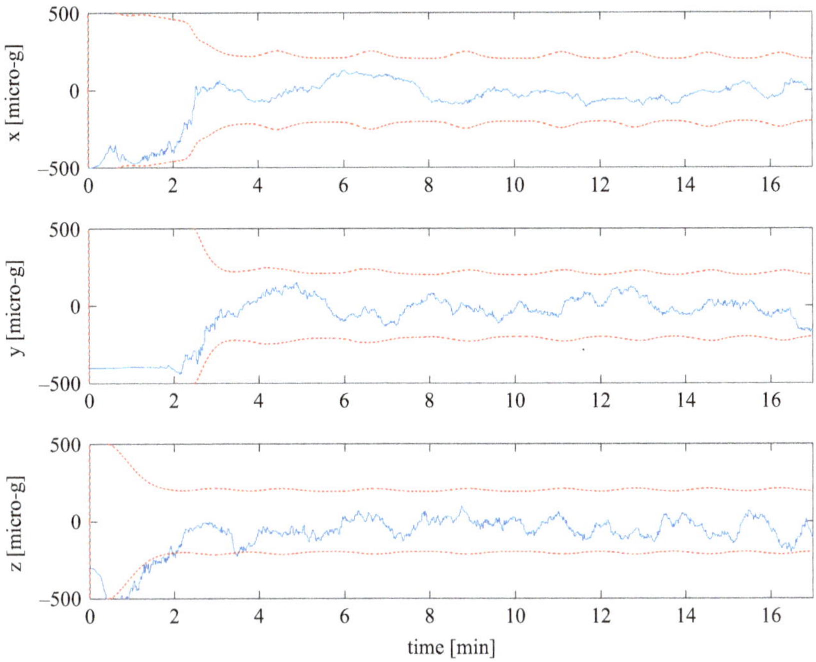

Figure 20.5 Error in accelerometer bias estimates. 18-State loosely coupled GPS/INS filter

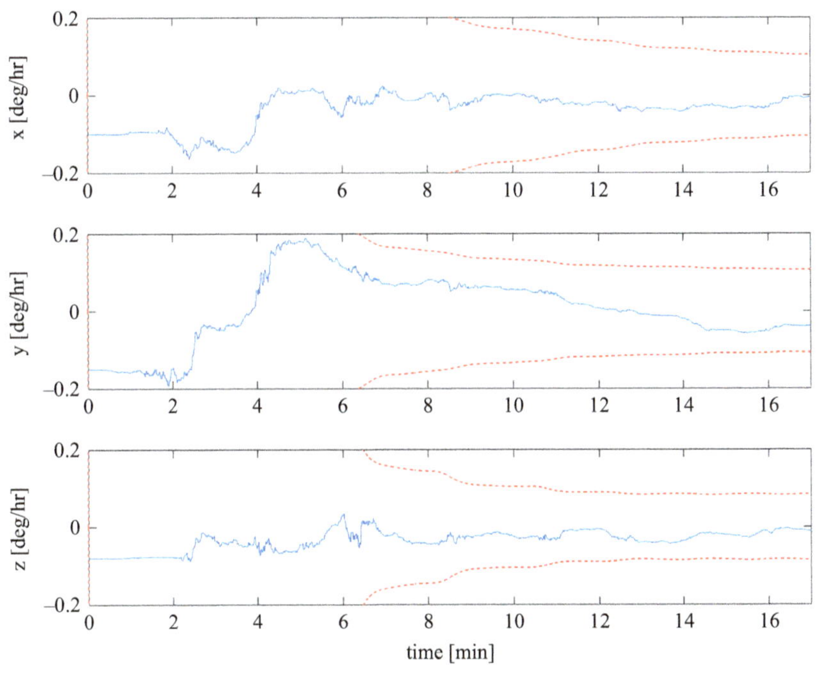

Figure 20.6 Error in gyro bias estimates. 18-State loosely coupled GPS/INS filter

20.4.2 Loosely coupled GPS/INS: variable number of visible satellites without compensation in the filter

For this second scenario, the number of visible satellites is reduced from six to four during the period from $t = 8$ minutes to $t = 9$ min. Figure 20.7 shows the east, north and up GPS position error components. In this scenario, we make no compensation for this effect in the Kalman filter. The results are shown in Figures 20.8–20.12. Since no compensation is being made, the Kalman filter erroneously takes the change in the observation ($\underline{z}$), weights it according to the Kalman gain, and applies it to all of the states. As Figures 20.8–20.12 show, the result is dramatic. Significant excursions occur in the position, velocity, attitude and sensor bias estimation errors.

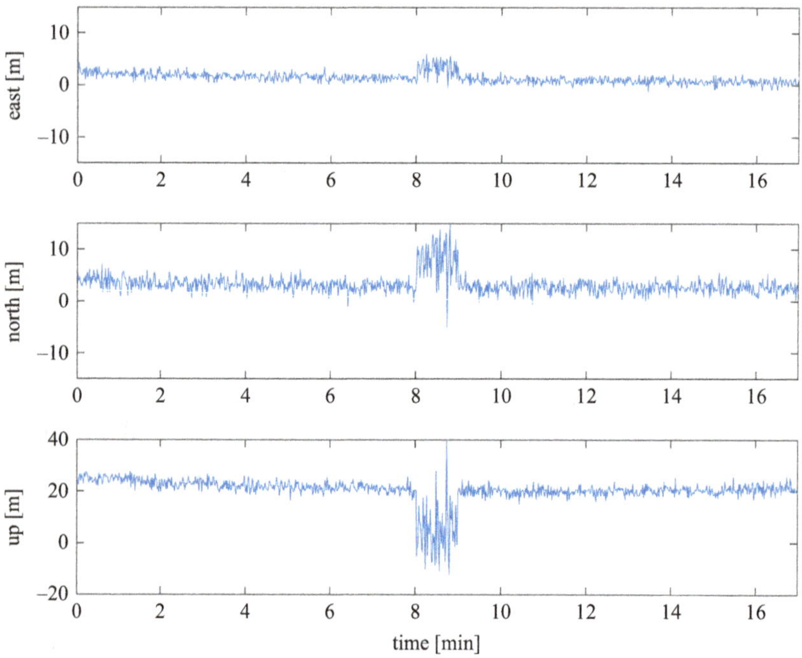

Figure 20.7 Error in the GPS-computed position. From t = 8 min to t = 9 min the number of visible satellites is reduced from 6 to 4

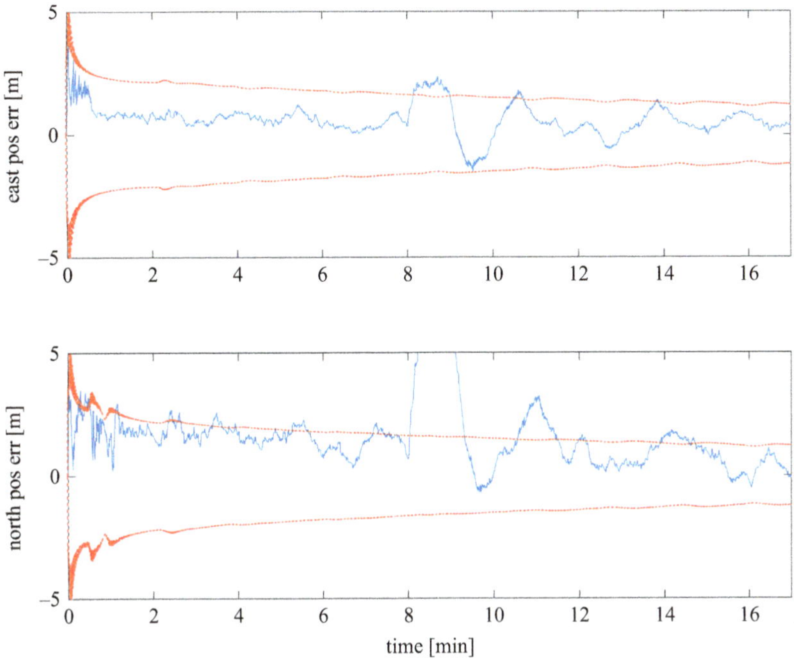

Figure 20.8 *Error in Kalman-corrected inertial position. 18-State loosely coupled GPS/INS filter. From t = 8 min to t = 9 min the number of visible satellites is reduced from 6 to 4 but without reset of the prediction error covariance*

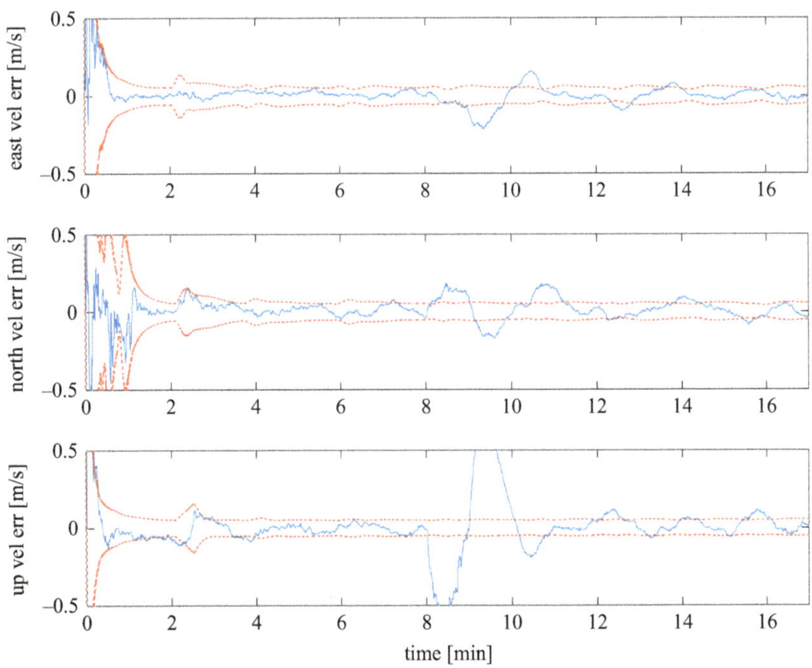

Figure 20.9 *Error in Kalman-corrected inertial velocity. 18-State loosely coupled GPS/INS filter without reset of the prediction error covariance*

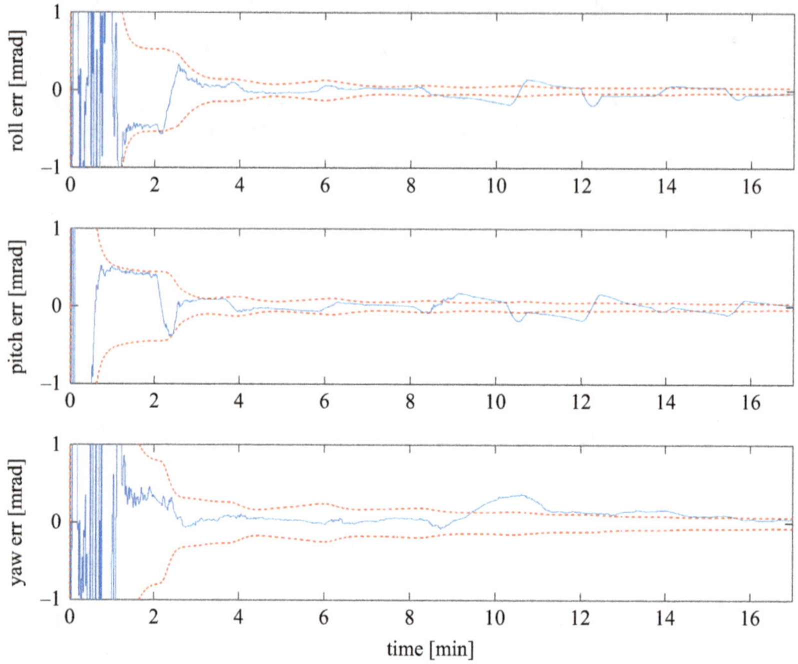

Figure 20.10 *Error in Kalman-corrected inertial attitude. 18-State loosely coupled GPS/INS filter without reset of the prediction error covariance*

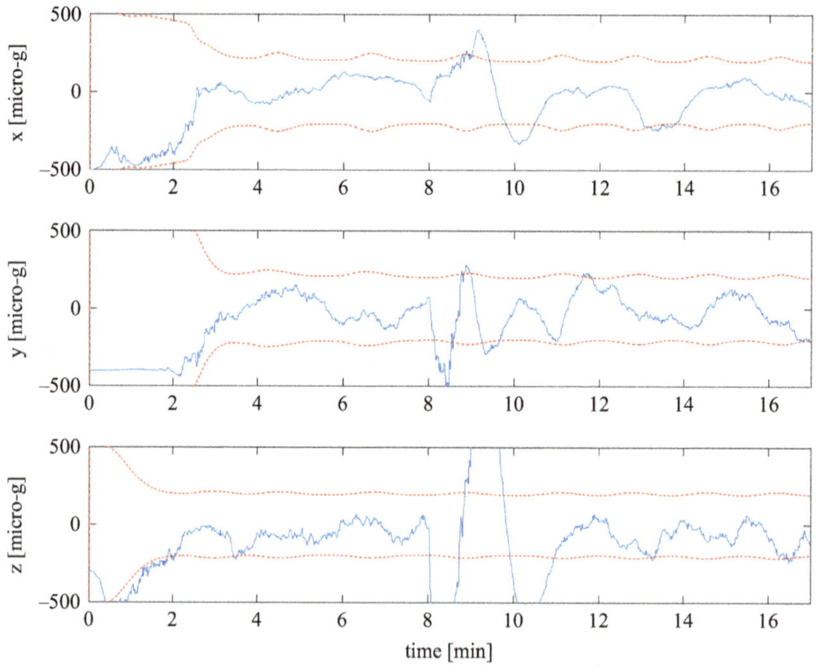

Figure 20.11 *Error in accelerometer bias estimates. 18-State loosely coupled GPS/INS filter without reset of the prediction error covariance*

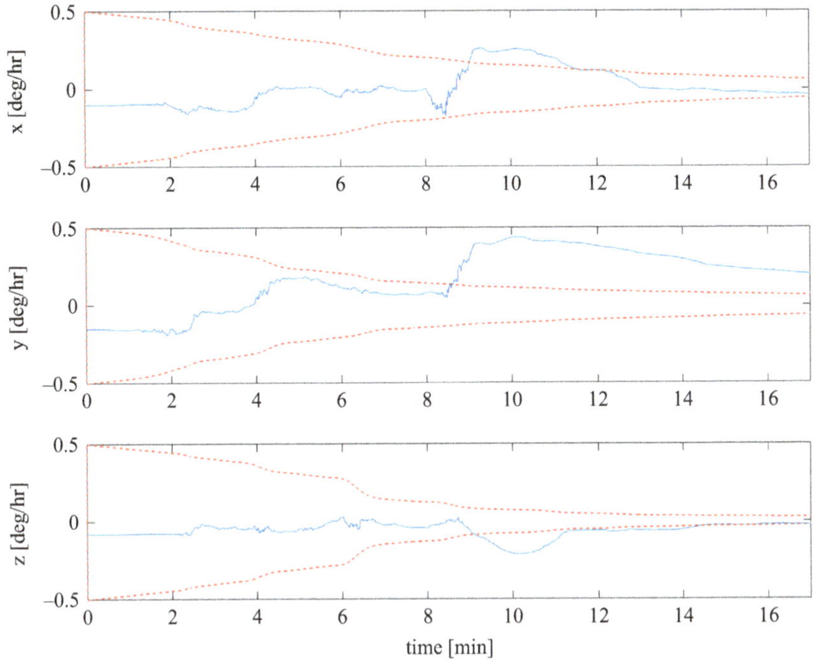

Figure 20.12 Error in gyro bias estimates. 18-State loosely coupled GPS/INS filter without reset of the prediction error covariance

20.4.3 Loosely coupled GPS/INS: variable number of visible satellites with compensation in the filter

Although the previously described jump in the observation ($\underline{z}$) is due solely to the GPS aiding data, the Kalman filter *does not know this* and hence does not deal with it appropriately. Although we have added three GPS position bias states to our filter, all the filter sees is a jump in the observation and it has no way to know which states the jump belongs to. What we want is to force the filter to put most, if not all, of the jump into the GPS position bias states and not in the rest of the states. We can do this by resetting the rows and columns of the prediction error covariance matrix that correspond to the GPS position bias states (rows/columns 16–18). Recall that when the prediction error covariance is large, the filter relies little on its own predictions and thus relies heavily on the measurements. By resetting only rows/columns 16–18 in the prediction error covariance matrix, and leaving the rest alone, we are telling the filter to place most of the observation into these states.

Resetting is straight-forward. Simply zero out the entire row/column and then reset the main diagonal element to its original large value. However, it must be done every time a satellite drops out of or is added into the computed solution. For the example here, the reset has to be performed when the number of visible satellites is reduced (at t = 8 min) and when the number of visible satellites is increased (at t = 9 min).

The results are shown in Figures 20.13–20.17 and a dramatic performance improvement may be seen when comparing against the previous scenario. Relatively minor glitches are observed in certain error components but the rest behave nearly identically to the case without a change in the number of visible satellites.

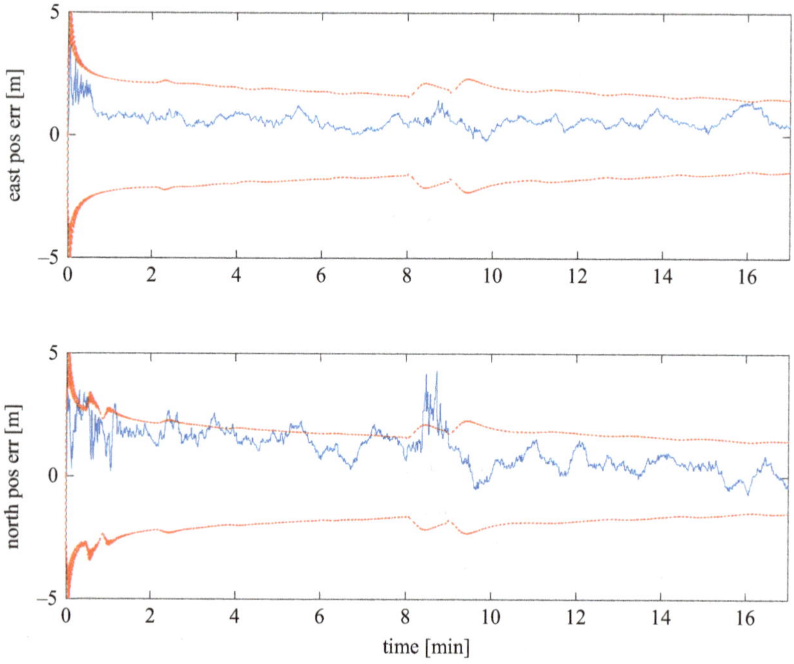

Figure 20.13 Error in Kalman-corrected inertial position. 18-State loosely coupled GPS/INS filter. From t = 8 min to t = 9 min the number of visible satellites is reduced from 6 to 4 and the prediction error covariance row/columns corresponding to the GPS position bias states are reset

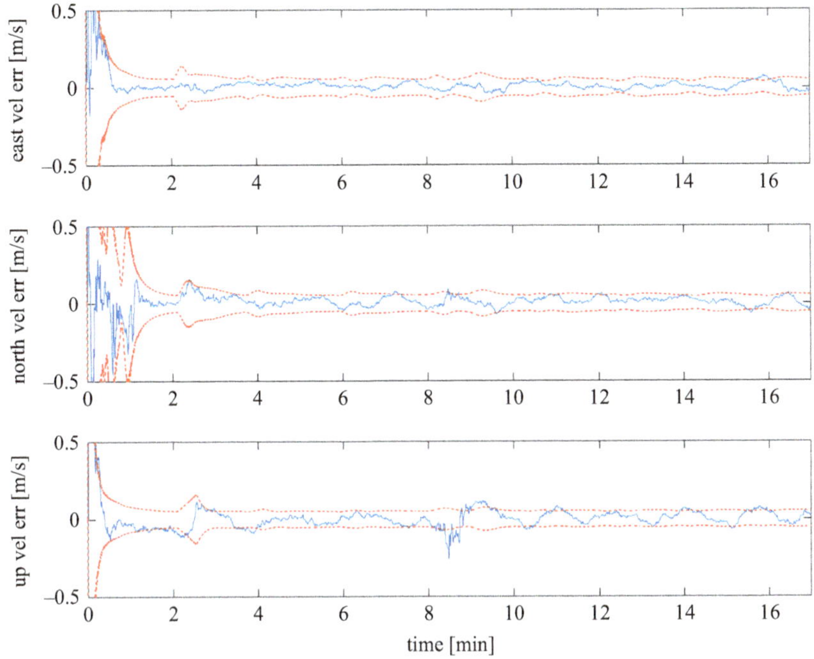

Figure 20.14 *Error in Kalman-corrected inertial velocity. 18-State loosely coupled GPS/INS filter with reset of the prediction error covariance*

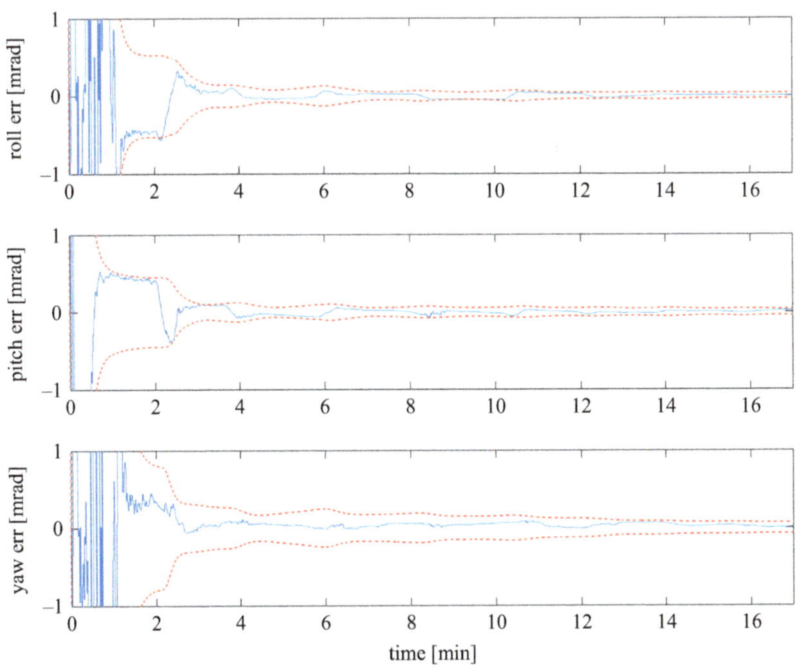

Figure 20.15 *Error in Kalman-corrected inertial attitude. 18-State loosely coupled GPS/INS filter with reset of the prediction error covariance*

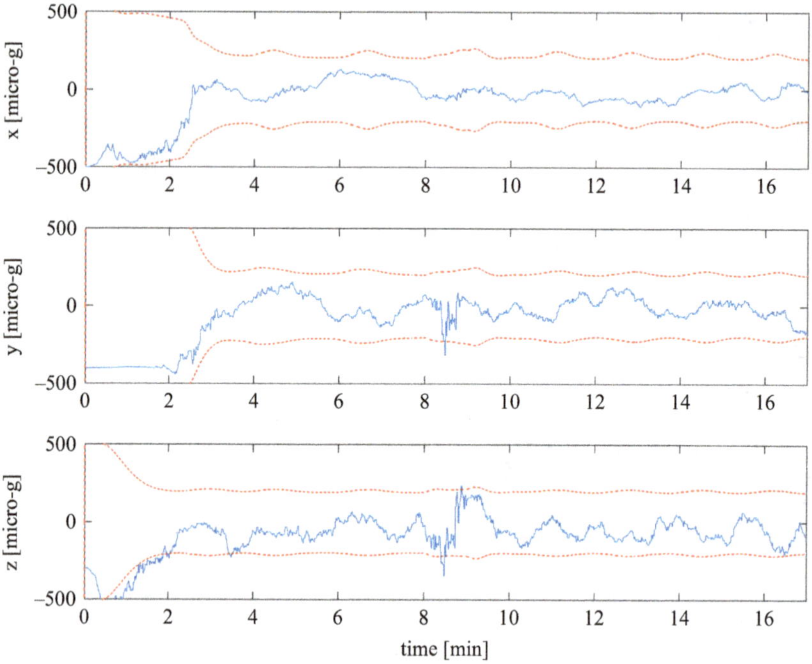

Figure 20.16 *Error in accelerometer bias estimates. 18-State loosely coupled GPS/INS filter with reset of the prediction error covariance*

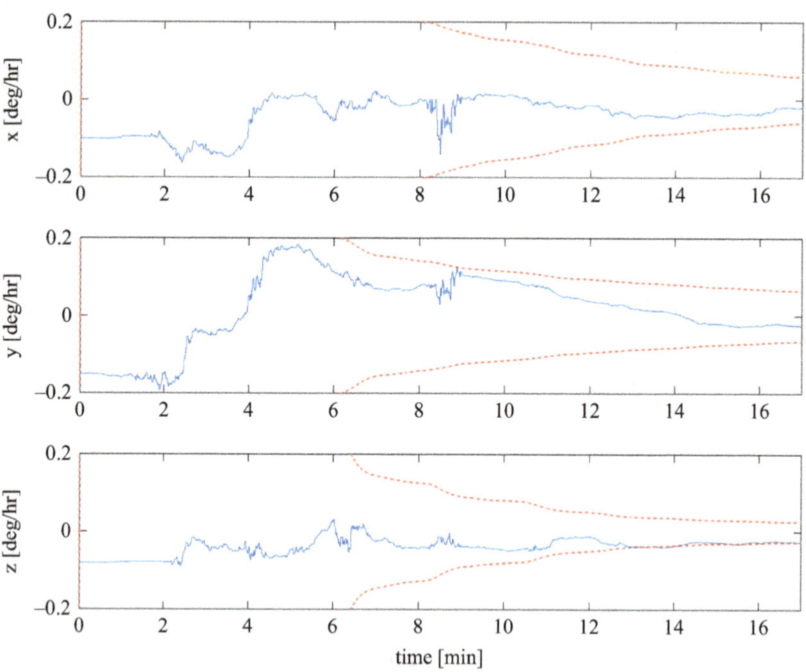

Figure 20.17 *Error in gyro bias estimates. 18-State loosely coupled GPS/INS filter with reset of the prediction error covariance*

20.5 Additional considerations

There are a number of practical considerations that must be taken into account when implementing a filter with real GNSS receivers and a real INS.

20.5.1 Cascaded filters

The Kalman filter assumes the measurement noise vector is independent white Gaussian. Put another way, the filter assumes it is being fed independent measurements. If the GNSS receiver is computing an ordinary least squares solution, this assumption is valid. However, most receivers instead utilize an internal Kalman filter to smooth the output data. The output is thus correlated with a time-constant set by the filter design. Thus, the GNSS receiver may be outputting data a rate of one sample per second but the independent data rate is actually much lower.

This leads to the so-called cascaded filter problem. The output of the GNSS receiver's Kalman filter is feeding the downstream GNSS-aided INS Kalman filter. If nothing is done to accommodate this situation, the integration filter will diverge and produce erroneous results since it is wrongly assuming the GNSS receiver data is uncorrelated. The solution is to determine the time-constant of the filter in the GNSS receiver and then decimate the data before feeding it to the integration filter. For example, if it is determined that the GNSS receiver's Kalman filter has a time-constant of 10 sec, then GNSS measurements should only be fed to the integration filter once every 20–30 sec to ensure data independence.

20.5.2 Measurement timing

In general, the GNSS receiver and INS are asynchronous. That is, the times of validity of the GNSS measurements/position solutions are not synchronized with the times of validity of the INS-determined position/velocity. Furthermore, there is latency between the time the GNSS measurements are formed and the time the data is received by the integration filter (this is known as transport delay).

Typically the GNSS receiver is outputting data at one sample per second whereas the inertial system may be outputting position, velocity, and attitude at tens to hundreds of times per second. This timing offset cannot be ignored if we want to get the best possible performance. The latency in the GNSS solution may be on the order of hundreds of milliseconds (possibly half a second or more depending on the receiver). Consider, as an example, an inertial system outputting a position at, say, 100 Hz. In one hundredth of a second, an aircraft flying at five hundred knots will travel more than two and half meters. Two and a half meters would be considered a non-trivial GNSS measurement error if it was not accounted for properly. First the INS outputs have to be saved and interpolated to the time of validity of the GNSS.

It is best to interpolate the INS to the time of validity of the GNSS since on these short time scales the inertial system essentially has no noise. The GNSS, by contrast, has considerably more noise than the INS. Nevertheless, once the interpolation is performed and the observable (the difference between the INS and GNSS position) is formed, it is old. That is, the time of validity of the observable significantly lags behind real time. There are at least three potential solutions. One is to perform the Kalman update at the time of validity of the observable, apply it to the stored INS solution at the time of validity, and then use the stored INS sensor data to extrapolate the inertial solution forward to real time. The second is to perform the Kalman update just mentioned but propagate the predicted state vector (i.e., estimated inertial errors) forward to real time and apply the corrections to the current INS solution. Third, we can revise the measurement equation in order to account for the time delay. This is significantly more complicated but has the advantage of performing the Kalman update with the most recent inertial solution.

20.5.3 Lever arm

Thus far, we have implicitly assumed the GNSS and INS are providing data for the same physical location. Since the GNSS antenna is not co-located with the INS sensors, this is not true. The GNSS receiver is forming a position that is valid at the GNSS antenna (generally on top of the fuselage for an airplane) whereas the INS is generating data with respect to the accelerometers and gyros that are in the INS which is located somewhere within the fuselage. There is a non-zero lever arm between the inertial unit inside the airplane and the GNSS antenna on top of the airplane. Thus the lever arm between the GNSS antenna and the INS must be taken into account. This involves using the relative positions of the GNSS antenna and the INS (fixed in body-frame coordinates), along with a lever arm computed using the INS-determined attitude, to convert the GNSS data to equivalent measurements/solutions valid at the location of the INS.

20.5.4 Q tuning

Up to this point, we have modeled the system noise to accommodate the accelerometer and gyro bias states (along with the GNSS position bias states). Noise was only modeled on the states that drive the position, velocity, and attitude errors. In reality we actually need to model some noise on all of the states in order to account for un-modeled error sources. For example, un-modeled scale factor errors can be accommodated by increasing the Q on the velocity and attitude states. It is possible to explicitly model the scale factor states, but at some point the state vector can no longer be expanded. The Kalman filter cannot be made artificially large because it will take an excessive amount of time to converge and there is a tradeoff known as the curse of dimensionality. Essentially the estimation error increases as the total number of states increases. Thus, in the end, there will be some amount of un-modeled error and some tweaking of the Q is needed in order to ensure stable filter performance.

20.5.5 R tuning

In a loosely coupled INS/GNSS, the measurement error covariance matrix is a function of the estimated pseudorange noise errors and the appropriate dilution of precision (DOP). The pseudorange noise errors can be estimated on the basis of receiver-measured carrier-to-noise ratios and satellite elevation angle. To first order, the pseudorange noise varies inversely with the sine of the elevation angle. DOP can be used to scale the noise from the range-domain to the position-domain. In practice, the values in the R matrix may need to be inflated somewhat to account for un-modeled/mis-modeled errors.

Reference

[1] Pue A. *Integration of GPS with Inertial Navigation Systems [Short Course Notes]*. NavtechGPS; 2007.

Chapter 21

GNSS-aided INS: tight coupling

21.1 Introduction

As we have discussed in previous chapters, tight-coupling utilizes the measurement data from the aiding source (as opposed to loose coupling that utilizes processed data). In the case of a GNSS receiver, pseudorange and delta-range (or possibly carrier-phase) measurements are processed by the integration filter rather than position and velocity. This requires information about the individual satellite position and velocity be fed to the filter since it has to form the inertial equivalent of the pseudorange in order to compute an observable corresponding to the difference between the INS and GNSS outputs.

21.2 Basic state model

Although it is possible to design the filter with fewer states, good performance can be achieved with at least the following:

- nine inertial error states (position, velocity and attitude error)
- three accel bias states
- three gyro bias states
- two GNSS receiver clock states
- N GNSS pseudorange bias states

The inertial error states and sensor bias states are the same as in the loosely coupled filter. The two GNSS receiver clock states are the same clock phase and frequency error states described in the earlier chapter about GNSS-only filtering. As with the GNSS position bias states of the loosely coupled filter, the pseudorange bias states of the tightly coupled filter serve to absorb the discontinuity of the measurements at the time of a constellation change (specifically when measurements from a new satellite are added). To a certain extent, these states can also partially estimate bias errors due to atmospheric delays and multipath. Again, as was the case for the position bias states, the range bias states use first-order Gauss-Markov models. The continuous-time system equation is thus given by:

$$\begin{bmatrix} \delta\dot{\underline{R}} \\ \delta\dot{\underline{V}} \\ \dot{\underline{\psi}} \\ \delta\dot{\underline{f}}^b \\ \delta\dot{\underline{\omega}}_{ib}^b \\ \frac{d}{dt}\Delta b \\ \frac{d}{dt}\Delta b^{\cdot} \\ \delta\dot{pr}_1 \\ \vdots \\ \delta\dot{pr}_N \end{bmatrix} = \begin{bmatrix} & & & 0_{3x3} & 0_{3x3} & 0 & 0 & 0 & 0 & 0 \\ & F_{INS} & & C_b^c & 0_{3x3} & 0 & 0 & 0 & 0 & 0 \\ & & & 0_{3x3} & -C_b^c & 0 & 0 & 0 & 0 & 0 \\ 0_{3x3} & 0_{3x3} & 0_{3x3} & \frac{-1}{\tau_{acc}}I & 0_{3x3} & 0 & 0 & 0 & 0 & 0 \\ 0_{3x3} & 0_{3x3} & 0_{3x3} & 0_{3x3} & \frac{-1}{\tau_{gyr}}I & 0 & 0 & 0 & 0 & 0 \\ 0_{1x3} & 0_{1x3} & 0_{1x3} & 0_{1x3} & 0_{1x3} & 0 & 1 & 0 & 0 & 0 \\ 0_{1x3} & 0_{1x3} & 0_{1x3} & 0_{1x3} & 0_{1x3} & 0 & 0 & 0 & 0 & 0 \\ 0_{1x3} & 0_{1x3} & 0_{1x3} & 0_{1x3} & 0_{1x3} & 0 & 0 & \frac{-1}{\tau_{pr}} & 0 & 0 \\ \vdots & \vdots & \vdots & \vdots & \vdots & \vdots & \vdots & 0 & \ddots & 0 \\ 0_{1x3} & 0_{1x3} & 0_{1x3} & 0_{1x3} & 0_{1x3} & 0 & 0 & 0 & 0 & \frac{-1}{\tau_{pr}} \end{bmatrix} \begin{bmatrix} \delta\underline{R} \\ \delta\underline{V} \\ \underline{\psi} \\ \delta\underline{f}^b \\ \delta\underline{\omega}_{ib}^b \\ \Delta b \\ \Delta b^{\cdot} \\ \delta pr_1 \\ \vdots \\ \delta pr_N \end{bmatrix} + \begin{bmatrix} 0_{3x1} \\ 0_{3x1} \\ 0_{3x1} \\ \underline{\eta}_{acc} \\ \underline{\eta}_{gyr} \\ w_{cf} \\ w_{ca} \\ \eta_{pr_1} \\ \vdots \\ \eta_{pr_N} \end{bmatrix} \tag{21.1}$$

where F_{INS} is the 9×9 core inertial error dynamics submatrix described in the previous chapter, I is a 3×3 identity matrix, δpr_i is the ith pseudorange bias state and the various η_s are the white noise inputs for the Gauss–Markov processes. Note the upper left 15×15 submatrix is identical to that of the loosely coupled filter. The total system dynamics matrix (F) in (21.1) has dimensions ($17 + N \times 17 + N$) where N is the maximum number of satellites tracked by the GNSS receiver. As discussed in the previous chapter, the state transition matrix is obtained using $\Phi = e^{F\Delta t}$ (or an approximation thereof) over short-time intervals such that F can be considered to be constant.

For short covariance propagation intervals, the system noise covariance matrix, Q, is frequently approximated by a diagonal matrix with non-zero elements specified according to the stochastic models of the associated state variables (see Section 14.2.6 and Equation (14.82) in [1]). As we have discussed in earlier chapters, the gyro biases and accel biases are all modeled as first-order Gauss–Markov processes. For tightly coupled filters, the range bias states may be modeled in this way as well.

As an example, we will consider the pseudorange bias state. Assume the combination of low-frequency GNSS pseudoranging errors (tropospheric delay, ionospheric delay, multipath and satellite clock, and ephemeris error) can be modeled with a first-order Gauss–Markov process with a variance of 18 m^2 (standard deviation of 4.24 m) and a time constant of one hour (3,600 sec). The variance of the continuous-time model white noise input is given by (see Appendix on stochastic processes):

$$q = \frac{2}{\tau}E[x^2] \tag{21.2}$$

Thus, for the current example:

$$q = \left(\frac{2}{3,600} \right) (18) = 0.01 \ (\text{m}^2/\text{s}) \tag{21.3}$$

Equations provided in the appendix can then be used to compute the associated Q matrix entry (i.e., q_k) for the pseudorange bias states for a given filter iteration interval (Δt).

The system noise covariance matrix for the two clock states was described in an earlier chapter. As discussed in Groves [1, Section 14.2.7], this matrix (Groves equation (14.88)) can also be approximated by a diagonal matrix (Groves equation (14.89)).

21.3 Measurement model

In order to derive the measurement model for the tightly coupled filter, we begin with the simple range formula for the distance from the ith satellite to the true user position:

$$r_{true_i} = \sqrt{(x_u - x_i)^2 + (y_u - y_i)^2 + (z_u - z_i)^2} \tag{21.4}$$

where the subscript u denotes true user position. Performing a Taylor Series expansion of (21.4) with respect to the position estimated by the INS and retaining only the zeroth and first-order terms results in:

$$\begin{aligned} r_{true_i} \approx\ & r_{INS_i} + \frac{x_{INS} - x_i}{r_{INS_i}}(x_u - x_{INS}) + \frac{y_{INS} - y_i}{r_{INS_i}}(y_u - y_{INS}) \\ & + \frac{z_{INS} - z_i}{r_{INS_i}}(z_u - z_{INS}) \end{aligned} \tag{21.5}$$

where:

$$r_{INS_i} = \sqrt{(x_{INS} - x_i)^2 + (y_{INS} - y_i)^2 + (z_{INS} - z_i)^2} \tag{21.6}$$

and the subscript *INS* denotes the user position as estimated by the INS. The components of the unit vector from the estimated INS position to the satellite are given by:

$$u_{x_i} = \frac{x_i - x_{INS}}{r_{INS_i}} \tag{21.7}$$

$$u_{y_i} = \frac{y_i - y_{INS}}{r_{INS_i}} \tag{21.8}$$

$$u_{z_i} = \frac{z_i - z_{INS}}{r_{INS_i}} \tag{21.9}$$

Substitution of Equations (21.7)–(21.9) into (21.5) yields:

$$r_{true_i} \approx r_{INS_i} - u_{x_i}(x_u - x_{INS}) - u_{y_i}(y_u - y_{INS}) - u_{z_i}(z_u - z_{INS}) \tag{21.10}$$

Note the change in sign on the unit vector terms as a result of the order of the terms in the numerators in Equations (21.7)–(21.9). For tight coupling, we need an observable defined in the measurement domain of the aiding source. In the present case that is the pseudorange domain. In (21.6), we have defined the range from the satellite to the INS-estimated position. In order to define an *INS-equivalent pseudorange*, let us first add the Kalman-filter estimated GNSS receiver clock offset ($\hat{b}_u$) to both sides of (21.10):

$$r_{true_i} + \hat{b}_u \approx r_{INS_i} + \hat{b}_u - u_{x_i}(x_u - x_{INS}) - u_{y_i}(y_u - y_{INS}) - u_{z_i}(z_u - z_{INS}) \tag{21.11}$$

Now we define the INS-equivalent pseudorange as:

$$PR_{INS_i} = r_{INS_i} + \hat{b}_u \tag{21.12}$$

which allows us to rewrite (21.11) as:

$$r_{true_i} + \hat{b}_u \approx PR_{INS_i} - u_{x_i}(x_u - x_{INS}) - u_{y_i}(y_u - y_{INS}) - u_{z_i}(z_u - z_{INS}) \tag{21.13}$$

The measured GNSS pseudorange from the ith satellite is related to the true range by:

$$PR_{meas_i} = r_{true_i} + b_u + \gamma_i + \nu_i \tag{21.14}$$

where b_u is the true receiver clock offset, γ is the sum of the bias-like errors (atmospheric delays and multipath) and ν is the measurement noise. Solving (21.14) for the true range and substituting into (21.13) yields:

$$PR_{meas_i} + (\hat{b}_u - b_u) - \gamma_i - \nu_i \approx PR_{INS_i} - u_{x_i}(x_u - x_{INS}) - u_{y_i}(y_u - y_{INS}) - u_{z_i}(z_u - z_{INS}) \tag{21.15}$$

The error in the estimated clock offset is defined as:

$$\Delta b_u = \hat{b}_u - b_u \tag{21.16}$$

Substituting (21.16) into (21.15):

$$PR_{meas_i} + \Delta b_u - \gamma_i - \nu_i \approx PR_{INS_i} - u_{x_i}(x_u - x_{INS}) - u_{y_i}(y_u - y_{INS}) - u_{z_i}(z_u - z_{INS}) \tag{21.17}$$

The range-domain observable for our tightly coupled filter can now be formed as the difference between the INS "pseudorange" and the GNSS measured pseudorange:

$$z_i = PR_{INS_i} - PR_{meas_i} \tag{21.18}$$

To be clear, (21.18) is the formula by which the actual observable is formed. To determine how it relates to the state vector, we need to form the theoretical measurement equation. Thus we *model* the observable by isolating the two pseudorange terms of (21.17) on the left hand side and substituting (21.18):

$$z_i = u_{x_i}(x_u - x_{INS}) + u_{y_i}(y_u - y_{INS}) + u_{z_i}(z_u - z_{INS}) + \Delta b_u - \gamma_i - \nu_i \quad (21.19)$$

If we define the components of the error of the estimated INS position as:

$$\delta R_x = x_{INS} - x_u \quad (21.20)$$

$$\delta R_y = y_{INS} - y_u \quad (21.21)$$

$$\delta R_z = z_{INS} - z_u \quad (21.22)$$

Then substitution of Equations (21.20)–(21.22) into (21.19) yields the desired linearized measurement equation:

$$z_i = -u_{x_i}\delta R_x - u_{y_i}\delta R_y - u_{z_i}\delta R_z + \Delta b_u - \gamma_i - \nu_i \quad (21.23)$$

Now let us recall the "pseudorange bias" states (δpr) from the continuous-time system equation (Equation (21.1)). We noted earlier that these states serve two purposes. One is to deal with the case of new satellites being added into the solution. The second is to account for the bias-like GNSS errors. By defining $\delta pr = \gamma$, we can rewrite (21.23) as:

$$z_i = -u_{x_i}\delta R_x - u_{y_i}\delta R_y - u_{z_i}\delta R_z + \Delta b_u - \delta pr_i - \nu_i \quad (21.24)$$

We can now write the Kalman measurement equation for a single observation ((21.24)) in terms of the state vector given in Equation (21.1):

$$z_i = \begin{bmatrix} -\underline{u}_i & 0_{1x12} & 1 & 0 & \cdots & 0 & -1 & 0 & \cdots & 0 \end{bmatrix} \begin{bmatrix} \underline{\delta R} \\ \underline{\delta V} \\ \underline{\psi} \\ \underline{\delta f^b} \\ \underline{\delta \omega_{ib}^b} \\ \Delta b \\ \Delta \dot{b} \\ \delta pr_1 \\ \vdots \\ \delta pr_i \\ \vdots \\ \delta pr_N \end{bmatrix} + \nu_i \quad (21.25)$$

where $\underline{u}_i$ is the unit vector from the user to the ith satellite: $\underline{u}_i = [u_{x_i}\ u_{y_i}\ u_{z_i}]$. The "1" (after the twelve zeros) multiplies the receiver clock offset (Δb) and the "-1" multiplies the ith pseudorange bias (δpr_i). Note the sign change on the measurement noise has been done to put (21.25) in the usual form and has no effect (since the relevant filter design variable is the measurement error covariance).

Equation (21.25) is the Kalman measurement equation for a single observation. If (21.25) is expanded to include a vector of N observations ($\underline{z}$), the full Kalman measurement equation is obtained. Note the full data matrix (H) consists of N rows where the ith row is given by the row vector in (21.25). The first three columns of H consist of the components of the unit vectors between the estimated INS position and the satellites. The 16th column consists of all ones. Columns 18 through 17+N are all zeros except for a single negative one in each column (located in the 17+i element of the ith row).

When the estimated receiver clock offset is incorporated into PR_{INS_i} as described above, the Kalman filter is estimating the error of the estimated receiver clock offset. The corrected total estimated receiver clock offset is thus:

$$(\hat{b}_u)_k = (\hat{b}_u)_{k-1} - \Delta\hat{b}_k \tag{21.26}$$

The Kalman measurement equation is completed by specifying the measurement error covariance matrix, R. See the earlier chapter on GNSS-only Kalman filtering.

Note that if we had delta-range measurements as well as pseudoranges, only a handful of changes would be needed. Delta-range bias states would be added to the state vector. The measurement vector would double in length to accommodate the additional measurements. The H matrix would be modified as follows: unit vector components would be added in the velocity error columns at the rows corresponding to the delta-range observables; 1s would also be added in these rows at the 17th column (i.e., the clock frequency error column) and then 1s would also be added in the elements corresponding to the delta-range bias states. Finally, the R matrix would be increased in size to model the noise of the delta-range measurements as well as the pseudoranges.

21.4 Simulation example: tightly coupled GPS/INS

To see an example of the tightly coupled filter performance, we will utilize the same trajectory and the same tactical grade IMU as was simulated in the previous chapter for the loosely coupled case. As before, ionospheric delay, tropospheric delay and noise are simulated on the GPS pseudorange measurements.

The simulation results are depicted in Figures 21.1–21.7. The results for position, velocity, and attitude errors along with accel and gyro bias estimation errors are comparable to the results shown previously for the loosely coupled simulation. Figures 21.6 and 21.7 compare the filter pseudorange bias estimates to the total pseudorange errors (consisting of atmospheric delays and measurement noise). With "well-behaved" atmospheric errors (e.g., slowly varying), the filter estimates are remarkably accurate. The noticeable drift between the actual pseudorange errors and the Kalman estimates can be traced in this simulation to a slowly increasing error in the Kalman's receiver clock bias estimate and thus tuning of the clock model would be needed to optimize performance.

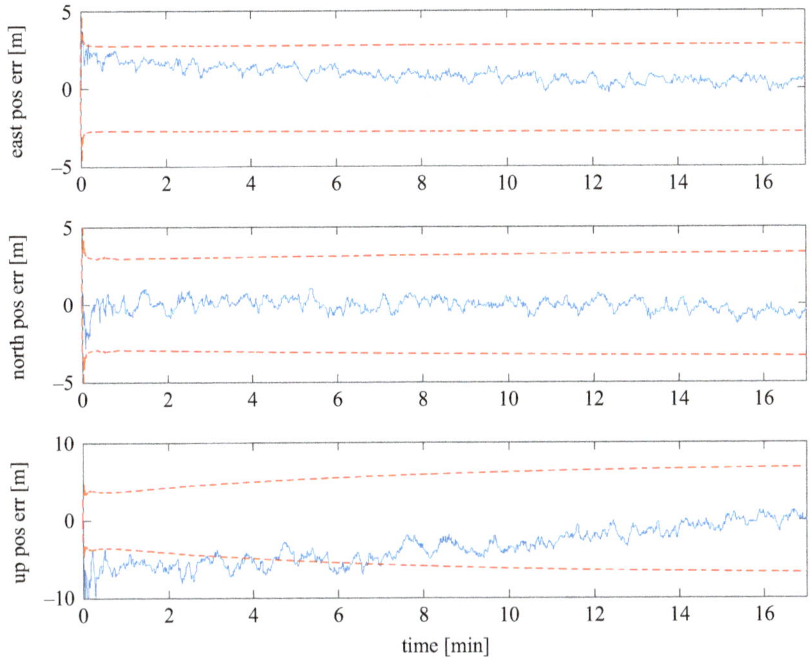

Figure 21.1 Error in Kalman-corrected inertial position. Tightly coupled GNSS/INS filter. The dashed lines depict the filter's estimated error covariance (specifically, the standard deviation of the given state)

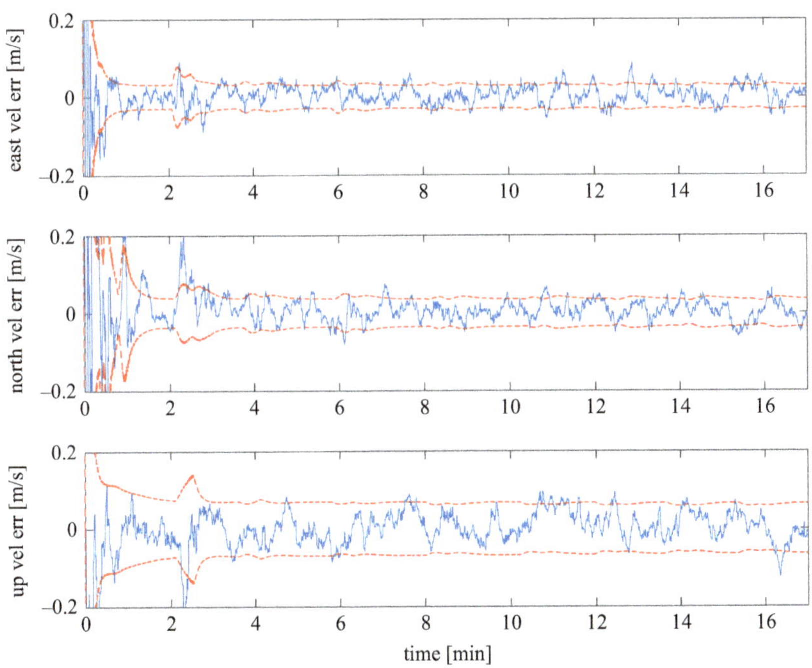

Figure 21.2 Error in Kalman-corrected inertial velocity. Tightly coupled GNSS/INS filter

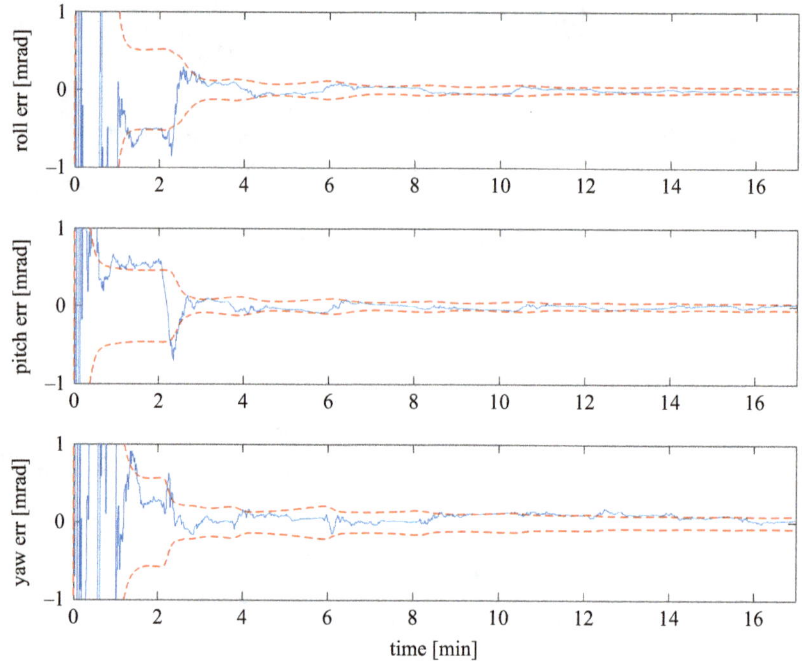

Figure 21.3 Error in Kalman-corrected inertial attitude. Tightly coupled GNSS/INS filter

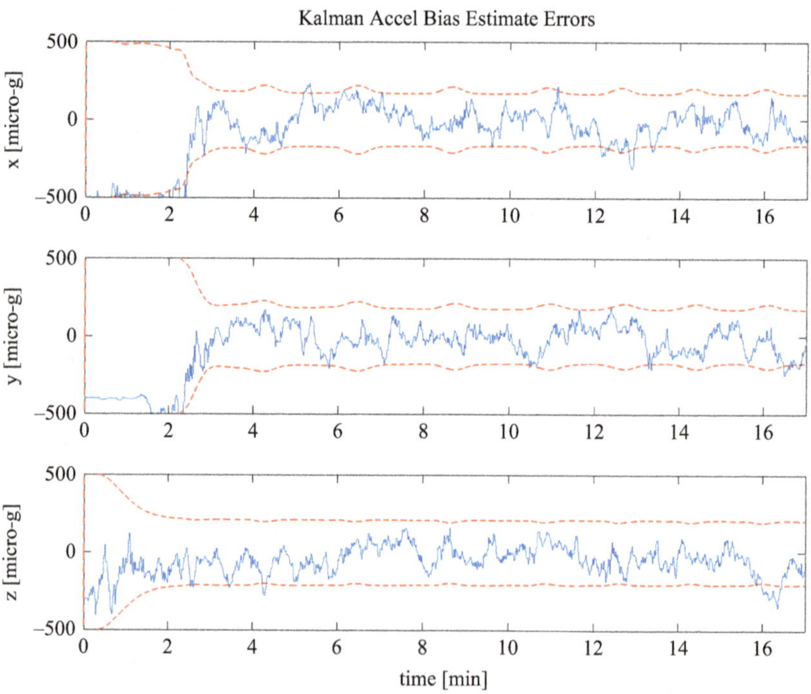

Figure 21.4 Error in accelerometer bias estimates. Tightly coupled GNSS/INS filter

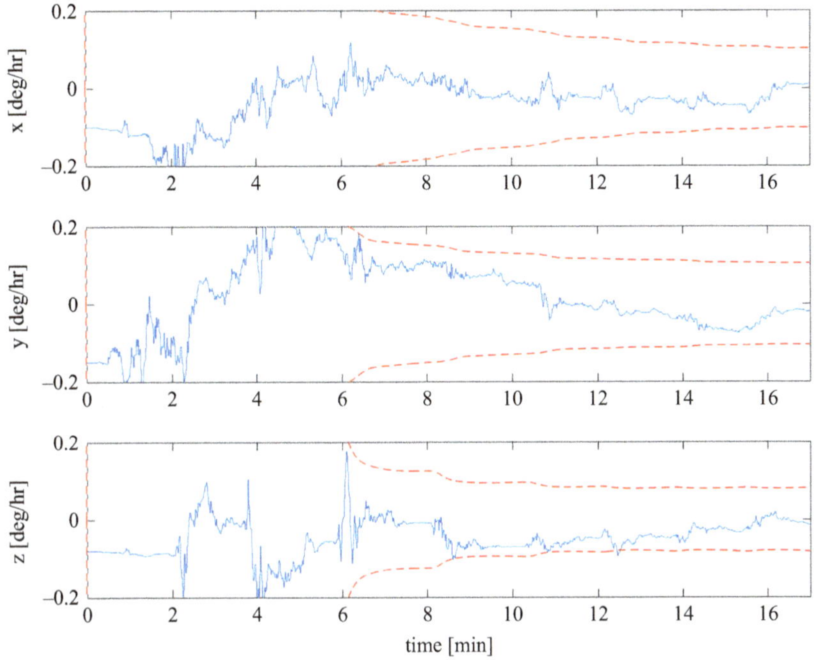

Figure 21.5 Error in gyro bias estimates. Tightly coupled GNSS/INS filter

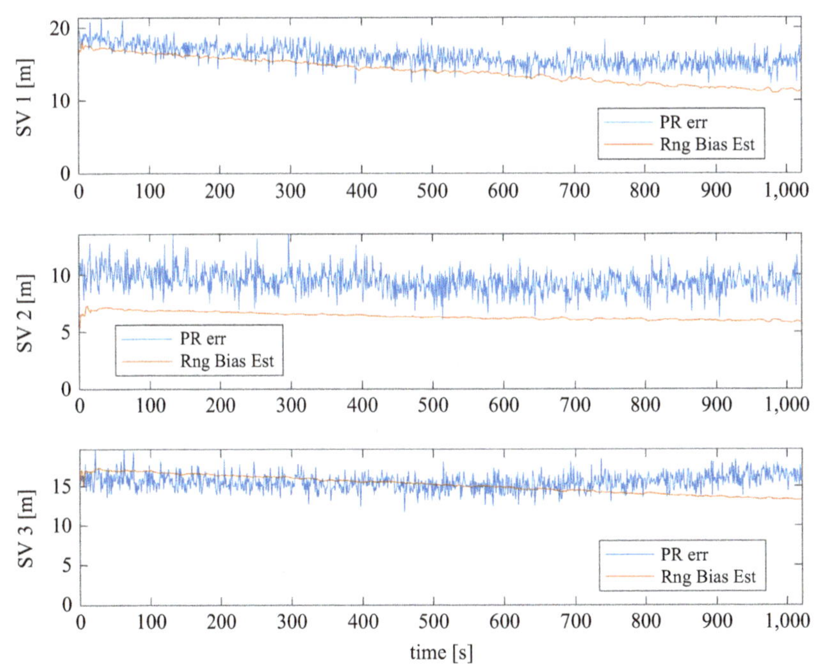

Figure 21.6 Pseudorange bias estimates for satellites 1–3. Tightly coupled GNSS/INS filter. The total pseudorange errors (PR err) consisting of atmospheric delays and measurement noise are depicted along with the filter estimate

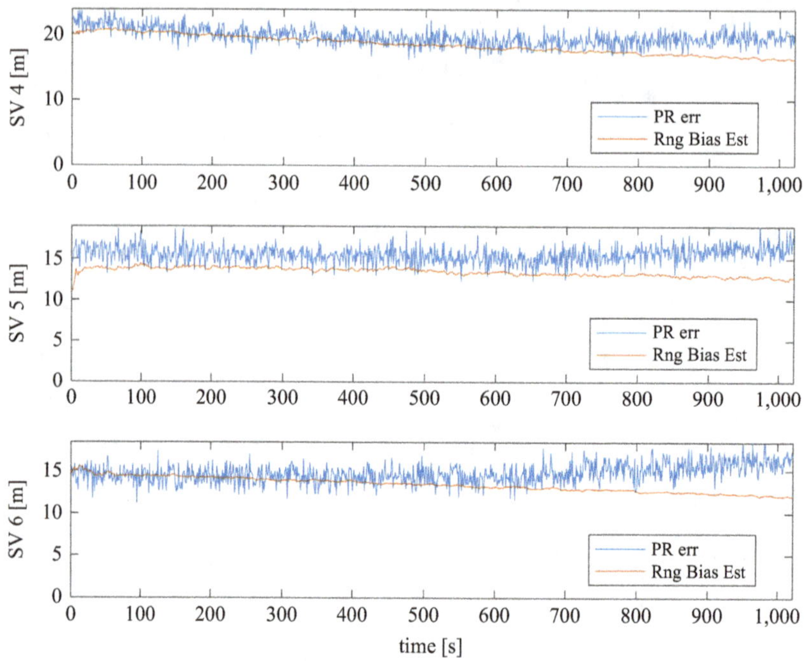

Figure 21.7 Pseudorange bias estimates for satellites 4–6. Tightly coupled GNSS/INS filter. The total pseudorange errors (PR err) consisting of atmospheric delays and measurement noise are depicted along with the filter estimate

21.5 Constellation changes

In the same manner as was discussed in the previous chapter for loose-coupling, when there is a change in the satellite constellation the prediction error covariance matrix must be modified. Specifically, when the receiver starts providing measurements on a new satellite (i.e., a satellite that has just risen above the mask angle), the row and column of P^- associated with the range bias state for that satellite must be zeroed out and the diagonal element reset to the initial value. In this way, the filter can gradually absorb the new information provided by the newly tracked satellite. If this was not done, the filter would instantaneously and incorrectly apply the measurement to the INS error state estimates and a discontinuity would result.

21.6 Lever arm

As with loose coupling, the lever arm between the INS and the GNSS antenna must be taken into account. For tight coupling, the INS-computed position is translated, through the lever arm, to the position of the GNSS antenna. The 'INS' terms in, for example, Equations (21.6)–(21.9) must be the computed INS position components translated to the position of the GNSS antenna. Since the vehicle attitude typically is not constant, the accuracy of this translation is limited in part by the accuracy of the INS attitude determination.

21.7 Data synchronization

As was discussed in the previous chapter, there is a data synchronization issue in the fact that by the time the GNSS data is received, it is "old." Specifically, there is a non-zero so-called "transport delay" of the data from the GNSS receiver to the processor that performs the Kalman filter computations. The inertial data has to be buffered so that it can be interpolated to the GNSS time of validity.

21.8 Q and R tuning

As with loose coupling, tuning of the Q matrix is needed to account for unmodeled errors. Since pseudorange noise and multipath errors are typically inversely proportional to elevation angle, the R values can be dynamically adjusted accordingly. It must be noted that the relatively benign conditions presented in the simulation results are not always present. Two items of particular note are the receiver clock and multipath-induced error. The two-state receiver clock model is adequate for clocks that have low phase-noise and jitter. For clocks that are not so well behaved, enhanced clock models (with additional states) may need to be built into the filter. Multipath-induced errors can degrade filter performance if their magnitude is sufficiently high (e.g., several meters or more). As shown earlier, slowly varying atmospheric errors, even with sizable magnitude, can be accommodated by the filter. However, if multipath errors with higher frequency components are also present, the range bias states will be unable to estimate both the atmospherics and multipath simultaneously. Attempts to add multipath error states along with the range bias states are generally not beneficial due to lack of observability. Conversely, if dual-frequency measurements are available along with a good tropospheric correction model, the remaining pseudorange error may be dominated by the multipath error. In such cases, the range bias states can be tuned for the multipath instead of the atmospheric errors.

21.9 A note about numerical instability and UD factorization

In the early days of the Kalman filter (e.g., 1960s), finite word-length effects in then state-of-the-art computers was a serious issue. The floating-point processor was still in its infancy and essentially unavailable for real-time embedded processors. One of the early uses of the Kalman filter was in the Apollo program where there was concern that the prediction/estimation error covariance matrices would become invalid due to finite word-length effects. Specifically, covariance matrices have the property of being symmetric and positive-semidefinite. If this property is lost, filter divergence can occur.

The first solution to this problem was invented by Dr James Potter in 1962 [2]. Potter had recently completed his doctorate in mathematics at MIT and joined MIT's Instrumentation Laboratory. Potter's ingenious solution took advantage of the fact that every positive semidefinite matrix has a square root. If M is a positive semidefinite matrix, then there exists a matrix B such that $M = BB = B^2$. Potter realized that if the square root of the covariance matrix were propagated then the covariance matrix could be constructed from the square-root and would be guaranteed to be positive-semidefinite. The implementation became known as the Potter Square Root Method.

This concept was later generalized by Bierman [4] who considered a wider variety of matrix factorization methods. One of the most popular is known as the UD Factorization method that relies on the decomposition of the covariance matrix into an upper-right triangular matrix (U) and a diagonal matrix (D). The idea is the same. Instead of propagating the covariance matrix, the U and D factors are propagated. More details may be found in [3,4].

References

[1] Groves P. *Principles of GNSS, Inertial, and Multisensor Integrated Navigation Systems*. 2nd ed. Boston, MA: Artech House; 2013.

[2] MIT News [homepage on the Internet]. Cambridge, MA: MIT; c2005-2022 [cited 2022 Aug 6]. Available from: https://news.mit.edu/2005/obit-potter.

[3] Brown RG, Hwang PYC. *Introduction to Random Signals and Applied Kalman Filtering: With MATLAB® Exercises*. 4th ed. Hoboken, NJ: John Wiley & Sons, Inc.; 2012.

[4] Bierman G. *Factorization Methods for Discrete Sequential Estimation*. San Diego, CA: Academic Press; 1977.

Chapter 22
Aided INS: observability and feedback

22.1 Introduction

We have dedicated a considerable number of pages to the description of the system and the measurement models in the Kalman Filter and, of course, we have applied that to GNSS/INS integration. We have developed the inertial error models and we have developed the GNSS measurement models. Recall the H matrix has a considerable number of zeroes in it and but yet it is the sole representation, within the filter, of the direct relationship between the measurements and the states. So the question is: How does the filter estimate all of the states and, in particular, those states that are *not* direct functions of the measurements? In a loosely coupled filter, for example, the input measurements consist of observed position differences. How does the filter use those observations to be able to estimate things like attitude error and gyro biases? What's up with that? How does that work?

22.2 H, P^- and the Magic K

The generic term is "observability." Specifically, how does the Kalman filter observe the states of interest based on the available measurements? In order to understand how it works we have to analyze not only the H matrix but also the prediction error covariance matrix (P^-) and what I like to call the "magic Kalman gain." The H matrix relates the measurements to the state vector. Recall that for the case of the 18-state loosely coupled filter the H matrix was given by:

$$H = \begin{bmatrix} 1 & 0 & 0 & 0 & \cdots & 0 & 1 & 0 & 0 \\ 0 & 1 & 0 & 0 & \cdots & 0 & 0 & 1 & 0 \\ 0 & 0 & 1 & 0 & \cdots & 0 & 0 & 0 & 1 \end{bmatrix} \tag{22.1}$$

H has 18 columns. The first three columns have non-zero elements for the inertial position error states and the last three columns have ones in them for the GPS position bias states. However, in between, there are 12 columns of zeros. Recall those columns of zeros correspond to the inertial velocity error, attitude error, accel bias, and gyro bias states. Thus, we have this interesting situation where we observe only position differences and yet we are trying to estimate a number of quantities that are not directly related to the observations.

Although H is filled mostly with zeros, the prediction error covariance matrix (P^-) contains not only the prediction error variances of the states, which are on the main diagonal, but also has the *covariances* on the off-diagonal elements. As we will see shortly, those covariances are the key to the observability question. So where do they come from? Recall that typically we initialize the prediction error covariance matrix as a diagonal matrix. Somehow the off-diagonal elements get populated. How does that happen? Recall the prediction error covariance equation:

$$P_{k+1}^- = \Phi_k P_k \Phi_k^T + Q_k \tag{22.2}$$

Although Q can have off diagonal elements, it is not uncommon, particularly for navigation-grade inertial systems, to approximate Q with a diagonal matrix. Thus the only way to obtain off-diagonal elements in the prediction error covariance matrix is through the state transition matrix and the estimation error covariance matrix. The state transition matrix is obtained from the system dynamics matrix where the off-diagonal elements embody the coupling between the states. Additionally, the estimation error covariance matrix is a function of the Kalman gain. Recall the Kalman gain equation:

$$K_k = P_k^- H_k^T (H_k P_k H_k^T + R_k)^{-1} \tag{22.3}$$

We know that H relates measurement space to state space. The first term in the denominator of (22.3) is the prediction error covariance converted from state space to measurement space. That is why we can add that term to R (which is obviously in measurement space). The numerator, since it does not have the leading H like the denominator, acts similarly but plays an additional role in allowing the total K to convert measurement space back into state space in the Kalman update equation. For the moment, just recall that the Kalman gain essentially is the ratio of the state prediction uncertainty (i.e., error covariance) to the sum of the state prediction uncertainty and the measurement uncertainty (i.e., error covariance).

We will focus on the example of a loosely coupled GPS/INS filter for the ease of explanation but the results are applicable to tight coupling as well. Recall that the GPS position bias states are primarily used to ease the transition that occurs when satellites are added to or dropped from the GPS position solution. Assuming the number of satellites in the solution has not changed in a while, we can thus ignore the non-zero elements in the last three columns of H (Equation (22.1)). With that in mind, recall the Kalman update equation:

$$\underline{\hat{x}}_k = \underline{\hat{x}}_k^- + K_k(\underline{z}_k - H_k \underline{\hat{x}}_k^-) \tag{22.4}$$

which can be rewritten as:

$$\underline{\hat{x}}_k = \underline{\hat{x}}_k^- + K_k(\underline{z}_k - \underline{z}_k^-) \tag{22.5}$$

where $\underline{z}_k^-$ is the predicted measurement vector (i.e., the measurement vector as predicted from the state prediction):

$$\underline{z}_k^- = H_k \underline{\hat{x}}_k^- \tag{22.6}$$

For the loosely coupled filter, the term inside the parentheses in equation (22.5) is a 3x1 vector and the Kalman gain is an 18×3 matrix. In order for the filter to estimate the 18 elements of the state vector, K must have non-zero elements throughout the 18 rows of its three columns. Now consider the numerator of the Kalman gain Equation (22.3) and remember that H is only non-zero in its first three columns (recall we have dropped the position bias states). Thus, it is the first three columns of the prediction error covariance matrix that yield the non-zero elements in K.

Let us now return to Equation (22.5). The term in parentheses is the actual measurement minus the predicted measurement. That term is called the "innovations" since it is the *new* information provided to the filter. It is not the whole measurement that is new but, rather, it is only the portion of the measurement that we were *not* able to predict that is the new information. Thus, the Kalman gain weights the new measurement information and adds it to its prediction to form the estimate. Therefore in order to take, for example, position differences, and then use them to estimate attitude error, gyro biases, accel biases, etc., there have to be non-zero elements in the Kalman gain relating the observations to all those different states. That is why I call it the "magic Kalman gain." It relates measurements to states that do not have direct relationships. Keep in mind, the key is the fidelity of the state models used to form the state transition matrix.

We can now better understand the temporal nature of the filter's initial transient (that is a fancy way to say the filter takes time to converge). It takes time for the off-diagonal elements in the prediction error covariance matrix to build up. Specifically, it requires a time-dependent relationship between the states as is described in the state transition matrix. Remember that we are estimating more states than we have measurements and that can only be done by exploiting the additional information obtained by processing multiple sets of measurements over time.

22.3 Observability

Furthermore, not only is there a requirement for a time-dependent relationship, but as we will see, the vehicle trajectory is a critical factor in the observability of some states. Recall the psi-angle inertial error model:

$$\underline{\delta \dot{R}}^c = -\underline{\omega}_{ec}^c \times \underline{\delta R}^c + \underline{\delta V}^c \tag{22.7}$$

$$\underline{\delta \dot{V}}^c = C_b^c \underline{\delta f}^b + \underline{\delta g}^c - \underline{\psi}^c \times \underline{f}^c - (2\omega_{ie}^c + \omega_{ec}^c) \times \underline{\delta V}^c \tag{22.8}$$

$$\underline{\dot{\psi}}^c = -\underline{\omega}_{ic}^c \times \underline{\psi}^c - C_b^c \underline{\delta \omega}_{ib}^b \tag{22.9}$$

With GNSS aiding, the measurement vector $\underline{z}$ is either position and velocity differences (loose coupling) or range and range rate differences (tight coupling). Inertial attitude and instrument errors are observed *through* the position and velocity errors *that they produce* in the inertial system solution. For the case of loose coupling, we are observing position and velocity errors and yet we want to be able to estimate attitude and inertial sensor errors. Without even looking at the equations, we know implicitly based on our undergraduate physics that accelerometer biases are going to produce linear velocity error growth, at least in the short term, and then quadratic position error growth. You knew that before you read this book. However, what about gyro biases and attitude errors?

Consider the error equations. A constant gyro bias produces linear growth in the psi-angle (Equation (22.9)) and the linear psi-angle growth produces quadratic growth in velocity error (Equation (22.8)) and thus cubic growth in the position error (all of this is in the short-term prior to Schuler effects).

Consider now the cross-product term in Equation (22.8). We see that not only is the velocity error growth a function of the attitude error (psi-angle), but it is also dependent on the specific force. Obviously if the specific force is zero, it does not matter what the psi-angle is, there will be no velocity error growth regardless. Now consider straight and level flight. In that case, the only sensed specific force is the reaction to gravity and that lies solely in the vertical direction (to first order). There is no horizontal component. Now remember the result of the cross-product is always *orthogonal* to the plane containing the two input vectors. Recall further that the cross-product of two collinear vectors is zero. This means that in straight and level flight, where the specific force is solely in the vertical direction, only the *horizontal* components of the attitude error (pitch and roll) will give rise to velocity error growth. Yaw error is *not* observable in straight-and-level flight since the yaw error is collinear with the specific force vector and thus the cross-product is zero.

Given this cross-product relationship, heading (and/or yaw) error will only cause velocity error growth when the vehicle is experiencing acceleration in the horizontal plane. Of course, in practice, significant acceleration in the direction of flight is impractical. When an aircraft is at cruise altitude the throttle is typically left alone. Sure, you could throttle back and forth but nobody is going to do that, in part, because it is unnecessarily hard on the engines. Thus, the only practical way to achieve heading observability at cruise altitude is to perform turns. Every once in a while the vehicle has to perform an s-turn. Navigation-grade INS manufacturers frequently have to educate their customers on this issue. For customers with long distance missions (e.g., oceanic crossings) who are unwilling to adapt their flight profiles (e.g., commercial flights), horizontal error growth due to poorly estimated heading error has to be accepted.

Vehicle maneuvers have beneficial effects on other states as well. For example, attitude errors and accel biases are not separately observable at constant vehicle attitude. Accel biases lead to erroneous velocities and transport rates and Schuler oscillations. However, attitude errors (specifically roll and/or pitch) do so also due to gravity coupling. This is clear from Equation (22.8). A constant accel bias and a constant roll or pitch error (manifested in a constant psi-angle) lead to linear velocity error growth (in the short term). Thus, if the true attitude is constant, there will be no observable difference between the attitude error and the accel bias. The filter will partially estimate the attitude error and partially estimate the accelerometer bias in accordance with the prediction error covariance associated with each state. Thus, neither the attitude errors nor the accelerometer biases are estimated perfectly during constant-attitude flight. However, they become separately observable when the vehicle attitude changes since this impacts the cross-product of the psi-angle with the specific force vector along with the projection of the accel biases into the navigation frame.

Given what we have discussed about straight and level flight, attitude error does not contribute to vertical velocity error (since the specific-force vector is vertical, attitude errors will only manifest velocity error in the horizontal direction). The remaining vertical velocity error sources are vertical accel bias, gravity model error, and Coriolis. Assuming the velocity errors are kept small via feedback of the Kalman correction (see the next section in this chapter), the Coriolis component of vertical velocity error will also be small. This leaves the combination of vertical accel bias and gravity modeling error. As a result, the vertical accelerometer bias state (absorbing also the gravity modeling error) is the most observable of the three accel biases and thus it is the most easily estimated. One consequence of this is that if the integrated system has to coast at some point, due loss of aiding, the system vertical error growth will be small relative to the horizontal error growth. This is non-intuitive since we know the free inertial solution has an unstable vertical channel. Nevertheless, it is the most observable.

A topic closely related to observability is state selection. Specifically, what states should we choose to put into our filter? The nine core inertial error states (three dimensions of position, velocity and attitude) are needed in most applications but after that the choice is dependent upon the application, the IMU sensors and the aiding sensors. For example, if the length of the primary "mission" is short (e.g., less than 15 min) and the gyros are of sufficient quality, estimation of accel biases is required but there may be little benefit to estimating the gyro biases. The reader may recall that we estimated scale factor error in one of the sled track examples in an earlier chapter but we did not do so in either the loosely coupled or tightly coupled examples of the previous two chapters. This was done partially to reduce the complexity of the new material. However, in a real scenario, the observability of scale factor error states will be a function of the dynamics of the trajectory. For example, gyro scale factor error on a slow-moving marine vehicle will be much less observable than it is on an aerobatic aircraft. Similarly, non-orthogonality errors (e.g., x-dimension acceleration being incorrectly sensed by the y-accelerometer) may not be observable if other sensor errors (e.g., bias and noise) are relatively larger and if the vehicle dynamics are low. In most cases, simulations are needed to determine which states should be included.

As we conclude this introduction to observability, the reader desiring to dig deeper is referred to two additional resources. Groves' book on integrated navigation [1] has an excellent discussion on this topic. The reader is also referred to the article by Rhee *et al.* [2]. It provides a very rigorous mathematical treatment of observability in GPS/INS.

22.4 Feedback of corrections

An aided-INS Kalman filter can be configured either in a feed forward or feedback manner. In a feed forward configuration, the errors estimated by the Kalman filter are subtracted from the stand-alone INS output. The limitation with this approach, however, is that the error model that we utilize to construct the system dynamics matrix is linear. As the inertial position, velocity and attitude errors grow over time, the linearity assumption breaks down. The filter's predictions thus become unreliable and the filter estimates will diverge from the actual values.

Therefore, in order to sustain operation over long periods of time (where "long" is a function of the quality of the IMU sensors), the error states have to be fed back as corrections to the basic inertial processing. In this way, the inertial errors are kept small and the linearity assumption in the error model is kept valid. In other words, we cannot just run the inertial all by itself and let it drift off indefinitely. The Kalman filter error estimates must be used to correct the position, velocity, and attitude update processing. We must also take the accelerometer and gyro error estimates and use them to correct the delta-thetas and delta-Vs that are going into the inertial position, velocity, and attitude updating.

This is all done in software. The hardware is not affected in any way by the feedback. In the first few decades of GPS/INS integration on military platforms, it was common practice to output three parallel solutions: INS only (e.g., "free inertial"), GPS only and GPS/INS. Both free inertial and aided inertial data could be output simultaneously, even with feedback, since the processing was all done in software.

After feedback, all of the Kalman filter state vector elements are reset to zeros. Thus the linearized complementary Kalman filter has been turned into an extended Kalman filter. Remember in an extended Kalman filter the linearization is based on the estimates obtained from the filter itself. That is exactly what is happening under feedback.

It should be noted that although the state vector is zeroed out after each Kalman update, the software must separately maintain the *total* values of the estimated sensor errors. These are needed to correct the gyro and accel measurements before they are used in the position/velocity/attitude updating. For example, when the filter estimates gyro bias, it is actually estimating the change in gyro bias over the update interval. This estimate is added to the total gyro bias estimate and the total is used to correct the raw gyro measurement (and obviously the same is needed for the accelerometers).

A couple of notes of caution: when implementing feedback, there is a risk of closed loop instability. It is generally safe to wait for the filter to converge before instituting feedback but feedback can be started earlier if great care is taken in tuning the initial prediction error and system noise covariance matrices to control the transient behavior of the filter output. In addition, if the filter is loosely coupled and if the GPS receiver has its own Kalman filter, then the updates must be performed at a sufficiently low rate such that the input observations can be considered uncorrelated.

22.5 Simulation example

We now consider a simulation example that highlights some of the key points discussed in this chapter regarding observability and feedback. A navigation-grade IMU is simulated on an F-16. The accelerometer biases range from 40 to 60 micro-g and the gyro biases range from 0.008 to 0.01 deg/hour. The ground track of the trajectory (Figure 22.1) shows the vehicle makes a few s-turns (the vehicle starts at the coordinate system origin) prior to a straight segment directly west followed by a 90 degree left turn and a straight segment to the south. East and north velocity components, shown in Figure 22.2, show that the straight west segment starts at approximately 10 min

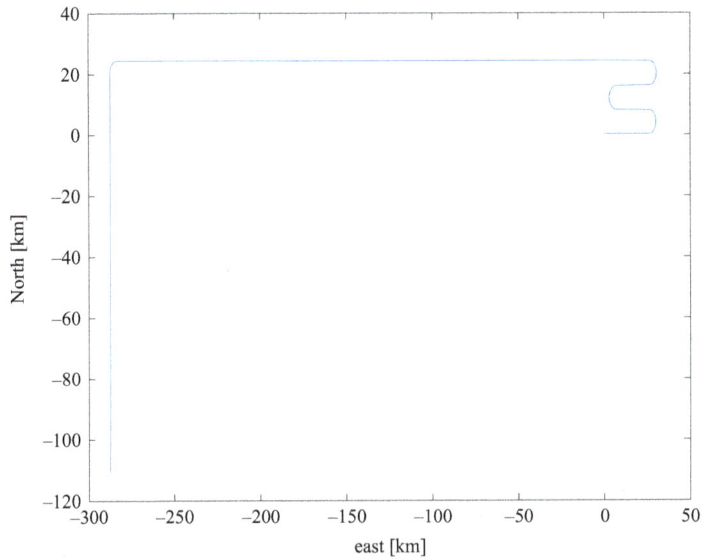

Figure 22.1 Ground track of simulated trajectory. Flight starts at the origin of the coordinate system

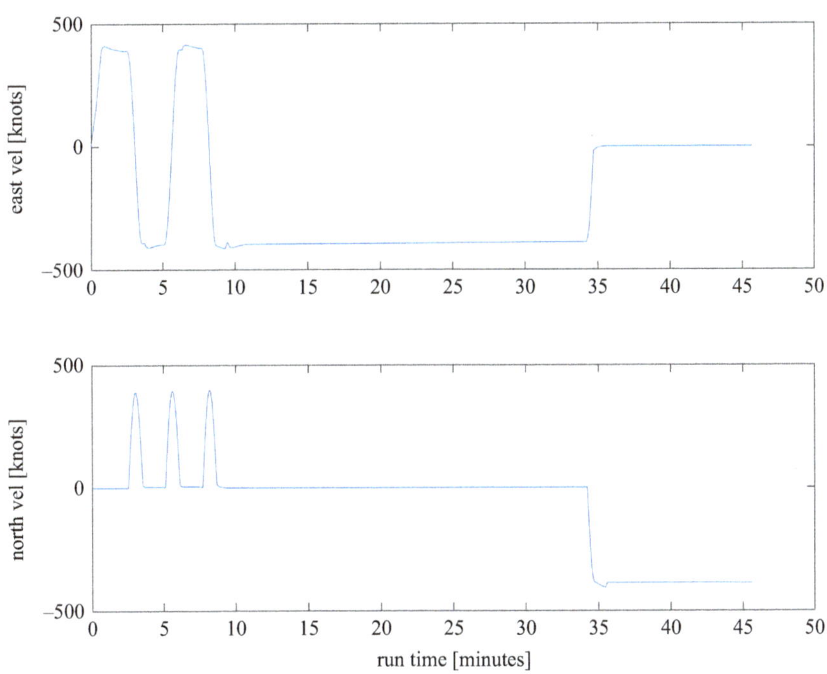

Figure 22.2 Velocity components of the simulated trajectory

and the left turn to the straight south segment starts at approximately 34 min into the flight. The altitude profile (Figure 22.3) shows the vehicle is executing step-climbs during the initial turns.

Figures 22.4–22.8 illustrate the performance obtained from a GPS-aided INS with a loosely coupled Kalman filter *without* the implementation of feedback. In order to eliminate any artifacts resulting from less-than-perfectly modeled/estimated GPS errors, the only GPS errors simulated were thermal noise. Since the IMU sensor biases were simulated as perfect constants, it is not surprising that the filter does well in most of the state estimates in this case. Nevertheless, the vertical components of most of the errors degrade significantly after the initiation of the aforementioned left turn to the south segment. The linearity assumptions in the filter have been violated and it is also not surprising that the effects are observed first in the vertical components (recall the instability of the vertical channel). If lower-grade IMU sensors had been simulated, the effects would have been observed sooner and in some of the horizontal components as well.

Figures 22.9–22.13 illustrate the performance obtained when feedback of the corrections is implemented. The previously observed excursions of the vertical components have been eliminated. Additionally, we can also note the impact of the previously described effects of vehicle dynamics on the state estimates and covariances. For example, in Figure 22.11, we see that although the Kalman's estimated error standard deviations are growing for all three components of attitude, they are an order of magnitude larger for the yaw component. Furthermore, close scrutiny

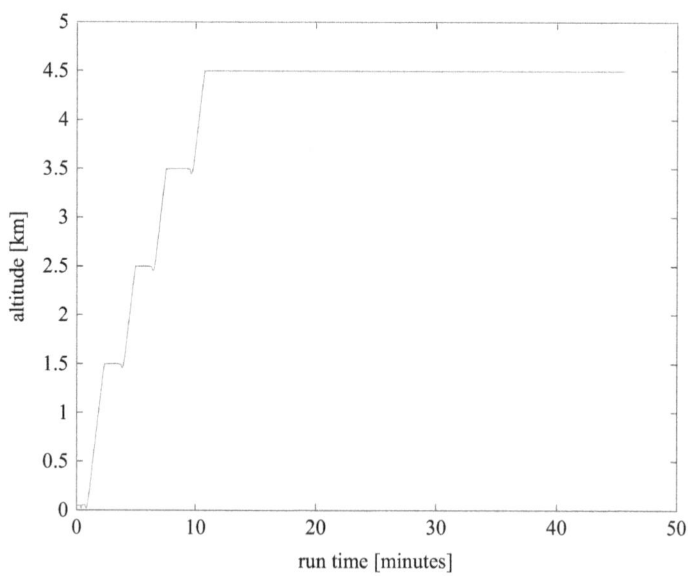

Figure 22.3 Altitude of the simulated trajectory

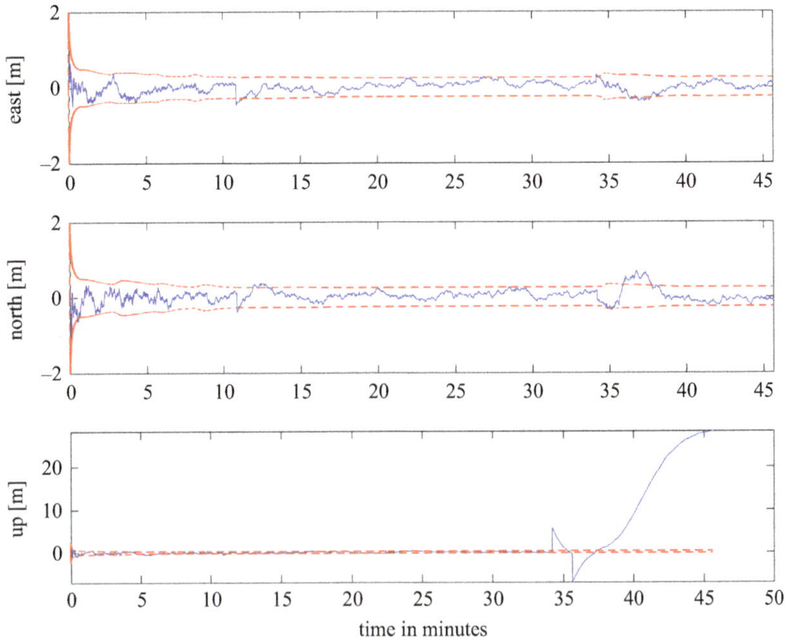

Figure 22.4 *Kalman-corrected position errors without implementation of feedback*

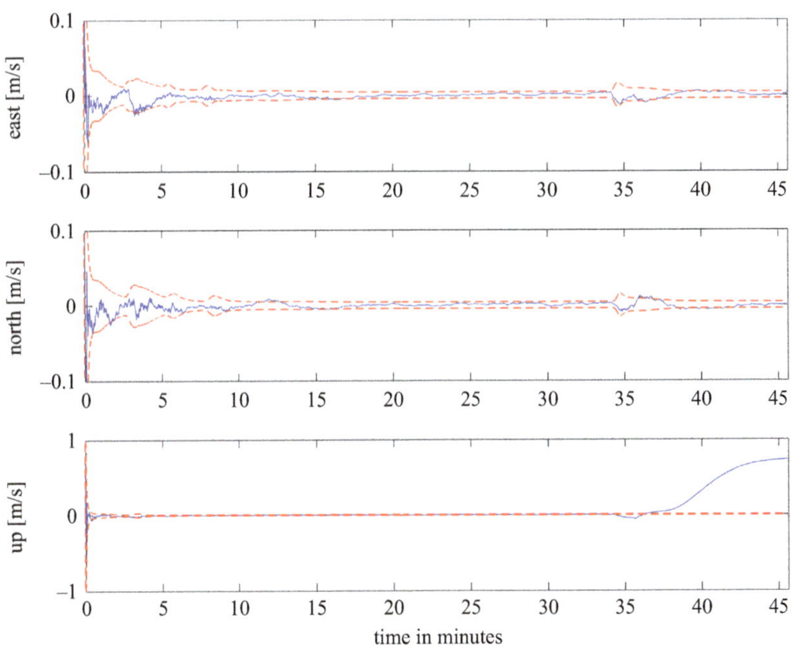

Figure 22.5 *Kalman-corrected velocity errors without implementation of feedback*

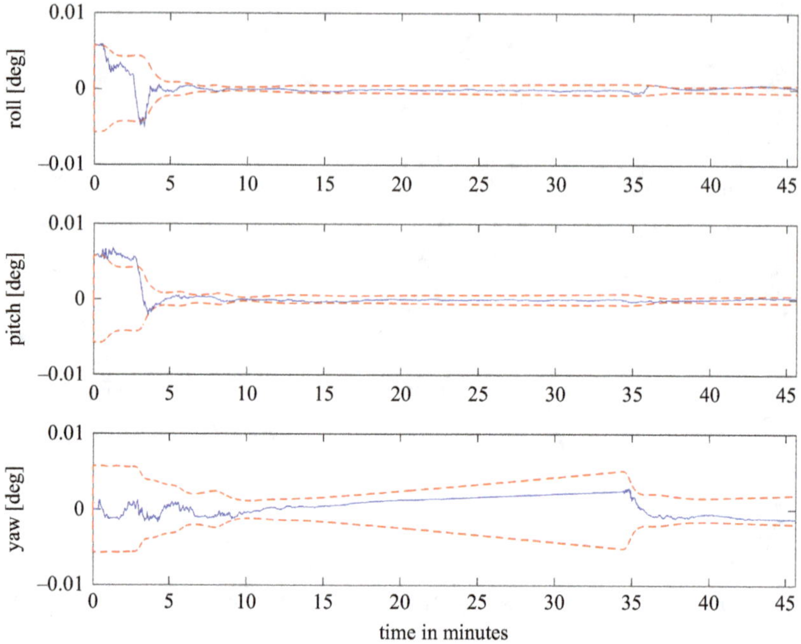

Figure 22.6 Kalman-corrected attitude errors without implementation of feedback

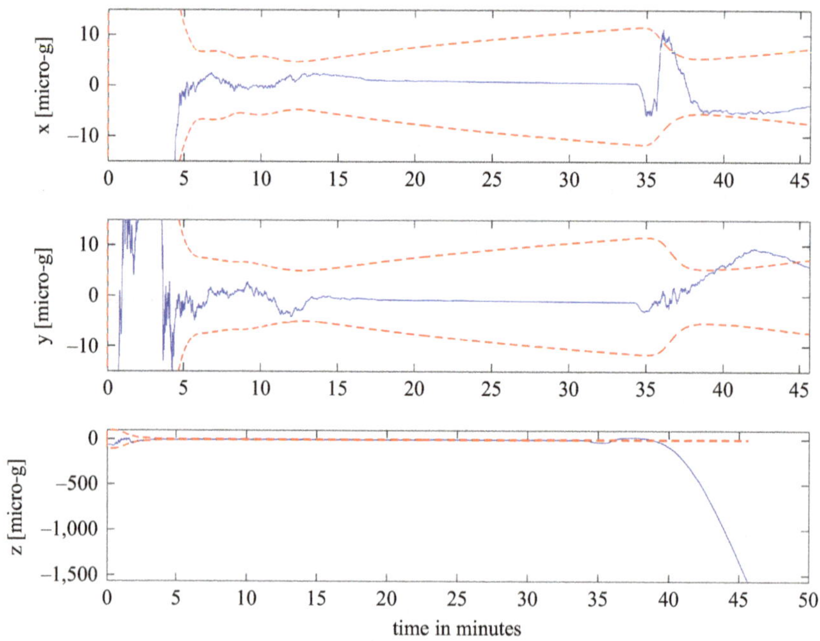

Figure 22.7 Kalman-corrected accelerometer bias estimation errors without implementation of feedback

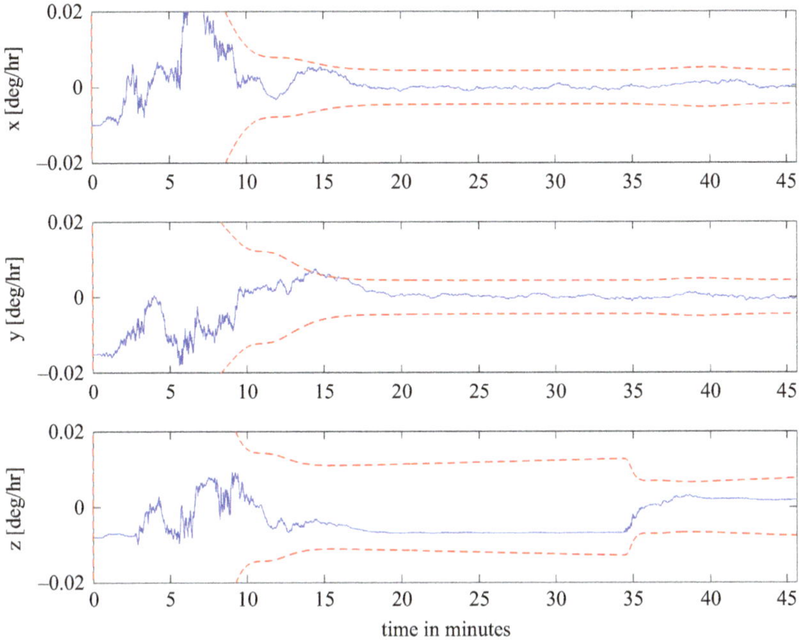

Figure 22.8 Kalman-corrected gyro bias estimation errors without implementation of feedback

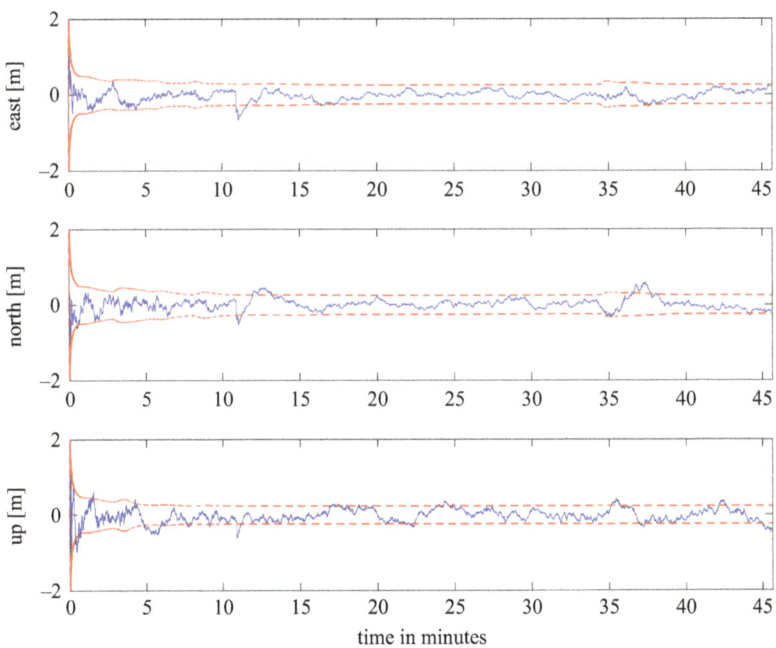

Figure 22.9 Kalman-corrected position errors with feedback of corrections

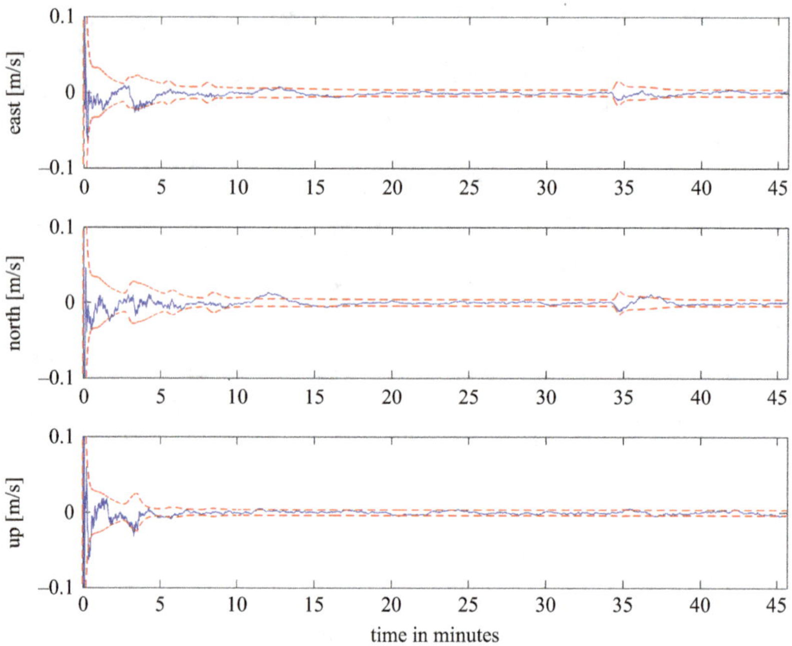

Figure 22.10 Kalman-corrected velocity errors with feedback of corrections

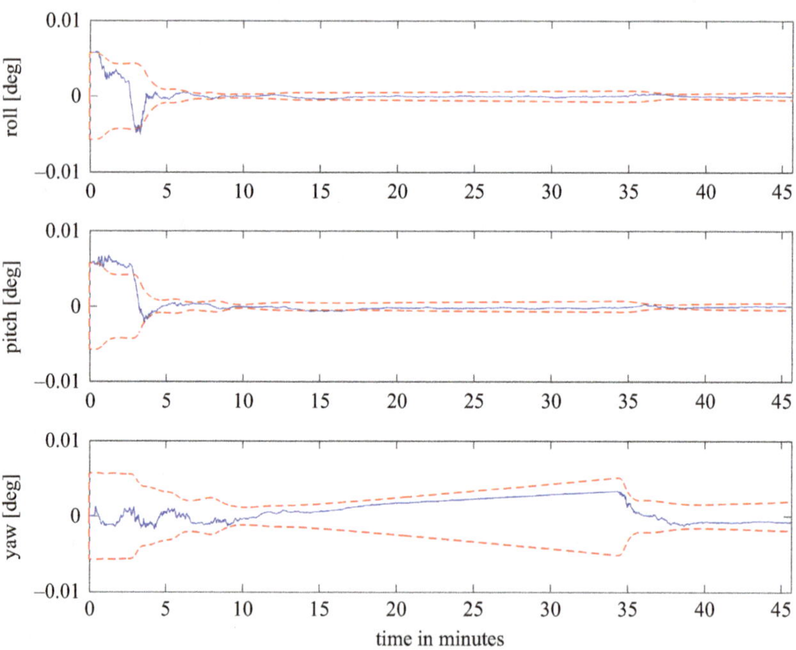

Figure 22.11 Kalman-corrected attitude errors with feedback of corrections

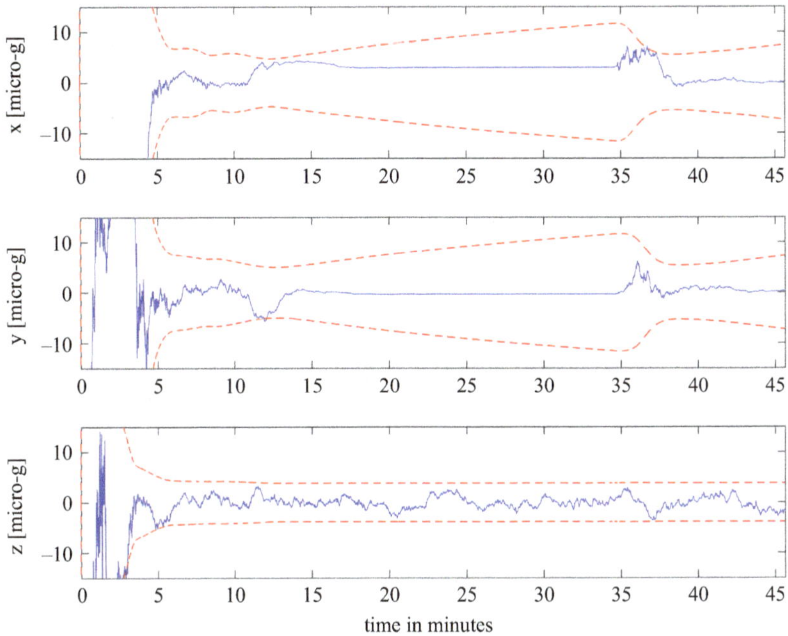

Figure 22.12 Kalman-corrected accelerometer bias estimation errors with feedback of corrections

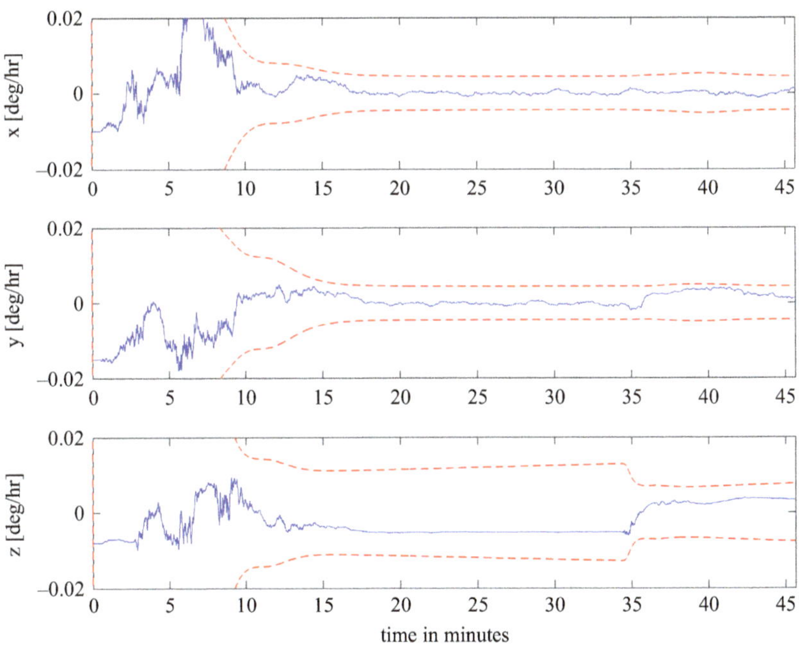

Figure 22.13 Kalman-corrected gyro bias estimation errors with feedback of corrections

of the errors shows that the pitch and roll errors are still being estimated during the straight west segment (from 10 min to 34 min) whereas the yaw error is not being estimated (notice the smoothness of the growth of the yaw estimation error during this segment). However, once the left turn to the south has been performed, the yaw error is quickly estimated and corrected by the filter.

Similarly, Figure 22.12 shows the horizontal accelerometer biases are not being estimated during the bulk of the straight west segment whereas the vertical accelerometer bias is being estimated the entire time. In particular, notice how the x-accelerometer bias is not quite estimated perfectly (note the offset during the straight west segment) but after the turn the filter has captured it. Recall that attitude errors and accelerometer biases are not separately observable during constant-attitude flight (e.g., straight-and-level flight). Lastly, the gyro bias estimation error plots (Figure 22.13) show horizontal gyro biases are estimated well once the filter has converged but that last turn is needed to improve the vertical gyro estimate.

References

[1] Groves P. *Principles of GNSS, Inertial, and Multisensor Integrated Navigation Systems*. 2nd ed. Boston, MA: Artech House; 2013.

[2] Rhee I, Abdel-Hafez MF, Speyer JL. Observability of an integrated GPS/INS during maneuvers. *IEEE Transactions on Aerospace and Electronic Systems*. 2004;40(2):399–416.

[3] Gelb A, editor. *Applied Optimal Estimation*. Cambridge, MA: The M.I.T. Press; 1974.

[4] Van Loan CF. Computing integrals involving the matrix exponential. *IEEE Transactions on Automatic Control*. 1978;23(3):491–493.

[5] Brown RG, Hwang PYC. *Introduction to Random Signals and Applied Kalman Filtering: With MATLAB® Exercises*. 4th ed. Hoboken, NJ: John Wiley & Sons, Inc.; 2012.

Chapter 23

Baro-inertial vertical channel

23.1 Introduction

As we have shown previously, the vertical channel of the inertial system is unstable. In fact, if you go back to legacy gimbal systems from the 1950s, it was not uncommon for there to be *no* vertical channel. They did not even have a vertical accelerometer. A gimbal platform was mechanized with two orthogonal accelerometers in the horizontal plane and that was it. It was thought that there was just no point in mechanizing the vertical channel.

Nevertheless, eventually it was realized that there *were* benefits to having an inertial vertical channel. Specifically, applications with a need for accurate vertical velocity drove this development. In the defense community, weapon delivery is one example where precise vertical velocity is important. There was thus a need to be able to stabilize the inertial vertical channel and the best available aiding sensor, prior to the advent of GNSS, was the barometric altimeter. Even to this day, aiding with a barometric altimeter is still utilized due to the fact that its availability is essentially 100% (limited only by the maintainability of the altimeter itself).

As was the case with satellite-based systems such as GPS, the main idea is that the barometric altimeter ("baro") provides long-term stability. Although the baro will in general exhibit bias-like error in its altitude estimate, that error does not grow as a function of time as do inertial errors. Note that as with the use of the term in inertial sensors, "bias" is understood to be a very low frequency, non-zero mean error component. Its variation is a function of weather changes. There is also some noise in the baro output. Finally, the baro exhibits lag in its output. Specifically, sudden changes in true altitude are not immediately reflected in the baro output. Since the inertial system has excellent dynamic response, proper integration will result in an aided vertical channel with stabilized vertical velocity, high data rate and low data latency.

Before discussing the Kalman filter mechanization, we will consider some legacy fixed-gain filters and will examine their strengths and weaknesses.

23.2 Simple fixed-gain complementary filter

Recall the complementary filter structure is applicable when combining two sensors wherein one of the sensors exhibits primarily low-frequency errors and the other sensor exhibits primarily high-frequency errors. In practice, this is never perfectly the case but we can nevertheless apply it where the assumptions are reasonable. In the present case, the inertial error is low-frequency whereas the baro error is a combination of high frequency error (noise) and a slowly changing bias.

Figure 23.1 [1] is an example of a second-order, fixed gain, complementary filter. The vertical accelerometer is compensated by gravity, any known accel bias and Coriolis corrections. Across the top of the figure, we see that the compensated vertical accelerometer measurement is integrated into vertical velocity and then integrated again to get altitude (along with the necessary initializations).

The bottom half of the figure depicts the aiding from the barometric altimeter. On the lower right, the indicated altitude and the baro altitude are differenced to form the observation. The difference is filtered by $G_1(s)$ to form an estimate of the inertial vertical velocity error and is also separately filtered by $G_2(s)$ to form an estimate of the inertial vertical acceleration error. Although it is not obvious at first glance, a mathematical analysis of this loop can be done to demonstrate that it is in fact a complementary filter.

The simplest forms of G_1 and G_2 are constants and in feedback loop parlance are referred to as "gains." As with any feedback control system, the gains are chosen based on the desired transient response (e.g., critically damped) that the designer is trying to achieve. A wide variety of compensation algorithms have been developed since baro stabilization of the inertial vertical channel was invented in the late 1950s [2]. Some algorithms were developed with gains that were dependent upon the magnitude of vertical velocity. For example, a military aircraft in a very steep dive may rapidly pass through atmospheric variations that are not accounted for by the standard barometric altitude computation. One approach to deal with this is to adapt the loop gain in accordance with the vertical velocity. In the open literature, one of the most advanced variable gain filters developed along these lines was described by Ausman ([3]).

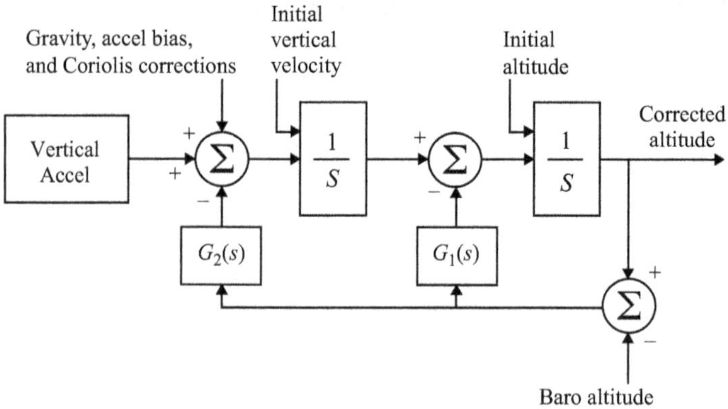

Figure 23.1 Simple fixed-gain inertial vertical aiding loop [1]

23.3 Three-state Kalman filter

The previously described fixed-gain filters provide acceptable performance for some applications. To improve, a first step is to be able to estimate and remove the baro bias. Figure 23.2 illustrates the classic complementary Kalman filter architecture (that we have described previously for radar and radionavigation system aiding of the inertial) applied to the vertical channel.

A three-state Kalman filter was described by Brown and Hwang [4] that estimates the baro bias (baro altitude error) along with the inertial vertical velocity and inertial altitude error:

x_1 = inertial vertical position error (altitude)
x_2 = inertial vertical velocity error
x_3 = barometric altimeter error

We start the filter design by determining the Kalman system and measurement equations. For this simple filter, the accelerometer error is assumed to be white noise. This is not realistic, of course, but we will address accel bias in a subsequent filter design. With the assumed accel error model, the inertial vertical velocity error is then random walk (i.e., integrated white noise) and the inertial altitude error is integrated random walk. The derivative of the altitude error is simply the vertical velocity error and the derivative of the vertical velocity error is the assumed accelerometer white noise:

$$\dot{x}_1(t) = x_2(t) \tag{23.1}$$

$$\dot{x}_2(t) = \varepsilon_2(t) \tag{23.2}$$

where $\varepsilon_2(t)$ is the driving white noise and the subscript "2" is used simply because the vertical velocity error is the second of the three state variables. Since the baro bias varies slowly (with change of latitude/longitude and/or weather) but does not grow without bound, a reasonable stochastic approximation is a first-order Gauss–Markov model:

$$\dot{x}_3(t) = -\frac{1}{\tau_{bb}}x_3(t) + \varepsilon_3(t) \tag{23.3}$$

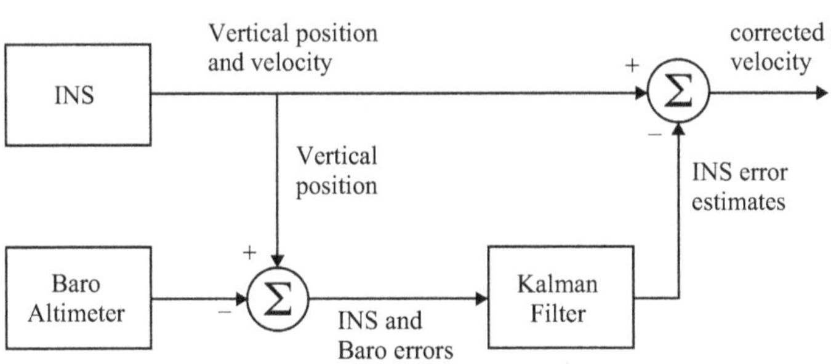

Figure 23.2 Vertical channel complementary Kalman filter

where τ_{bb} is the time-constant of the Gauss–Markov process that is simulating the slowly varying baro bias (in practical terms this may be tens of minutes) and $\varepsilon_3(t)$ is the driving white noise. The state equation is thus:

$$\begin{bmatrix} \dot{x}_1 \\ \dot{x}_2 \\ \dot{x}_3 \end{bmatrix} = \begin{bmatrix} 0 & 1 & 0 \\ 0 & 0 & 0 \\ 0 & 0 & -\frac{1}{\tau_{bb}} \end{bmatrix} \begin{bmatrix} x_1 \\ x_2 \\ x_3 \end{bmatrix} + \begin{bmatrix} 0 \\ \varepsilon_2 \\ \varepsilon_3 \end{bmatrix} \tag{23.4}$$

The discrete-time solution of (23.4) gives us the Kalman system equation for this example:

$$\begin{bmatrix} x_1 \\ x_2 \\ x_3 \end{bmatrix}_{k+1} = \begin{bmatrix} 1 & \Delta t & 0 \\ 0 & 1 & 0 \\ 0 & 0 & e^{-\frac{1}{\tau_{bb}}\Delta t} \end{bmatrix} \begin{bmatrix} x_1 \\ x_2 \\ x_3 \end{bmatrix}_{k} + \begin{bmatrix} 0 \\ w_2 \\ w_3 \end{bmatrix}_{k} \tag{23.5}$$

where the 3×3 matrix is the state transition matrix and covariance matrix of the system noise vector is given by [4]:

$$Q = \begin{bmatrix} A\frac{\Delta t^3}{3} & A\frac{\Delta t^2}{2} & 0 \\ A\frac{\Delta t^2}{2} & A\Delta t & 0 \\ 0 & 0 & \sigma^2\left(1 - e^{-\frac{2\Delta t}{\tau_{bb}}}\right) \end{bmatrix} \tag{23.6}$$

where A is the power spectral density of the accelerometer white noise process $\varepsilon_2(t)$ and σ is the standard deviation of the first-order Gauss–Markov process being used to model the baro bias.

Turning our attention now to the Kalman measurement equation, the observation is formed by differencing the baro and inertial altitude estimates:

$$z = h_{INS} - h_{baro} \tag{23.7}$$

In order to determine the Kalman measurement equation, we first expand (23.7) by recognizing that each estimate is the sum of the truth and the sensor error:

$$z = (h_{true} + \delta h_{INS}) - (h_{true} + \delta h_{baro}) \tag{23.8}$$

Which, of course, simplifies to:

$$z = \delta h_{INS} - \delta h_{baro} \tag{23.9}$$

Given our state vector, the Kalman measurement equation may now be written by inspection:

$$z_k = \begin{bmatrix} 1 & 0 & -1 \end{bmatrix} \begin{bmatrix} x_1 \\ x_2 \\ x_3 \end{bmatrix} + v_k \tag{23.10}$$

where the 1×3 matrix is H. v_k models the baro measurement noise. The baro bias is being estimated in x_3 so v_k accounts for whatever high frequency error exists in the baro altitude measurement. For the Kalman filter, R is a scalar equal to the variance of v_k.

23.4 Five-state Kalman filter

The previously described three-state filter estimated the baro bias as well as the inertial vertical velocity and altitude error. However, as was admitted, it is inadequate in the fact that it modeled accelerometer error as white noise. Ausman [5] has described a five-state Kalman filter for the inertial vertical channel that improves on the three-state model by explicitly estimating the vertical accelerometer bias and estimating a baro scale factor error and a sea-level baro bias. The five-state model was originally proposed in 1969 by Barham and Manville [6] with baro scale factor error modeled as a first-order Gauss–Markov process with a time constant of 10,000 seconds. Ausman improved the design with a higher fidelity baro scale factor error model. Specifically, the National Oceanographic and Atmospheric Administration [7] noted that air density deviations from the Standard Atmospheric model tended to reverse sign at an altitude of approximately 26,000 feet (≈ 8 km). Ausman proposed the following model for scale factor error:

$$k = k_0\left(1 - \frac{h_t}{h_{rev}}\right) \tag{23.11}$$

where k is the scale factor error, k_0 is the sea level scale factor error, h_t is the true altitude, and h_{rev} is the altitude at which the scale factor error changes (reverses) sign. The barometric altimeter error is then the sea-level bias plus the integral of the scale factor error from sea-level up to the true altitude:

$$\delta h_{baro} = \delta h_0 + \int_0^{h_t} k d\lambda \tag{23.12}$$

where δh_0 is the sea-level baro bias. Substituting (23.11) into (23.12) yields:

$$\delta h_{baro} = \delta h_0 + \int_0^{h_t} k_0\left(1 - \frac{\lambda}{h_{rev}}\right) d\lambda \tag{23.13}$$

Evaluation of the definite integral results in:

$$\delta h_{baro} = \delta h_0 + h_t k_0\left(1 - \frac{h_t}{2h_{rev}}\right) \tag{23.14}$$

The state equations for the five states are conceptually simple. The derivative of inertial altitude error is equal to the inertial vertical velocity error:

$$\delta \dot{h}_{INS} = \delta v_z \tag{23.15}$$

The derivative of inertial vertical velocity error is the sum of the vertical accelerometer error (bias) and the positive feedback term resulting from gravity being imperfectly compensated due to altitude error:

$$\delta \dot{v}_z = \delta a_z + \left(\frac{2g}{R}\right)\delta h_{INS} \tag{23.16}$$

where δa_z is the vertical accelerometer bias, g is the gravity and R is the radius of the earth.

The three remaining states (accel bias, sea-level baro scale factor error and sea-level baro bias) are all modeled as first-order Gauss–Markov processes. The time constant of the accel bias should be set in accordance with the performance of the sensor. For nav-grade units, this is generally on the order of an hour. Generically, we can simply state:

$$\delta\dot{a}_z = -\frac{1}{\tau_{acc}}\delta a_z + \varepsilon_{acc} \tag{23.17}$$

where τ_{acc} is the accel bias time constant and ε_{acc} is the driving white noise.

Setting the time constants for the two baro states is arguably as much art as it is science since their variation is, of course, a function of the local weather. Ausman suggested values on the order of 10,000 seconds (approximately 2.78 hours) but longer or shorter periods could be justified depending upon the expected atmospheric conditions. Ausman also suggested making the time constant a function of horizontal velocity. His rationale was that high speeds (e.g., jet aircraft) would likely move the vehicle through weather systems more quickly. Thus, his state equations for the two baro states are:

$$\dot{k}_0 = -\frac{1}{\tau_{atm}}\left(\frac{1}{2} + \frac{v_h}{v_{ref}}\right) + \varepsilon_{k0} \tag{23.18}$$

$$\delta\dot{h}_0 = -\frac{1}{\tau_{atm}}\left(\frac{1}{2} + \frac{v_h}{v_{ref}}\right) + \varepsilon_{h0} \tag{23.19}$$

where τ_{atm} is the nominal time-constant of the sea-level baro scale factor error and bias states, v_h is the horizontal velocity, v_{ref} is the reference horizontal velocity, and ε_k and ε_{h0} are the driving white noise. Ausman suggested setting v_{ref} to approximately 300 knots (e.g., approximately 150 m/s). Note the terms in parentheses simply modify the magnitude of the nominal time constant. For horizontal speeds less than v_{ref}, the time constant will be proportionately longer and conversely for speeds higher than v_{ref}.

Having defined the state equations for each individual state, we can gather them as follows:

$$\begin{bmatrix} \delta\dot{h}_{INS} \\ \delta\dot{v}_z \\ \delta\dot{a}_z \\ \dot{k}_0 \\ \delta\dot{h}_0 \end{bmatrix} = \begin{bmatrix} 0 & 1 & 0 & 0 & 0 \\ \frac{2g}{R} & 0 & 1 & 0 & 0 \\ 0 & 0 & -\frac{1}{\tau_{acc}} & 0 & 0 \\ 0 & 0 & 0 & F_{k_0} & 0 \\ 0 & 0 & 0 & 0 & F_{h_0} \end{bmatrix} \begin{bmatrix} \delta h_{INS} \\ \delta v_z \\ \delta a_z \\ k_0 \\ \delta h_0 \end{bmatrix} + \begin{bmatrix} 0 \\ 0 \\ \varepsilon_{acc} \\ \varepsilon_{atm} \\ \varepsilon_{atm} \end{bmatrix} \tag{23.20}$$

where:

$$F_{k_0} = F_{h_0} = -\frac{1}{\tau_{atm}}\left(\frac{1}{2} + \frac{v_h}{v_{ref}}\right) \tag{23.21}$$

The 5×5 matrix in (23.20) is the system dynamics matrix (F) and since it is constant over a Kalman update interval we can determine the state transition matrix as we have done before:

$$\Phi = e^{F\Delta t} \tag{23.22}$$

where Δt is the Kalman update interval.

Ausman recommended specifying the Q matrix as follows. For the inertial altitude error state, a small amount of noise is modeled to account for "computer resolution." However, this suggestion was made based upon embedded hardware available in the 1980s. Nevertheless, even with modern computational capability, a small amount of modeled noise is prudent simply to keep the filter "awake" in estimating these states. For example:

$$Q(1,1) = \Delta t(0.3)[m^2] \tag{23.23}$$

where Δt is the filter update interval. For the inertial vertical velocity error, again, a small amount of noise is modeled but an additional term, based on velocity change, is suggested to account for unmodeled scale-factor error:

$$Q(2,2) = \Delta t\left[10^{-6} + 10^{-8}\left(v_z[k] - v_z[k-1]\right)\right]\left[\left(\frac{m}{s}\right)^2\right] \tag{23.24}$$

For the vertical accel bias state, system noise is needed to account for bias shifts, vibration effects and gravity modeling error. Ausman suggests a constant value to account for the bias shifts and vibration effects along with a horizontal velocity-dependent term to account for unmodeled changes in gravity vector as the vehicle moves over the surface of the earth:

$$Q(3,3) = \Delta t\left[2 + \frac{v_h}{v_{ref}}\right] \times 10^{-10}\left[\left(\frac{\text{m}}{\text{s}^2}\right)^2\right] \tag{23.25}$$

It should be noted that the length scales of weather systems and gravity fields are arguably different (gravity variations tend to happen over shorter lengths than weather systems) so a different v_{ref} may be needed in, for example, (23.21) than in (23.25).

For the baro scale factor error state, a small amount of noise is suggested to account for temporal variation and an additional vertical velocity-dependent term is suggested to account for error in the assumed linear model of altitude dependency.

$$Q(4,4) = \Delta t\left[2 \times 10^{-8} + \left(\frac{v_z}{v_{ref}}\right)^2 \times 10^{-6}\right] \tag{23.26}$$

Note that since $Q(4,4)$ is a scale-factor term, there are no units. Finally, for the baro bias state, there is again a constant term to account for temporal variation and a horizontal velocity-dependent term to account for spatial variation:

$$Q(5,5) = \Delta t(0.1)\left[1 + \frac{v_h}{v_{ref}}\right]\left[m^2\right] \tag{23.27}$$

Having defined the state vector, the state transition matrix and the system noise covariance matrix, we are done with the Kalman system equation and can now turn our attention to the Kalman measurement equation. We start by modeling the inertial altitude as the sum of the truth and the inertial altitude error:

$$h_{INS} = h_t + \delta h_{INS} \tag{23.28}$$

In like manner, the baro altitude model is:

$$h_{baro} = h_t + \delta h_{baro} \tag{23.29}$$

Substitution of (23.14) into (23.29) yields:

$$h_{baro} = h_t + \delta h_0 + h_t k_0 \left(1 - \frac{h_t}{2h_{rev}}\right) \tag{23.30}$$

The observation is formed by differencing the inertial and baro altitudes:

$$z = h_{INS} - h_{baro} \tag{23.31}$$

Equation (23.31) is the quantity that is input to the filter. To determine the measurement equation, we first expand it:

$$z = h_{INS} - h_{baro} = \delta h_{INS} - \delta h_0 - h_t k_0 \left(1 - \frac{h_t}{2h_{rev}}\right) \tag{23.32}$$

In real-time operation, however, the true altitude is unknown so we approximate it with the Kalman-corrected inertial altitude:

$$z = \delta h_{INS} - \delta h_0 - \tilde{h}_{INS} k_0 \left(1 - \frac{\tilde{h}_{INS}}{2h_{rev}}\right) \tag{23.33}$$

where $\tilde{h}_{INS}$ is the Kalman-corrected inertial altitude. From (23.20), the state vector is:

$$\underline{x} = \begin{bmatrix} \delta h_{INS} \\ \delta v_z \\ \delta a_z \\ k_0 \\ \delta h_0 \end{bmatrix} \tag{23.34}$$

Recall the generic form of the Kalman measurement equation is:

$$\underline{z} = H\underline{x} + \underline{v} \tag{23.35}$$

In the current case, we have a scalar measurement. By comparing (23.33) with (23.34) and (23.35), it follows that:

$$H = \begin{bmatrix} 1 & 0 & 0 & h4 & -1 \end{bmatrix} \tag{23.36}$$

where

$$h4 = -\tilde{h}_{INS} \left(1 - \frac{\tilde{h}_{INS}}{2h_{rev}}\right) \tag{23.37}$$

The Kalman measurement noise covariance matrix (R, a scalar in the current case) models a number of factors. As mentioned in the three-state design, R models the high frequency baro measurement noise. For high dynamic vehicles and for additional precision, a term can also be added to R to account for additional noise during periods of high rates of climb or descent. Finally, a term can also be added to R to account for lag in the baro output. Specific numbers are dependent upon the particular barometric altimeter being used and the expected vertical dynamics but the high frequency noise component is generally on the order of 1 m^2 assuming a 1-sec update interval.

The filter design is completed by specifying the initial prediction for the state vector and the associated initial prediction error covariance matrix. Since the states are initially unknown and assumed to be zero mean, the initial predicted state vector is simply all zeros. As is typically done in actual practice, the initial prediction error covariance matrix is assumed to be diagonal (almost, see below). Ausman recommended the following values:

$$P_1^-(1,1) = (150\ [\mathrm{m}])^2 \tag{23.38}$$

$$P_1^-(2,2) = \left(0.3\ \left[\frac{\mathrm{m}}{\mathrm{s}}\right]\right)^2 \tag{23.39}$$

$$P_1^-(3,3) = \left(0.002\ \left[\frac{\mathrm{m}}{\mathrm{s}^2}\right]\right)^2 \tag{23.40}$$

$$P_1^-(4,4) = (0.05)^2 \tag{23.41}$$

$$P_1^-(5,5) = (150\ [\mathrm{m}])^2 \tag{23.42}$$

$$P_1^-(1,5) = (150\ [\mathrm{m}])^2 \tag{23.43}$$

where $P_1^-(1,1)$ and $P_1^-(1,5)$ have been set equal to each other with the assumption that the inertial altitude has been initialized to the baro altitude. All of the aforementioned values (Q, P, R) must be adjusted in accordance with the actual hardware being used (the accelerometer, in particular).

To aid the filter during the initialization transient, it can be assumed that the aircraft is initially parked at a location with a known altitude. Similarly, since baro aiding of the vertical channel is frequently considered to be a back-up mode in case of loss of GNSS, the filter can be initialized assuming GNSS altitude is available for some initial period. As noted by Ausman [5], the only changes in the filter during this initialization are in the observation, data matrix, and measurement variance. Assuming an aircraft parked at a known location (that closely approximates the truth), the observation is:

$$z = h_{INS} - h_t \tag{23.44}$$

Comparing (23.44) with (23.34), we see that:

$$H = \begin{bmatrix} 1 & 0 & 0 & 0 & 0 \end{bmatrix} \tag{23.45}$$

Since in reality, the altitude of the "known" location is not actually known perfectly (and this is especially true for something like GNSS), we need to model the measurement error/noise. Given nominal GNSS performance, a reasonable value is:

$$R = (3\ [\mathrm{m}])^2 \tag{23.46}$$

23.4.1 Simulation results

For the simulation to be realistic, a non-linear scale factor error truth model is needed. Ausman suggested starting with the following error model for the baro altitude as being reasonably representative:

$$\varepsilon_{baro_{alt}} = \left(k_{0t} + 0.04\frac{\beta}{h_{rev}} \right) \beta + \delta h_0 \tag{23.47}$$

where k_{0t} is the true sea-level scale factor error and:

$$\beta = h_t \left(1 - \frac{h_t}{2h_{rev}} \right) \tag{23.48}$$

The scale factor error at a given true altitude can be derived [5] from this and the result is:

$$k_t = \left(k_{0t} + 0.08\frac{\beta}{h_{rev}} \right) \left(1 - \frac{h_t}{h_{rev}} \right) \tag{23.49}$$

where k_t is the true scale factor error. A plot of the scale factor error truth model and the linear model implemented in the Kalman filter is given in Figure 23.3.

In order to demonstrate the performance of the filter at varying altitudes, a series of climbs and descents were simulated as shown in Figure 23.4. In order to assess the performance of the filter, a tactical-grade INS was simulated. The accelerometer biases were simulated in the range of 300–500 micro-gs and the gyro biases were

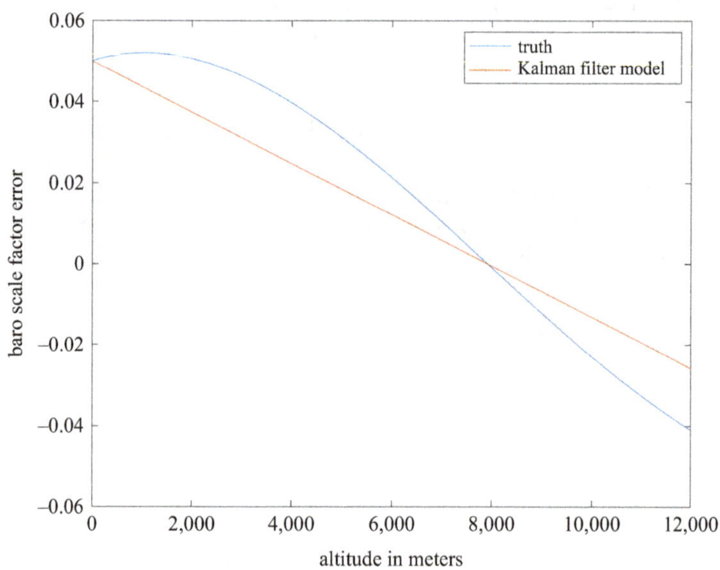

Figure 23.3 Baro scale factor error truth model and the linear model implemented by the Kalman filter

on the order of 0.1 deg/hour. The inertial position/velocity/attitude updating was simulated at a 100 Hz rate and the Kalman filter was updated at 1 Hz. Since transient performance was not the focus, the filter was initialized by assuming true altitude was known for the first 100 s (with $R = 1\ [\mathrm{m}]^2$). Since weather system changes were not simulated (nor gravity model errors), the velocity-dependent terms in F and Q were dropped. The results are given in Figures 23.5–23.10. Note that in the filter error plots, the dashed red lines are computed from the square root of the relevant diagonal element of the estimation error covariance matrix and thus represent the filter's one-sigma estimates of its own error.

The baro altimeter error is shown in Figure 23.5. It varies in accordance with the simulated sea level baro bias of 150 m and the scale factor error truth model. The error in the Kalman-corrected inertial altitude is shown in Figure 23.6. It takes a couple of climbs and descents to converge fully but remember that the critical performance parameter in the baro-inertial loop is the corrected inertial vertical velocity error. Figure 23.7 shows the inertial vertical velocity is well-stabilized throughout the 40 min of the flight after initialization.

Figure 23.8 shows the filter's vertical accel bias estimation error. It performs reasonably well but the relatively loose bounds and the divergence near the end of the run indicates the filter could use additional tuning. The error in the estimated sea level baro bias is very small as shown in Figure 23.9. Finally, the error in the sea level baro scale factor error (Figure 23.10) shows convergence after the first climb and descent.

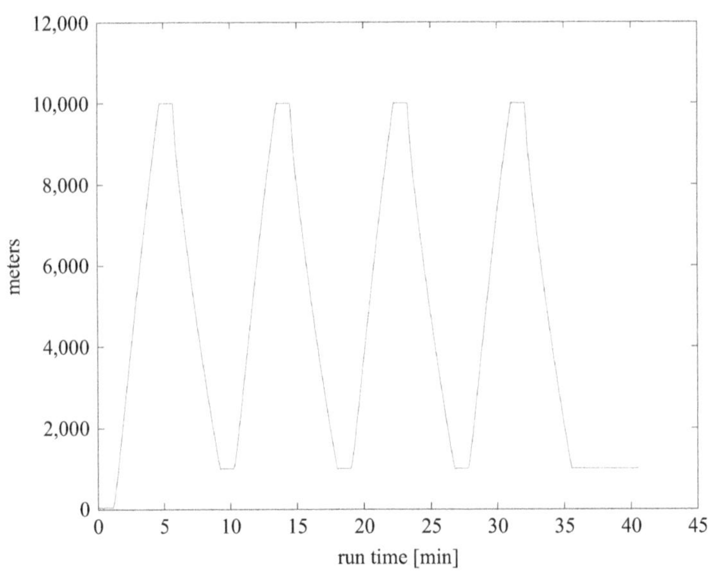

Figure 23.4 Altitude profile for the simulated flight path

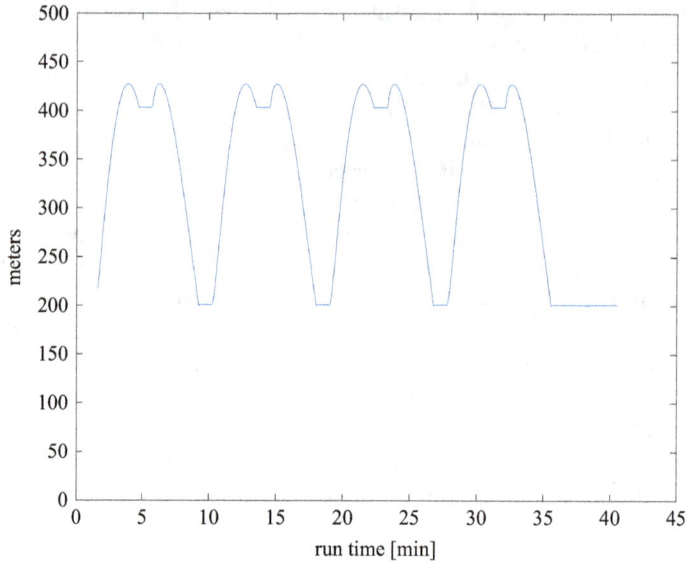

Figure 23.5 Barometric altimeter error

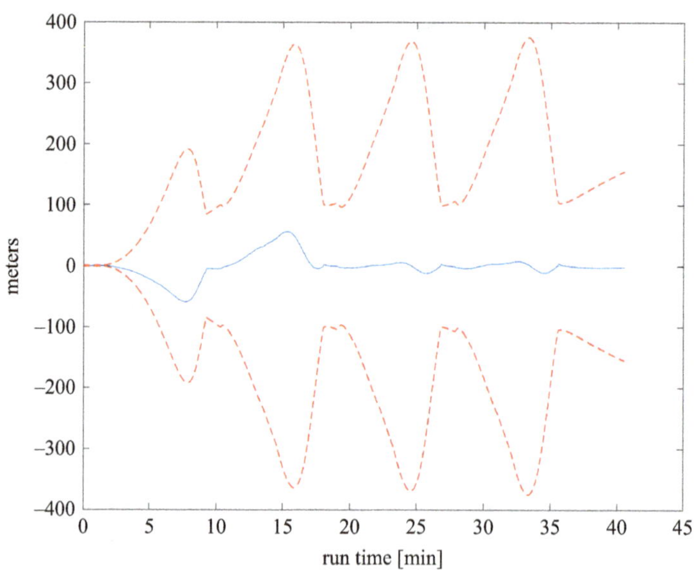

Figure 23.6 Error in the Kalman-corrected inertial altitude and one-sigma covariance bounds

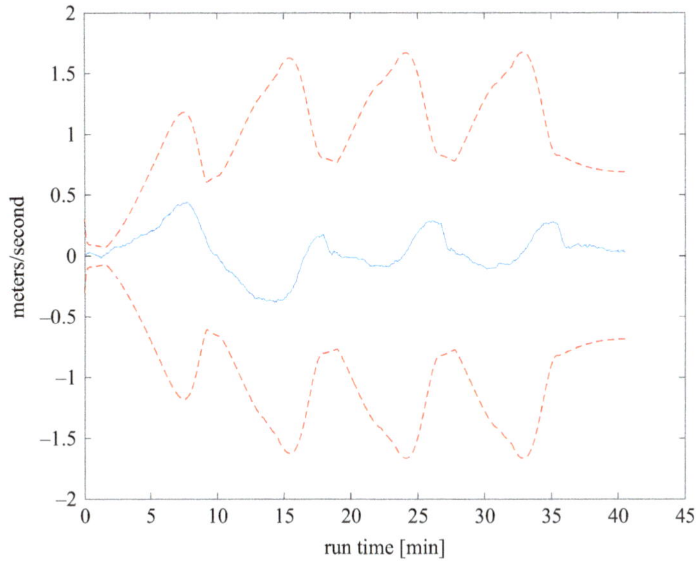

Figure 23.7 Error in the Kalman-corrected inertial vertical velocity and one-sigma covariance bounds

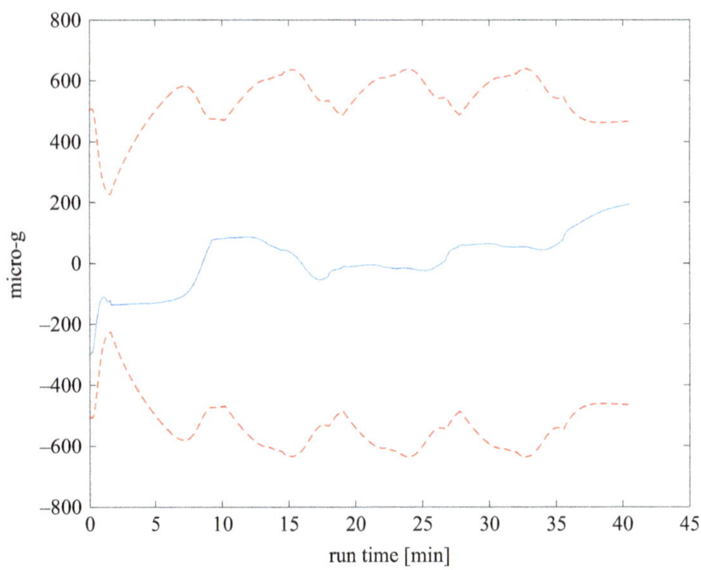

Figure 23.8 Vertical accel bias estimation error and one-sigma covariance bounds

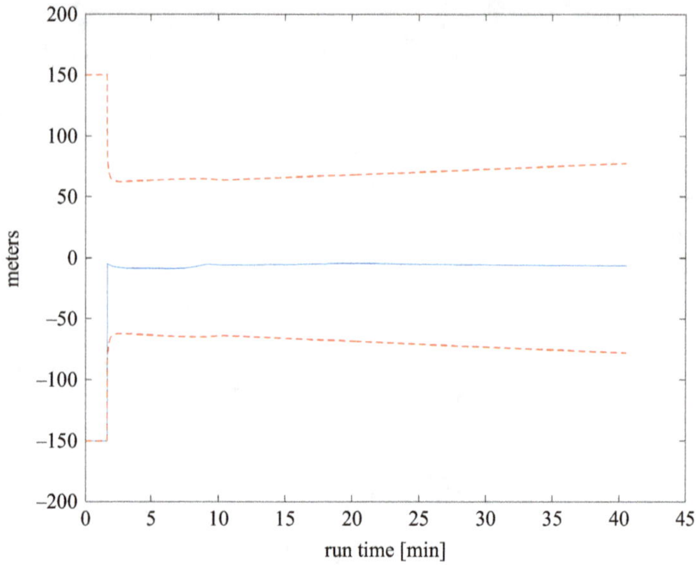

Figure 23.9 Sea level baro bias estimation error and one-sigma covariance bounds

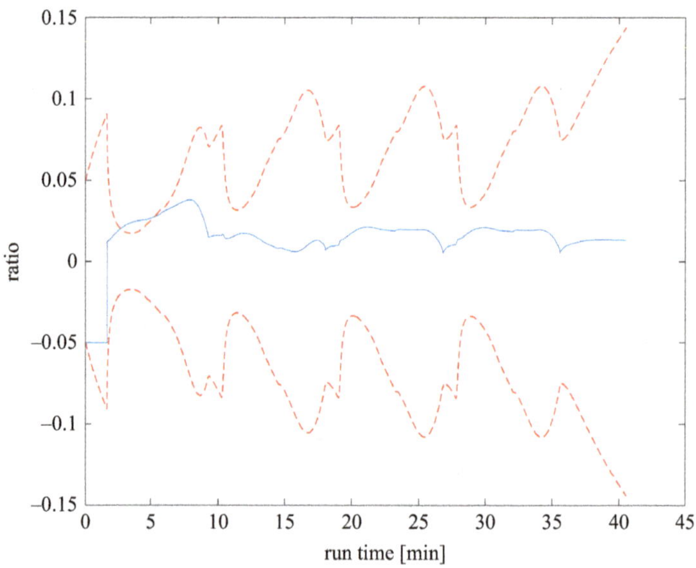

Figure 23.10 Sea level baro scale factor error estimation error and one-sigma covariance bounds

23.5 Additional considerations

The simulation results presented in the previous section should only be considered as a start to the design of a robust baro-inertial Kalman filter. As mentioned earlier, in reality weather systems can present significant changes from the standard atmospheric model and additional inertial sensor errors (accelerometer scale factor error, vibration sensitivity) must be considered. Both [3] and [5] discuss at length the effects that variable rates of climb and descent have on filter performance.

References

[1] Kayton M, Fried W. *Avionics Navigation Systems*. 2nd ed. New York, NY: John Wiley & Sons; 1997.

[2] Perkins KL, Palmer RR, Ausman JS, inventors; Inc. North American Aviation, assignee. Vertical velocity measuring system. United States Patent US 3005348; 1961 Oct 24.

[3] Ausman JS. Baro-inertial loop for the USAF Standard RLG INU. *NAVIGATION: Journal of the Institute of Navigation*. 1991;38(2):205–220.

[4] Brown RG, Hwang PYC. *Introduction to Random Signals and Applied Kalman Filtering: with MATLAB® Exercises*. 4th ed. Hoboken, NJ: John Wiley & Sons, Inc.; 2012.

[5] Ausman JS. A Kalman filter mechanization for the baro-inertial vertical channel. In: *Proceedings of the 47th Annual Meeting of the Institute of Navigation*; 1991.

[6] Barham PM, Manville P. Application of Kalman Filtering to Baro/Inertial Height Systems. Royal Aircraft Establishment; 1969. Technical Report 69131.

[7] U.S. Standard Atmosphere, 1976. National Oceanographic and Atmospheric Administration; 1976. NOAA-S/T 76-1562.

Chapter 24
Inertial initialization—Part B

24.1 Introduction

In an earlier chapter, we introduced the topic of inertial initialization. We showed how with a stationary vehicle the accelerometer outputs could be processed to determine rough initial estimates of pitch and roll and, having done so, the gyro outputs could be processed to determine a rough estimate of yaw or heading (or wander angle depending on the chosen convention). Although it is common for the entire process to be referred to as "alignment," it is more precisely "leveling" and "alignment." Although enhanced versions of this kind of processing (with fixed gain filters) were used in early strapdown systems (i.e., 1970s), it is more common in modern practice to use a Kalman filter.

24.2 Ground alignment and zero velocity updates

In normal aircraft operation, the inertial navigation system is initialized while the vehicle is parked on the ramp and thus the initialization is referred to as "ground" alignment. More generically this operation may be referred to as "stationary alignment" but Groves [1] notes that it is more properly referred to as "quasi-stationary alignment" since the vehicle may be experiencing some buffeting while the initialization is being performed. Wind gusts and loading operations (baggage, catering and passengers) all can cause some low frequency vibration of the airframe.

Initializing the position is straight-forward if the vehicle is parked at a known/surveyed location and initial velocity is effectively zero. As discussed previously, rough estimates of attitude can be formed directly from the accel and gyro measurements and, with a stationary vehicle, the measurements can be averaged over some interval to reduce noise. For the Kalman filter implementation, however, an aiding observable can be formed by exploiting the known stationary condition of the vehicle.

Specifically, the inertial system position/velocity/attitude updating is performed as normal (the so-called "free inertial" processing) starting with the known initial position and velocity, the rough initial estimates of roll and pitch and an arbitrary initial value for yaw/heading (assuming no a priori information). Even if we considered a completely theoretical scenario in which there were no sensor errors, the inertial system would still start to build up non-zero velocity values (and thus change in position) due to erroneous earth-rate compensation of the attitude. Recall that earth rate has north and vertical components that are dependent upon latitude. Latitude is assumed to be known since the vehicle is at a known location. However, since yaw/heading is not known, the guessed (and certainly erroneous) initial yaw will cause the earth rate compensation to be incorrect on all three axes. This will eventually cause the estimated nav-frame to be not locally level and gravity will couple into the estimated horizontal axes inducing Schuler oscillations.

We saw this in the earlier chapter on Inertial Error Simulation. In that chapter we saw that an initial azimuth error of only 1 milli-radian would induce velocity errors of more than 0.6 m/s given enough time for the Schuler oscillation to reach its maximum value. In the current situation, the initial azimuth error is, by definition, not small. Figure 24.1 shows, as an example, the position and velocity error growth over 60 sec given an initial azimuth error of $\frac{\pi}{4}$ radians.

The combination of the known stationary condition of the vehicle and the non-zero computed velocity (and non-zero computed displacement) of the vehicle provides the observable for the Kalman filter. The simplest example is the so-called "zero velocity update" sometimes abbreviated as ZUPT. The velocity computed by the "free inertial" processing is the observed velocity error and thus is the input to the filter. The H matrix consists of zeros except for "1"s in the relevant rows/columns (i.e., those corresponding to the velocity error states). Recall from the Observability chapter that the filter exploits the known coupling between the states (encapsulated

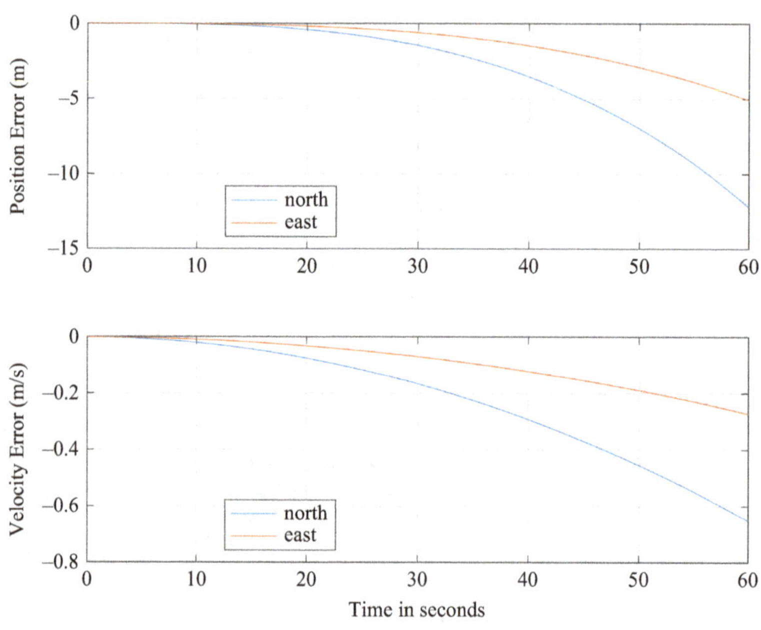

Figure 24.1 Position and velocity error growth due to a large ($\frac{\pi}{4}$ radians) initial azimuth error. A stationary vehicle was simulated at a latitude of 45 degrees. No sensor or other errors were simulated

by the state transition matrix) to populate off-diagonal elements of the prediction error covariance matrix which the Kalman gain uses to estimate those states that are not directly observable. A different implementation, described by Groves [1], utilizes the observed change of position (position displacement or integrated velocity, which should also be zero) rather than the observed velocity.

Some inertial systems have a mode control with "align" and "navigate" modes. The unit assumes that when it is in "align" mode the vehicle is stationary. The operator (pilot) selects the align mode to permit the alignment process to start. In practice this is somewhat inadvisable. Operators frequently are very busy with their own start-up procedures and the workload is typically quite high even without additional buttons to push and modes to worry about on one of the many pieces of equipment in their vehicle. Thus, it is better if the inertial unit itself can determine if the vehicle is stationary or not. "Motion detection" or "motion detect" is the name of this process. More detail on this topic is provided in [1].

24.3 Wide-angle or coarse alignment

The accel and gyro processing discussed in the earlier chapter is a form of coarse leveling and the so-called wide-angle or coarse alignment. In normal circumstances, coarse leveling yields estimates of pitch and roll that are accurate to a few degrees. The so-called "fine leveling" may then proceed in parallel with the coarse alignment Kalman filter processing.

The main challenge with no a priori knowledge of yaw is that the error equations are non-linear and thus not directly amenable to a conventional Kalman filter design. To see this, assume that yaw/heading is initially predicted to any random value. In the worst case, the difference between the initial prediction and the true value is π radians. This is much too large to maintain linearity in the error equations. One of the techniques utilized, instead of estimating the yaw error, is to estimate the error of the *sine* and *cosine* of the yaw error [1]. Since both the sine and cosine have a zero mean (given a random input that is uniform on the interval from 0 to 2π), both states can have their initial predictions set to zero.

The choice to replace the yaw error state with the error of the sine and cosine of the yaw error has a non-trivial ripple effect on the filter design. New error equations must be derived and then used in the formation of a new system dynamics matrix (F) and finally a new state transition matrix (Φ) and system noise covariance matrix (Q).

In practice, coarse leveling requires only a few seconds of accel data (given nav-grade sensors) and fine leveling is typically completed in less than a minute. Coarse alignment is typically also completed in less than a minute (assuming the vehicle is not close to the Earth's poles). This leaves fine alignment as the remaining task.

24.4 Fine alignment

When coarse alignment is complete, yaw/heading is known to within a few degrees. This permits the "in-flight" Kalman filter to be utilized since the linearity assumptions are valid. “Fine alignment” is simply the use of the in-flight filter aided by ZUPTS. The fine alignment process continues at least until the filter has determined that the attitude error covariances are sufficiently small to permit normal operation within the specification values of the system. The system may make an annunciation that the inertial system is ready for the vehicle to start moving. Even after this the system will continue to process ZUPTS (and thus refine the alignment even further) until the threshold of the motion detection algorithm is triggered. At low and mid-latitudes, 3–5 min are typically required for fine alignment.

Reference

[1] Groves P. *Principles of GNSS, Inertial, and Multisensor Integrated Navigation Systems*. 2nd ed. Boston, MA: Artech House; 2013.

Chapter 25
Conclusion: inertial+

I have been in the field of navigation for over 35 years at the time of this writing and along the way it has been my very great privilege to be able to work with some truly outstanding navigation researchers. Two of them had the very same perspective on navigation but worded it slightly different. One lamented the fact that the modern phrase "GPS/INS," or more recently "GNSS/INS," was a misnomer since the purpose of the GNSS is to aid the INS, not the other way around (and thus it should be INS/GPS or INS/GNSS). The second colleague very eloquently stated that the future of navigation is "inertial-plus." Plus what? Plus whatever the current aiding source happens to be. In the 1960s, it was fly-over fixes. In the 1970s, it was a Doppler navigator or TACAN. Since the early 1990s, it has been GPS/GNSS. With concerns over GNSS vulnerabilities, the future of aiding will likely be a combination of GNSS, signals-of-opportunity, vision systems and who knows what else. Nevertheless, the core is, and will be, INS.

The laws of physics provide the input and thus there is no need to rely on radio-frequency signals that can be disrupted. Even the vertical channel can be stabilized with a barometric altimeter which is also essentially immune to external disturbances (particularly for aircraft at altitude). Inertial systems provide high data rates, low data latencies, and high-quality attitude determination as a by-product of determining velocity and position. Their prime technical weakness is long-term drift, hence the need for external aiding. At the time of this writing, the only non-technical *problem* with navigation-grade inertial systems is SWAP-C (size, weight and power, and cost). Significant research and development efforts have been expended toward developing chip-scale navigation-grade IMUs and it seems likely that production units will become available well within the careers of the younger readers of this book.

Furthermore, quantum-based IMUs are also a hot topic of research. These technologies hold the promise of several orders of magnitude improvement in residual bias errors. However, these technologies also have weaknesses regarding bandwidth and will almost certainly need to be integrated with a "traditional" IMU to achieve acceptable performance. Since the high bandwidth unit is the core, we are talking about a strategic-grade IMU *aiding* a navigation-grade IMU! No matter how you look at it, the future of navigation is inertial-plus!

Appendix A
Two important stochastic processes

The random walk and first-order Gauss–Markov processes are two of the most important stochastic processes used to model inertial sensor errors. They are both used extensively in aided-inertial Kalman filters.

A.1 Random walk

The random walk process is simply a running integral of Gaussian white noise. As such it serves as the model for velocity error due to accelerometer white noise and for angular error due to gyro rate white noise. These two errors are thus known, respectively, as "velocity random walk" and "angle random walk." As shown in Figure A.1, the Gaussian white noise is indicated by $w(t)$ and the output of the integrator is the random walk process $x(t)$.

In discrete time, the process is given by:

$$x_{k+1} = x_k + w_k \tag{A.1}$$

where w_k is a zero-mean white Gaussian sequence.

The random walk process is not stationary since its statistical characteristics are functions of time. The ensemble mean and variance of the continuous-time random walk process are given by [1]:

$$E[x] = 0 \tag{A.2}$$

$$E[x^2] = q \cdot t \tag{A.3}$$

where q is the variance of the continuous-time white noise input $w(t)$.

Thus, we note that the variance of the random walk process grows *linearly* with time. To illustrate this, Figure A.2 depicts an example of discrete-time Gaussian white

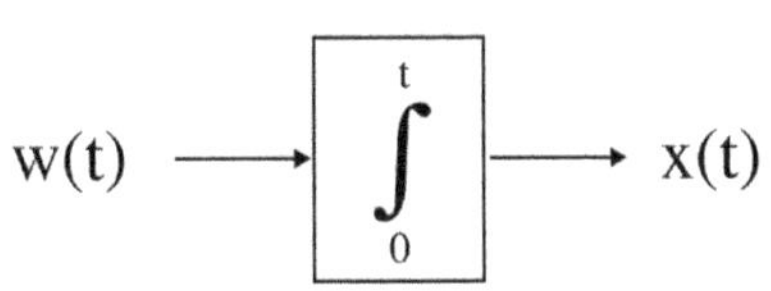

Figure A.1 The continuous-time random walk process. The input, w(t), is Gaussian white noise

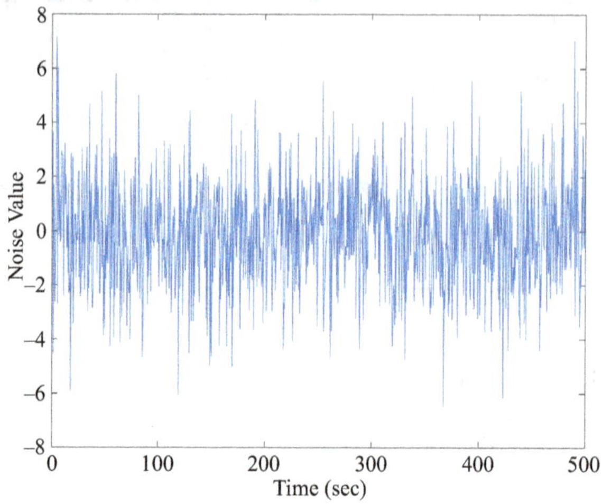

Figure A.2 Discrete-time white noise that subsequently will be used to generate random walk and first-order Gauss–Markov processes. Data rate is 2 Hz

noise with a standard deviation of 2. The running integral of this white noise is given in Figure A.3. You will note that, although it is a random process, it appears to be drifting over time. The term "random walk" comes from the idea of trying to model the path of somebody who is going down a very narrow alley and had too much to drink so they have a random chance of taking one step forward or one step back at every moment in time. As you can imagine, as time goes on the person gets farther and farther away from the starting point.

This is certainly the case in the particular run shown in Figure A.3. What happens if we do it all over again with a fresh set of random numbers? The path may not trend negative. It may go positive or it may hang around the middle. It just depends on how the random numbers play out. Figure A.4 shows the results for 1,000 realizations of the random walk process. You can see that sometimes the process ends up going negative, sometimes positive, and sometimes it stays in between. Figure A.4 shows clearly that when looking over the entire ensemble, the standard deviation (or variance) of the process grows with time. Note however that, when looking over the entire *ensemble*, the mean value is zero. A brief look back at Figure A.3 shows, however, this is clearly *not* the case for a *particular* realization.

If the previous statements seem a bit puzzling, remember that in the data shown in Figure A.4, computing the ensemble mean or variance involves processing all of the realizations for a particular *moment in time*. Conceptually, when looking at Figure A.4, we are taking vertical slices of the data and then computing the mean and variance. We observe that a vertical slice at, say, 100 sec, has a narrower range of values than does a vertical slice at 500 sec. Thus, the variance is increasing with time. The mean value, however, remains zero. Since the process is not stationary, statistics computed over

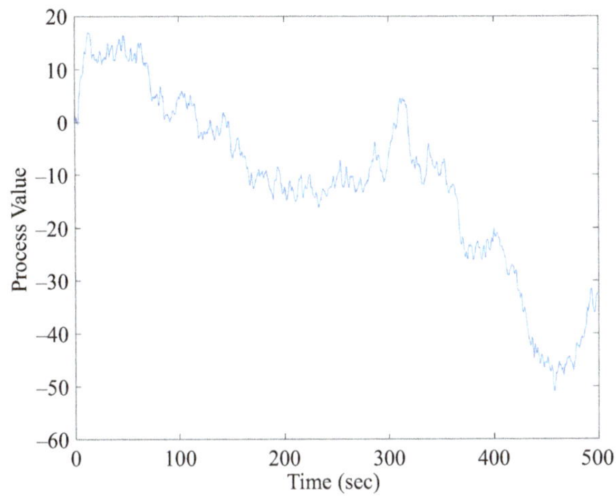

Figure A.3 An example of a random walk process. This example was created using the white noise shown in Figure A.2

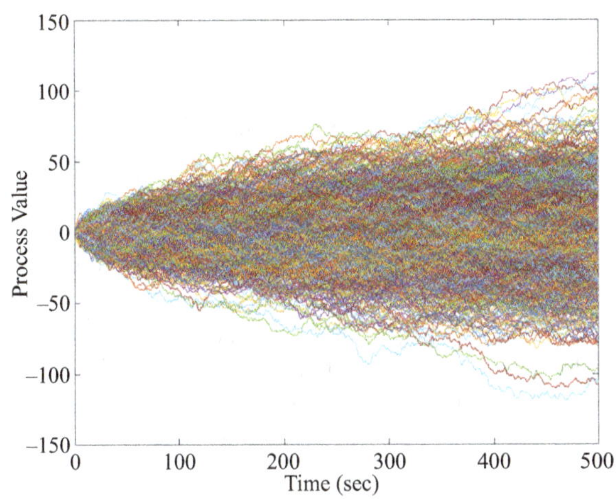

Figure A.4 One thousand realizations of a discrete-time random walk process. The input discrete-time white noise was generated at a 2 Hz sampling rate with a standard deviation of 2

time *will not* yield the same result as ensemble statistics. For example, the mean value (computed over time) of the random walk example shown in Figure A.3 clearly is not zero even though the ensemble mean (as illustrated in Figure A.4) is. For a discrete-time random walk process, the variance expression given by (A.3) is modified by the sampling interval [2]:

$$E[x_k^2] = (q \cdot \Delta t)t_k \tag{A.4}$$

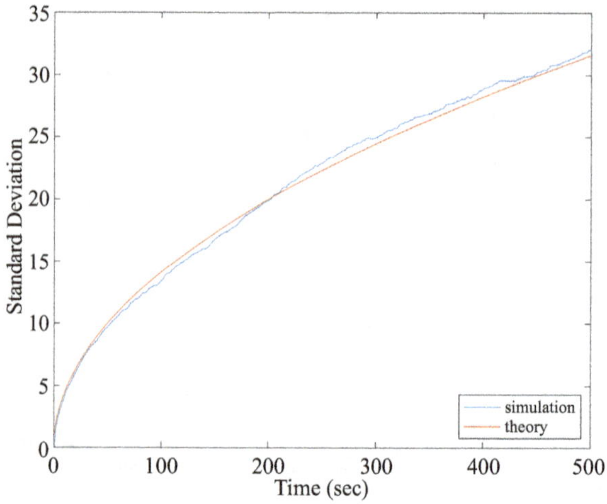

Figure A.5 Ensemble standard deviation of the data illustrated in Figure A.4 along with the theoretical value of a discrete-time random walk process generated from 2 Hz discrete-time white noise with a standard deviation of 2

where $E[x_k^2]$ is the variance of the random walk process at time index k, q is the variance of the discrete-time white noise input, Δt is the sampling interval and t_k is the discrete-time sampling time at time index k. Note the standard deviation is simply the square-root of the expression given in (A.4).

Figure A.5 illustrates the ensemble standard deviation computed from the data shown in Figure A.4 along with the theoretical value given by the square-root of Equation (A.4). There is clearly good agreement and it would be even better if we used a larger number of runs. This growth in standard deviation in proportion to the square-root of time explains why, for example, gyro angle random walk is expressed in terms of "degrees-per-root-hour." Since the variance of a random walk process grows linearly with time, the standard deviation of a random walk process grows with the square-root of time.

A.2 First-order Gauss–Markov process

The first-order Gauss–Markov process is obtained quite simply by applying negative feedback to the random walk process (see Figure A.6). This negative feedback, as we will see, ends up providing stability and prevents the output from growing without bound.

Recall that a state equation is the differential equation that describes a state variable. If we take the output of the first-order Gauss–Markov process, $x(t)$, as our state variable, then we see by inspection of Figure A.6 that the derivative of $x(t)$ is

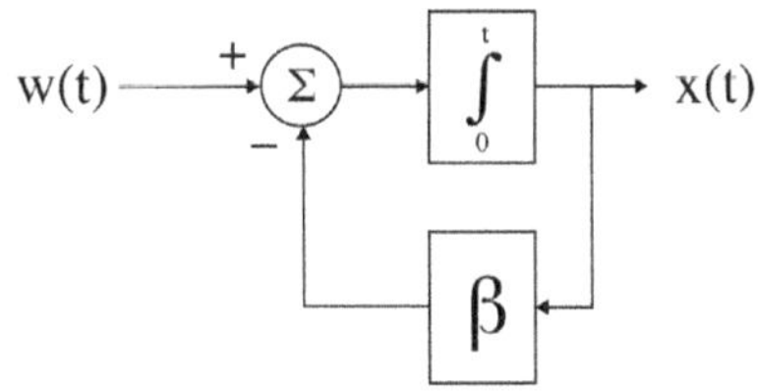

Figure A.6 The first-order Gauss–Markov process. The input, w(t), is Gaussian white noise and the time-constant is the inverse of β

given by the output of the summing junction. The continuous-time state equation is thus:

$$\dot{x}(t) = -\beta x(t) + w(t) \tag{A.5}$$

The mean and variance of the first-order Gauss–Markov process are given by [2]:

$$E[x] = 0 \tag{A.6}$$

$$E[x^2] = \frac{q}{2\beta} \tag{A.7}$$

where q is the variance of the continuous-time white noise input $w(t)$. Note that, unlike the random walk process, the variance of the first-order Gauss–Markov process is *not* a function of time and is, in fact, a constant. It is a stationary random process.

In discrete time, the first-order Gauss–Markov process is given by [2]:

$$x_{k+1} = e^{-\beta\Delta t} x_k + w_k \tag{A.8}$$

where Δt is the time-step in the discrete-time process and the time-constant of the process is:

$$\tau = \frac{1}{\beta} \tag{A.9}$$

Solving (A.7) for q yields:

$$q = 2\beta E[x^2] = \frac{2}{\tau} E[x^2] \tag{A.10}$$

The relationship between the continuous-time model white noise input and the discrete-time model white noise input is given by [2]:

$$q_k = \frac{q}{2\beta}[1 - e^{-2\beta\Delta t}] \tag{A.11}$$

where q_k is the variance of the white noise input of the discrete-time model and Δt is the time difference between the current sample and previous sample. Equation (A.11) can be simplified by expanding the exponential term with a Taylor Series and retaining only zeroth and first-order terms (valid for small Δt):

$$q_k \approx \frac{q}{2\beta}[1 - (1 - 2\beta\Delta t)] = q\Delta t \tag{A.12}$$

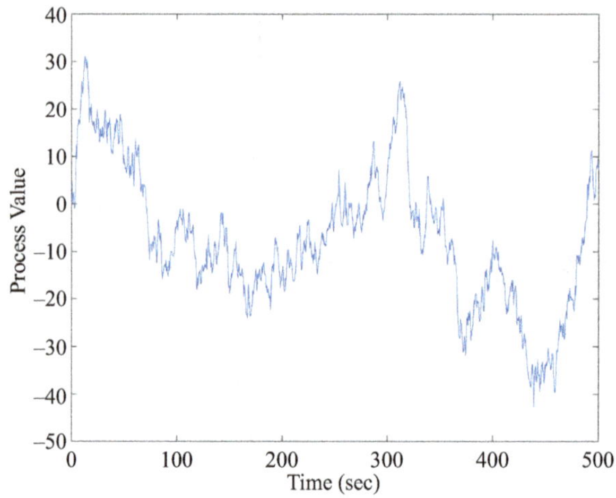

Figure A.7 A discrete-time first-order Gauss–Markov process generated using the white noise depicted in Figure A.2 with the time-constant set to 50 sec

The discrete-time variance can thus be approximated by the multiplication of the continuous-time variance by the time interval.

Figure A.7 shows the output of Equation (A.8) when the white noise depicted in Figure A.2 is the input and the time-constant is set to 50 sec. Although there appears to be an overall negative slope to the process, this is not actually the case as is shown in Figure A.8 where 10 realizations of the process are plotted simultaneously. As mentioned earlier, the ensemble variance of the first-order Gauss–Markov process is constant.

For a discrete-time first-order Gauss–Markov process, the variance is given by [2]:

$$E[x_k^2] = q_k \cdot (1 - e^{-2\beta\Delta t})^{-1} \tag{A.13}$$

Figure A.9 illustrates the ensemble standard deviation computed from the data in Figure A.8 along with the theoretical result provided by the square-root of the expression in (A.13). Since the data depicted in Figure A.8 was initialized at zero, it takes a while for the process to reach steady-state. Examination of Figure A.9 seems to indicate that it takes about two time-constants ($2 \cdot 50 = 100$ sec in this example) for the process to reach steady-state. The noise in the estimated standard deviation (computed from the data) is somewhat large due to the fact that only 10 data sets were utilized.

At this point, you may still be scratching your head wondering what exactly is the time-constant of the first-order Gauss–Markov process. The time-constant, also known as the "correlation time" of the stochastic process, indicates the so-called "serial correlation" of the process. Serial correlation is an indication of the predictability of the stochastic process. If we are given a few data points of a stochastic process, and we can predict the next data point with high probability, then

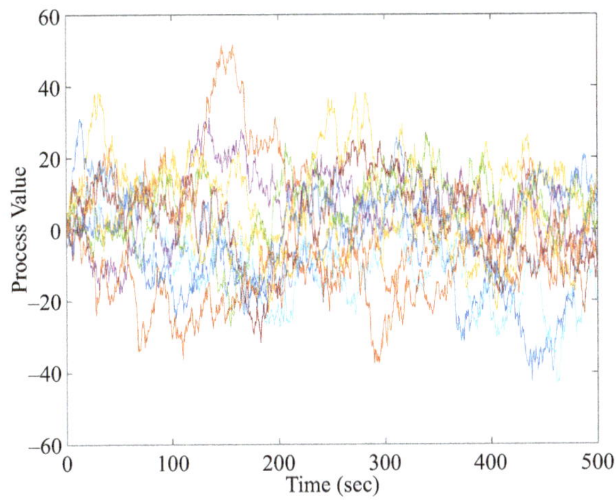

Figure A.8 *Ten realizations of a first-order Gauss–Markov process with a 50-sec time-constant and input discrete-time white noise with a standard deviation of 2*

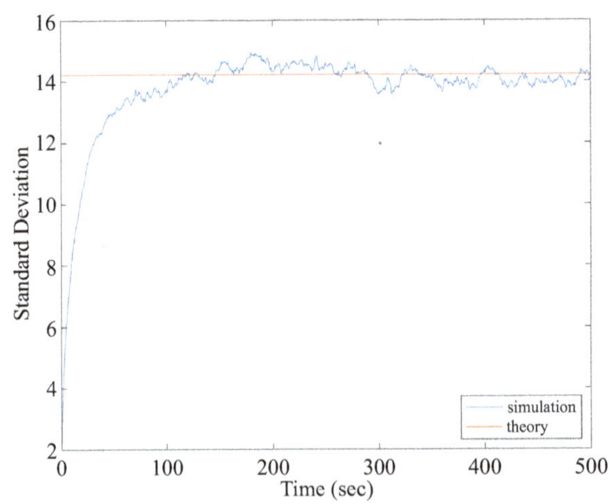

Figure A.9 *Ensemble standard deviation of the data illustrated in Figure A.8 along with the theoretical value of a discrete-time first-order Gauss–Markov process with a 50-sec time-constant generated from 2 Hz discrete-time white noise with a standard deviation of 2*

we would say that the process has high serial correlation. White noise, on the other hand, is completely unpredictable and therefore we would say that the serial correlation is zero. In general, if we have a process that has some serial correlation then, given a few data points in the process, it is possible to predict the next point with some degree of accuracy. If we look at the data in Figure A.7, there is clearly a positive trend in the data from $t = 180$ to $t = 300$ sec. If we observed the process from $t = 200$ to $t = 250$ sec, we could say with some certainty that the process would likely trend positive over the next few tens of seconds. The time-constant (or correlation time) provides an indication of the limit of our ability to predict ahead.

The time-constant can be estimated from the data itself by computing the autocorrelation function. For a discrete-time process, x_k, with N data points, the autocorrelation function is given by:

$$R(\tau) = \frac{1}{R_{max}} \sum_{k=1}^{N} x_k x_{k+\tau} \tag{A.14}$$

where R_{max} is the maximum value of $R(\tau)$ and simply normalizes the function. τ is known as the "lag" value. The autocorrelation function thus involves taking the data and point-by-point multiplying it with a shifted version of itself and then summing the result.

The theoretical normalized autocorrelation function of the first-order Gauss–Markov process is given by [2]:

$$R(\tau) = e^{-\beta|\tau|} \tag{A.15}$$

The average of the normalized autocorrelation functions of the ten realizations shown in Figure A.8 is plotted in Figure A.10 along with the theoretical value.

The time-constant is defined by the lag at which the normalized autocorrelation function decays to a value of e^{-1} (approximately 0.368). This value of e^{-1} is also depicted in Figure A.10 and we see that the theoretical normalized autocorrelation

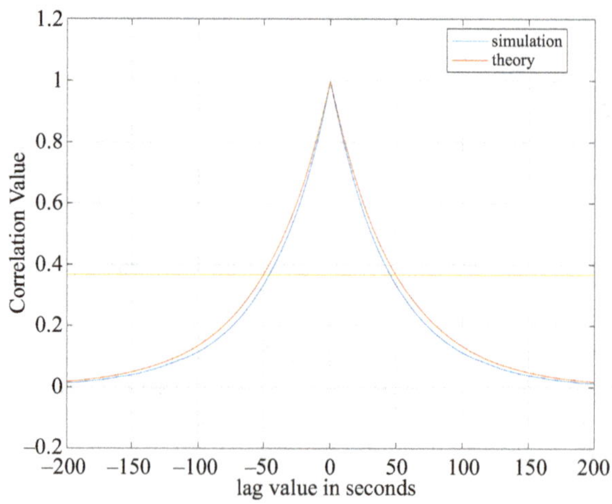

Figure A.10 Average of the normalized autocorrelation functions computed from the data shown in Figure A.8 along with the theoretical result. The time-constant is the lag at which the function decays to a value of e^{-1} *(depicted by the horizontal line)*

function decays to this value at a lag of 50 sec. This should not be surprising since we set $\beta = \frac{1}{50}$ in Equation (A.15). However, note that the result obtained from the data closely matches this theoretical value.

A.3 A note about uncorrelated samples

If, for some reason, we want to obtain uncorrelated samples from a time series of a correlated random process, we generally need an interval of approximately two to three time-constants. For example, the trends shown in the example of Figure A.7 have durations of one to two time-constants. If the samples are far enough apart that the next sample cannot be predicted with any certainty, then they are uncorrelated. That is important when data with serial correlation is input to the Kalman filter (e.g., data that has previously been filtered generally has some serial correlation). The Kalman filter inherently assumes that the input measurements are independent (uncorrelated). If you have correlated measurement data then you have to subsample (i.e., decimate) in order to ensure that the data is uncorrelated before it can be input to a Kalman filter.

References

[1] Brown RG, Hwang PYC. *Introduction to Random Signals and Applied Kalman Filtering: With MATLAB® Exercises*. 4th ed. Hoboken, NJ: John Wiley & Sons, Inc.; 2012.

[2] Gelb A, editor. *Applied Optimal Estimation*. Cambridge, MA: The M.I.T. Press; 1974.

Appendix B
One-at-a-time measurement processing

"One-at-a-time measurement processing" is an alternative formulation to the Kalman filter that performs the updating with scalar processing instead of matrix processing. It eliminates the need for a matrix inverse and was essential for legacy processors that were unable to handle the computational burden. Of course, this is certainly less of a concern today but there is still a plenty of operational software in the field built on this principle. As a result, it is useful to be familiar with it.

Let us take the example of GPS/INS integration where the GPS receiver is providing measurements at a data rate of 1 Hz and thus the Kalman update is also being executed at 1 Hz. At a given moment, the GPS receiver could be tracking and forming measurements on, say, eight satellites. The conventional Kalman filter has to process the entire observation vector (z) at the same time.

Figure B.1 illustrates a procedure that eliminates the need for the matrix inverse in the computation of the Kalman gain. The processing assumes the measurement error covariance matrix is a diagonal matrix. This assumption implies that all measurements are statistically independent. This is frequently the case or at least is taken as a reasonable approximation. Furthermore, there are procedures to diagonalize a matrix but that is beyond the scope of what we will consider herein.

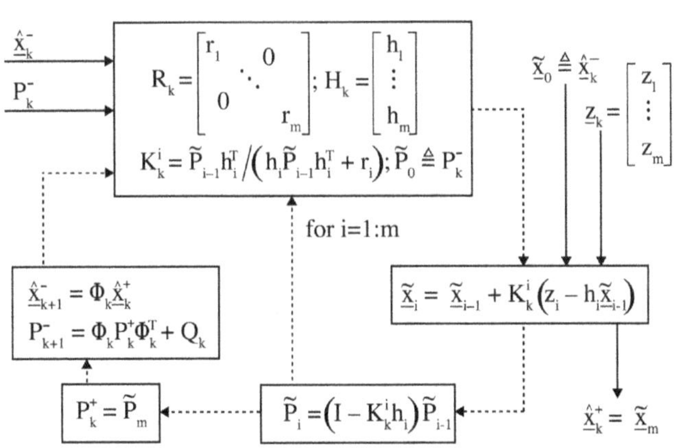

Figure B.1 Block diagram for Kalman filter one-at-a-time measurement processing [1].

Given a diagonal measurement error covariance matrix, we can reformulate the Kalman filter. The process starts conventionally with initialization and ends conventionally with the computation of the state prediction and its associated prediction error covariance matrix. The one-at-a-time processing only affects how the Kalman gain, state estimate and estimation error covariance are computed.

At a given time index, k, we have a state prediction ($\hat{\underline{x}}_k^-$) and its prediction error covariance matrix (P_k^-). [Note the hat symbol is explicitly distinguishing the state vector as being *not* the truth; it is either a prediction or an estimate.] As mentioned above, the measurement error covariance matrix is assumed to be diagonal:

$$R_k = \begin{bmatrix} r_1 & 0 & \cdots & 0 \\ 0 & r_2 & 0 & \cdots \\ \cdots & 0 & \ddots & 0 \\ 0 & \cdots & 0 & r_m \end{bmatrix} \tag{B.1}$$

We will denote the ith *row* of the data matrix as h_i and thus H_k may be written as:

$$H_k = \begin{bmatrix} h_1 \\ h_2 \\ \vdots \\ h_m \end{bmatrix} \tag{B.2}$$

The measurement (observation) vector is defined conventionally as:

$$\underline{z}_k = \begin{bmatrix} z_1 \\ z_2 \\ \vdots \\ z_m \end{bmatrix} \tag{B.3}$$

The state vector estimate and the estimation error covariance matrix are built up incrementally as the measurements are processed one at a time. Variables that are internal to the one-at-a-time measurement processing loop are initialized as:

$$\tilde{\underline{x}}_0 = \hat{\underline{x}}_k^- \tag{B.4}$$

$$\tilde{P}_0 = P_k^- \tag{B.5}$$

At a given time index, k, the loop over the measurements in the vector ($\underline{z}_k$) is given by:

for $i = 1{:}m$,

$$K_k^i = \tilde{P}_{i-1} h_i^T / \left(h_i \tilde{P}_{i-1} h_i^T \right)$$
$$\tilde{\underline{x}}_i = \tilde{\underline{x}}_{i-1} + K_k^i \left(z_i - h_i \tilde{\underline{x}}_{i-1} \right)$$
$$\tilde{P}_i = \left(I - K_k^i h_i \right) \tilde{P}_{i-1}$$

end

Once the loop is completed, final estimation results are given simply by:

$$\hat{\underline{x}}_k^+ = \tilde{\underline{x}}_m \tag{B.6}$$

$$P_k^+ = \tilde{P}_m \tag{B.7}$$

where the superscript "+" explicitly denotes an estimate (versus a prediction which we denote by superscript "−"). As mentioned earlier, the computation of the state prediction and prediction error covariance matrix are unchanged.

It should be noted that modern tools such as MATLAB® are so extremely efficient in matrix processing (including inverses) that the computational advantages of the one-at-a-time processing are negligible in some cases. However, the scalar processing is still used in existing software bases due to holdover from legacy embedded processors and compilers.

Reference

[1] Levy L. *Integration of GPS with Inertial Navigation Systems [Short Course Notes]*. NavtechGPS; 2000.

Index

www.ingramcontent.com/pod-product-compliance
Lightning Source LLC
LaVergne TN
LVHW011832200926
848484LV00002B/347